Lake Ecology in Kashmir, India

Mubashir Jeelani

Lake Ecology in Kashmir, India

Impact of Environmental Features on the Biodiversity of Urban Lakes

Springer

Mubashir Jeelani
Department of Environmental Sciences
Government Degree College Bemina
Srinagar, Jammu and Kashmir, India

ISBN 978-3-319-40879-8 ISBN 978-3-319-40880-4 (eBook)
DOI 10.1007/978-3-319-40880-4

Library of Congress Control Number: 2016947429

Printed on acid-free paper

This Springer imprint is published by Springer Nature
The registered company is Springer International Publishing AG Switzerland

Acknowledgments

I express my deep sense of gratitude to my supervisor, Dr. Harbhajan Kaur, Professor, Department of Zoology and Environmental Sciences, Punjabi University, Patiala. I am and will always remain thankful to her for her invaluable guidance, constant encouragement, affectionate attitude, and healthy criticism during the inception and execution of this study, and without her continual inspiration, it would have not been possible to complete this study.

I equally express my sincere thanks to my co-supervisor Dr. S. G. Sarwar for his inspiring and priceless guidance during the course of this research work and for providing me all necessary facilities for carrying out my research work uninterrupted in Hydrobiology Research Laboratory, S. P. College, Srinagar Kashmir.

I take this opportunity and privilege to express my deep sense of gratitude and respectful regards to Prof. (Dr.) M. Johal, head, Department of Zoology and Environmental Sciences, Punjabi University, Patiala, for providing all necessary facilities. I also express my courteous gratitude to Prof. (Dr.) D. S. Sidhu and Prof. (Dr.) S. S. Grewal (Retd.), former heads of the department for their unwavering support and help.

Words fail to express my heartfelt gratitude and sincere thanks to Dr. M.A. Shah, former head, Department of Environment and Water Management, S. P. College, Srinagar, for his generous help during different stages of this study.

My special gratitude, respect, and regards go to Mr. G.S. Ahluwalia (Xen.) and Mrs. S.G. Sarwar for their innumerable kindness and sympathetic attitude during the course of this study.

I am thankful to my fellow scholars and friends for their timely help and encouragement during the course of this study.

I am deeply grateful to my parents for their cooperation and help during the course of this work. My special thanks are extended to my sister for giving me unending support and encouragement needed to complete this thesis.

Mubashir Jeelani

Contents

Introduction

For centuries biologists have been interested in the diversity of life forms, their evolution and extinction. The role of biological diversity in the functioning of different ecosystems has also been investigated for several decades. However, the convention on biological diversity adopted at the Rio Earth summit in June, 1992, has led to a resurgence of interest in the subject of biodiversity and its various human dimensions. The convention requires *interalia*, an assessment of the current status of biodiversity in all countries and formulation of integrated strategies for its conservation. India, the largest nation in the Indian subcontinent, and the second most populous country in the world, has also great geological, geomorphological, climatic, biotic and cultural diversity (Gopal 1990). A large diversity of aquatic habitats is ensured by a high average annual rainfall of about 1000mm, with very large spatial and temporal variability. The lentic habitats mainly comprise natural lakes, reservoirs and impoundments. The natural lakes are rather few and mostly confined to the Himalayan belt. They are estimated to cover a total area of 0.72 million ha. (Jhingran 1991b). However, there are over 4290 large (>15m maximum depth or > 1 million m storage) and innumerable smaller reservoirs covering an area of over 3.15 million hectares (Sugunan 1995 and Suryanarayanan 1996). Besides, there are several thousands of temple tanks, particularly in peninsular India and village ponds.

The floristic component of aquatic life is dominated by algae in open deep waters whereas higher plants (macrophytes) dominate the shallow water areas and wetlands. The total algal flora of India has never been estimated although a preliminary report puts the Indian algal flora (excluding marine forms) at about 1800 species of which about two-thirds are the blue green algae. A few systematic accounts of some representative groups were published long ago (Desikachary 1959; Randhawa 1959; Venkataraman 1961; Ramanathan 1964 and Philipose 1967). Similarly total species richness of higher aquatic vegetation in India has not been estimated. Biswas and Calder (1937) described about 170 species of aquatic and marsh plants from the then British India but excluded all woody species. Cook (1996) described about 800 species from India at elevations below 1000m while total number of aquatic species to exceed 1250 in the Himalayan and higher altitudes in Peninsular India.

M. Jeelani, *Lake Ecology in Kashmir, India*,
DOI 10.1007/978-3-319-40880-4_1

According to the report published by the Zoological Survey of India (Anonymous 1991), about 16,500 species which is more than 20% of the total fauna (78,000 species) in India, are aquatic and majority of these inhabit freshwater. Among the major and better known freshwater groups, there are about 300 Rotifers, 285 Molluscs, 100 Cladocera 100 Ostracods, more than 300 copepoda, 742 fishes, and about 1000 avian species (cf. Michael and Sharma 1988); (Subba Rao 1989). A large proportion of insecta, which constitute more than 40% of the Indian fauna, have their larval stages in aquatic habitats.

Whereas considerable attention is being paid to the decline in the diversity of fishes, avifauna and a few other vertebrates, as well as to their conservation, the importance of diversity of planktonic and benthic organisms and aquatic vegetation is generally overlooked. It is these groups of aquatic biota which are crucial to the sustenance of fisheries and all other animals in the aquatic systems (Gopal 1990).

The planktonic communities due to their small sizes and short generation periods are quick to respond to environmental changes (APHA 1989). They are able to sense a wide variety of disturbances including nutrient loading (Pandit 1980), acidification (Bist 1986) and contamination (Vijver et al. 1997). In response to these disturbances, their standing crop and species composition get greatly affected. Therefore, changes in their abundance, species diversity and community composition can provide important indications of environmental changes and disturbances.

Kashmir Valley situated in the northern part of India within the Himalayan region, is bestowed with a vast array of fresh water bodies. These water bodies exhibit a great diversity primarily owing to variations in their origin, altitudinal situation and the nature of biotic pressures. Water bodies, ranging from 1505m to 4000m above sea level, include Dal lake, Wular lake and Anchar lake while other lakes of valley located in the flood plains of river Jhelum include Naranbagh, Manasbal, Trigam, Nilnag and Alipather.

Increasing population has been responsible for introducing many undesirable changes to aquatic environment. The Dal and the Anchar lakes, over the years, have been subjected to over exploitation for economic purposes. Open water areas of the Dal lake have been converted into floating gardens to enhance agricultural production while lake peripheries have been encroached upon to construct residential houses, hotels and house boats. A large number of illegal houseboats have encroached within the lake area. House hold wastes and sewage from these are drained into the lake indiscriminately at several sites. Agricultural run-off from the surrounding fields, especially the fertilizer and pesticide wastes from the floating gardens find their way into the lake water. The Anchar lake, too, is contaminated by ample amount of sewage coming from the bordering human settlements and agricultural run-off added from the surrounding fields. Above all, S.K. Institute of Medical Sciences hospital situated North-East of lake empties its effluence into the lake.

With this background, it becomes imperative to study the impact of various anthropogenic activities on the nutritive status and biodiversity of these two lakes viz. the Dal and the Anchar lake. This is precisely the main objective of the present study which has been undertaken to investigate physico-chemical and biological parameters of these lakes to understand the magnitude of threat imposed by discharges from urban human settlements to the ecology of these two lakes so that possible conservative measures could be undertaken to restore the natural lake ecology. The present study would prove useful in understanding the conservative planning and management of polluting factors.

Review of Literature

A lot of work has been carried out in the field of limnology throughout the globe and as such a voluminous literature is available on the subject. In view of the objectives of the present research, a critical survey of literature was carried out to gather information on various aspects of limnology of various water bodies. Forel (1901) of Switzerland may be considered as the founder of the modern limnology. A detailed study on physico-chemical and biological characteristics of a lake was carried out by Juday and Birge (1933); Welch (1952) and Edmondson (1959). The most elaborate treatise on all aspects of limnology was given by Hutchinson (1941, 1957 and 1969).

Barnes (1965) carried out his work on ice-water and reported the growth of certain lower organisms. Lund (1951) in his study found that winter diatom populations grew rapidly without adding nutrients if they were artificially exposed to higher irradiances. George et al. (1966) carried out a study on the River Kali and studied the impact of wastes of various industries discharged into this river. Saxena et al. (1966) while working on the river Ganga at Kanpur, depicted water quality of the river water. Munawar (1970a, b, 1974b) carried out limnological studies on freshwater ponds of Hyderabad and recorded the periodicity of various taxonomic groups such as volvocales, chlorococcales, desmids, diatoms and blue-greens and their correlations with fluctuations in the physico-chemical parameters of the ecosystem. Khan and Siddiqui (1974) recorded seasonal fluctuations in gross and net production values in surface waters of Aligarh fish pond. Unni (1985) investigated 24 reservoirs in central India for their physical, chemical and biological features. Bilgrami et al. (1994) carried out studies on different physico-chemical and biological characteristics of the River Ganga. Sharma and Rathore (2000) carried out limnological studies on all the water bodies of Berach river system. The changing nature of Indian rivers through changes in channels, hydrologic function, habitats and species composition has been investigated by Johal et al.(2005).

The valley of Kashmir existing in the greater Himalaya abounds a vast array of freshwater bodies viz., lakes, ponds, wetlands, springs, streams and rivers. These varied freshwater ecosystems which are of great aesthetic, cultural, socio-economic

M. Jeelani, *Lake Ecology in Kashmir, India*,
DOI 10.1007/978-3-319-40880-4_2

and ecological value and play an important role in the conservation of genetic resources of both plants and animals, remained unexplored ecologically until Hutchinson (1941) highlighted the movement of phosphorus from mud to water in small stratified lakes. Later on Kaul and Zutshi (1965) provided an extensive data on the ecology, genesis and development of floating islands while Zutshi (1968) provided data on the morphometry, transparency, thermal structure, water and sediment chemistry and macrophytic communities in four lakes of Kashmir valley. Das and Subla (1970) put forth a theory on the origin and evolution of Ichthyofauna of Kashmir. Das et al. (1969) while investigating the organic production and general ecology of some high altitude lakes of Kashmir reported 42 phytoplankton genera. Kant and Kachroo (1972, 1973, 1974 and 1975) investigated seasonal dynamics of phytoplankton and their distribution. Zutshi and Vass (1978) carried out studies on the morphometry and physico-chemical features of the Dal lake. Khan and Zutshi (1979) explored phytoplankton production of the lower Siwalik lakes and Kashmir Himalayan lake. Kaul and Handoo (1980) surveyed some freshwater bodies of Kashmir and found them as hard water types with alkaline pH and high concentration of divalent cations. Zutshi and Wanganeo (1984) investigated phytoplanktons and their primary production in the Lake Manasbal. Kundanger and Zutshi (1985) while working on two rural lakes, registered 127 species of phytoplankton. Sarwar (1989) worked out the limnological features of Kashmir lakes and found diatoms to be responsible for low silicate content. In another study, Sarwar et al. (1996) investigated the impact of floating gardens on the limnological features of the Dal lake. Pandit (1996) carried out work on various ecological aspects of Kashmir lakes and highlighted the factors leading to their degradation. Rather et al. (2001) carried out a limnological study on Hokarsar wetland.

The effects of physical forces such as light and heat and chemicals discharged into water are of great significance as they are solely responsible for certain phenomena such as thermal stratification, chemical stratification and seasonal qualitative and quantitative variations in the biota. These changes also alter the quality of water. The ever increasing populations in the present era are contributing to a maximum extent in influencing the physico-chemical properties and biological features of most water bodies. Review of each parameter taken in the present study is dealt with separately as follows.

Water Temperature

Love and Goodwin (1959) studied the effect of natural water temperature on the development of certain organisms in Georgia while Bagnoulus and Homji (1959) classified general climatic conditions of Kashmir on the basis of temperature and subsequently this classification was followed by Zutshi et al. (1980), Patra and Nayak (1982) and Yousuf et al. (1986). Water temperature has been observed to follow the general atmospheric temperature. This has widely been observed by Rice (1938); Welch (1952); Rao (1955); Zafar (1955); Macan (1963); Venkateshwarlu and Jayanti (1968); Munawar (1970a); Hannan and Young (1974); Qadri and Yousuf (1978); Swarup and Singh (1979); Sharma et al. (1981); Harshey et al. (1982); Mahadevan and Krishnaswamy (1983); Bagde and Varma (1985a, b); Palharya and Malviya (1988); Shyam Sunder (1988). Munawar (1970a) while carrying out limnological studies on freshwater ponds in Hyderabad elaborated that smaller water bodies reacted quickly to the changes in atmospheric temperature.

Jolly and Chapman (1966) made a preliminary study of effects of pollution on Farmers Creek and Cox's river with respect to temperature. Works on the similar lines have been done by Singh et al. (1982); Bisht (1986); Sarwar (1987); Palharya and Malviya (1988); Shah (1988); Prakash (1990); Adholia et al. (1991); Jindal and Kumar (1993); Bath (1996); Syal (1996); Batcha (1997) and Bath and Kaur (1998). Kumar and Sharma (1999); Katti et al. (2002) and Kaur et al. (2003) found higher temperature with contamination of sewage.

Depth

Depth which influences the physical and chemical properties of water, has been recorded in various lakes of Kashmir by Kaul (1977), and Zutshi et al. (1980). Naumann (1919); Thieneman (1925) and Rawson (1955) described trophic status of some lakes on the basis of depth.

Turbidity and Water Transparency

Turbidity, an expression of optical property of water, is a major determinant of conditions and productivity of an aquatic ecosystem. David (1956); Sengar and Sing (1986), Wanganeo et al. (1996); Pagioro et al. (1997); Kumar and Sharma (1999) and Maruthi et al. (2000) observed levels of turbidity in different aquatic ecosystems. Thorat and Sultana (2000); Das et al. (2003) and Sharma and Verma (2003) while investigating hydrobiological parameters of different water bodies, found increase in turbidity due to suspended solids discharged from sewage and silt.

Khan and Zutshi (1980); Zutshi et al. (1980) and Wanganeo et al. (1996) recorded transparency for Kashmir lakes. David (1956); Senger and Sing (1986); Kumar and Sharma (1999); Thorat and Sultana (2000); Maruthi et al. (2000); Das et al. (2003); Verma (2003) and Esmaeili and Johal (2005) observed remarkable changes in the transparency of water under the impact of various types of pollutants.

pH

Mohanty (1975) while studying physico-chemical features of the outer channel of the Chilka lake, recorded seasonal variations in pH of water. Seasonal variations in pH values have also been observed by Zutshi and Vass (1978) and Khan and Zutshi (1980) for Kashmir lakes and Sharma and Dheneshwar (1986); Mishra (1988) and Bandhopadhyay and Gopal (1991) for other water bodies.

Wanganeo et al. (1996) recorded variations in pH across the water column of Nilnag lake of Kashmir. Singh et al. (1982); Sharma and Dheneshwar (1986); Mishra (1988); Gopal (1990); Jindal and Kumar (1993); Khalique (1995); Bath (1996); Islam and Islam (1996); Syal (1996); Sarwar (1999); Narian and Chauhan (2000); Saha et al. (2000) and Valermathi et al. (2002) recorded change in pH values with addition of sewage and agricultural effluents.

Qadri and Yousuf (1978); Khan (1979); Zutshi et al. (1980); Sarwar (1987 and 1991a, b) investigated the impact of macrophytic vegetation on pH values of water. Sarwar and Majid (1997) while studying abiotic features of Wular lake, attributed alkaline pH to the presence of calcium rich rocks while Sarwar (1999) attributed it to higher silicate content in the rocky beds. Singh (1960); Mohanty (1975); Zutshi and Vass (1978); Khan and Zutshi (1980); Singh et al. (1982); Bisht (1986); Sharma and Dhaneshwar (1986); Mishra (1988); Abel (1989); Prakash (1990); Bondhopadhyay and Gopal (1991), Kundanger and Ahmad (1991); Bath (1996); Jameson and Rana (1996); Wanganeo et al. (1996); Khabade et al. (2003) and Esmaeili and Johal (2005) described the nature of water bodies based on pH values.

Dissolved Oxygen

Dissolved oxygen which is a parameter of primary importance in the aquatic ecosystems by virtue of its role in both chemical as well as biological reaction, has been studied by Matthews 1998 with respect to water fluctuations under various environmental conditions. Effect of temperature on dissolved oxygen has been observed by Hutchinson (1957); Chakraborty et al. (1959); Reid (1961); Jolly and Chapman (1966); Pahwa and Mehrotra(1966); Ray et al. (1966); Venkateshwarlu and Jayanti (1968); De Smet and Evens (1972); Cairns et al. (1975); Vasisht and Sharma (1975); King (1981); Badola and Singh (1981); Woodward (1984); Bhowmick and Singh (1985); Bhatt et al. (1988); Palharya and Malviya (1988); Rana and Palria (1988); Bandopadhyay and Gopal (1991); DattaMunshi and Singh (1991); Goel and Chavan (1991); Jhingran (1991a, b); Shastree et al. (1991); Tripathi et al. (1991); Mathuthu et al. (1993); Anjana and Kanhere (1998); Kumar (1995a, b); Bath (1996); Kaur et al. (1996a); Syal (1996); Bath and Kaur (1998); Jameel (1998); Shivanikar et al. (1999); Valarmathi et al. (2002); Kumar et al. (2003); Kaur et al. (2003) and Prasannakumari et al. (2003) carried out investigation on effects of sewage effluents on dissolved oxygen contents. Sculthrope (1967), Sarwar (1987); Roy (2000); Kaur et al. (2001) and Khatri and Dhankhar (2003) carried out studies on the impact of various anthropogenic activities on dissolved oxygen contents of water.

Pahwa and Mehrotra (1966); Zutshi and Vass (1971); Badola and Singh (1981) and Bhatt et al. (1988) recorded changes in the dissolved oxygen contents in a water body having a surface cover of *Salvinia* sp.

Total Alkalinity

Alkalinity of water is its capacity to neutralize acid and is a measure of productivity as has been suggested by Moyle (1946). Its relationship with pH of water was been investigated by Freiser and Fernando (1966); Qadri and Yousuf (1980a, b) and Zutshi et al. (1980). Kaul and Handoo (1980) and Wanganeo et al. (1996) recorded alkalinity across the water column of lakes. An attempt was made to classify lake water on the basis of alkalinity values by Phillipose (1960); Spence (1964) and Sarwar and Rifat (1991).

Gibbs (1970) suggested that alkalinity is contributed by carbonaceous salts present in rocks. Water (1957); Zutshi et al. (1980); Patra and Nayak (1982); Trivedy and Goel (1986); Sarwar and Zutshi (1987a, b); Weiler (1988); Sarwar and Zutshi (1989); Sarwar (1991a, b); Chapman and Kimstach (1992); Meybeck et al. (1992); Syal (1996); Sarwar et al. (1996); Sarwar (1999); Valarmathi et al. (2002); Kumar et al. (2003) and Prasannakumari et al. (2003) recorded changes in alkalinity values in relation to levels of organic wastes discharged into water.

Total Hardness

Mairs (1966) suggested total hardness to be a complex mixture of cations and anions while Cole (1975) recorded calcium and magnesium to account for most of the hardness.

Zutshi (1968); Vass (1973); Zutshi et al. (1980); Kundanger and Zutshi (1985); Sarwar (1987 and 1991a, b); Sarwar and Wazir (1988); Sarwar and Rifat (1991) and Sarwar et al. (1996) discussed the role of rocks in contributing hardness to water of various Kashmir lakes.

Thomson (1952); Singh et al. (1969); Bhuyan (1970); Upadhyaya and Roy (1982); Badge and Verma (1985a, b); Trivedy and Goel (1986); Sinha (1987); Sinha and Prasad (1988); Gopal (1990); Sinha et al. (1991); Chapman and Kimstach (1992) and Meybeck et al. (1992). Syal (1996) and Kaur et al. (2003) investigated the impact of sewage on the hardness values of water. Kannan (1991) has provided the standards depicting the nature of water on the basis of total hardness values.

Calcium

Pearsall (1930) while studying the phytoplankton in English lakes, recorded seasonal trends in calcium values of water. Similarly Ganapati (1940); Zafar (1964); Hussainy (1965); Boznaik and Kennedy (1968); Mac Crimmon and Kelso (1970); Munawar (1970a, b); Daborn (1976); Zutshi and Vass (1978); Swarup and Singh (1979); Qadri and Yousuf (1980a, b); Bagde and Verma (1985a, b); Sarwar (1987); Sinha and Prasad (1988); Chapman and Kimstach (1992); Saradhamani and Sivakumar (1995) and Syal (1996) also recorded seasonal fluctuations in calcium ions in various lakes and rivers.

Thomson (1952) observed dissolution of calcium in river water from the rocks along its course. Zutshi (1968); Bhuyan (1970); Vass (1973); Zutshi et al. (1980); Upadhyaya and Roy (1982); Kundanger and Zutshi (1985); Trivedy and Goel (1986); Sarwar (1987, 1991), Sarwar and Wazir (1988); Sarwar and Rifat (1991); Chapman and Kimstach (1992); Meybeck et al. (1992) and Sarwar et al. (1996) while investigating the levels of calcium in water also found rocks to be the major source of calcium in water..

Calcium levels in water have been found to change under the impact of sewage and agricultural wastes. Studies in this respect have been done by Singh et al. (1969), Badge and Verma (1985a, b); Trivedy and Goel (1986); Sinha (1987); Sinha and Prasad (1988); Gopal (1990); Sinha et al. (1991); Syal (1996); Sarwar (1999) and Kaur et al. (2003).

Magnesium

Rawson (1939a, b) described the productivity of Okanagan lake on the basis of magnesium concentration. Bhuyan(1970) while studying physico-chemical qualities of the water of some ancient tanks in Assam suggested rocks to be the major source of magnesium in water. Zutshi et al. (1980) also suggested rocks to be the major source of magnesium in Kashmir lakes.

Zafar (1964); Singh et al. (1969); Munawar (1970a, b); Zutshi and Vass (1978); Zutshi et al. (1980); Bagde and Varma (1985a, b); Trivedy and Goel (1986); Sarwar and Rifat (1991); Sinha et al. (1991) and Saradhamani and Sivakumar (1995) investigated the seasonal trends in magnesium values.

Contribution of magnesium by sewage is reported by Singh et al. (1969) while studying the Algal flora of sewage in relation to physico-chemical variables. Munawar (1970a, b); Bagde and Verma (1985a, b); Sinha et al. (1991); Syal (1996); Sarwar (1999) and Kaur et al. (2003) also recorded changes in magnesium levels in water with addition of sewage discharges.

Iron

Wanganeo et al. (1996) in their studies on limnological studies on a dimictic Himalayan lake, recorded iron concentration in surface water and bottom waters. Sivakumar et al. (2000) also recorded the iron concentration across the water column. Hasan and Pandey (1985); Wanganeo et al. (1996); Suvarna and Somashekar (2000) and Sajeev (1999) recorded seasonal trends in iron concentration.

Sarwar and Wazir (1988); Sarwar and Zutshi (1989); and Sarwar (1991a, b); Sarwar (1999) recorded iron concentration in lakes adjoining the vegetable fields while Trivedy and Goel (1986); Israili (1991), Sajeev (1999) studied the iron concentration in water bodies receiving sewage.

Chloride

Chlorides in water are generally due to the salts of sodium, potassium and calcium as has been described by Kaushik and Saksena (1999). Seasonal variations in the chloride values of water have been observed by Ownbey and Kee (1967); Zutshi and Vass (1978); Ajmal et al. (1985) and Shastree et al. (1991).

Thresh et al. (1944); Gonzalves and Joshi (1946); Blum (1957); Sawyer (1960); Zafar (1964); Lakshminarayan (1965); Singh (1965); Venketeshwarlu (1969); Munawar (1970a, b); Cole (1975); Goel et al. (1980); Mehra (1986); Trivedy and Goel (1986); Das (1989); Sarwar and Zutshi (1989); Shastree et al. (1991); Bath (1996); Sarwar (1999); Das et al. (2003) and Prasannakumari et al. (2003) recorded the impact of sewage contamination on chloride contents. James et al. (1990) made assessment of chloride concentration in water of Uyyakondan channel, Tamilnadu after it receives hospital waste.

ISI (1986) have prescribed maximum permissible limits of chloride in water.

Ammonical Nitrogen

Bruce (1958) suggested the values of ammonical nitrogen as a sign of organic pollution. Bandela et al. (1999) recorded the ammonical nitrogen across the water column in a lake. Harold (1934); Zutshi and Vass (1978); Ramakrishnaiah and Sarkar (1986); Adoni and Joshi (1987) and Bandopadhyay and Gopal (1991) recorded seasonal variations in ammonical nitrogen. Thresh et al. (1944); Seenayya (1971); King (1981); Das (1989); James et al. (1990); Sarwar (1991a, b); Bhatt et al. (1999) and Narian and Chauhan (2000) studied the contribution of household effluents and residues of detergents on ammonical-nitrogen. Ammonical nitrogen likely to be due to microbial decomposition of macrophytes and excretory products of aquatic biota was suggested by Wetzel (1975) and Bhatt et al. (1999). Trivedy and Goel (1986) suggested that high concentration of ammonical nitrogen in a water body indicated discharge of fresh faecal matter into it. Hospital wastes also add ammonical nitrogen to water as suggested by James et al. (1990). ISI (1986) has recommended permissible limits of ammonical nitrogen in water.

Nitrate Nitrogen

Nitrogen is one of the major constituent of cellular protoplasm of photosynthetic organisms. Occurrence of nitrate form of nitrogen in various water bodies has been reported by Thrash et al. (1944); Sylvester (1961); Jolly and Chapman (1966); Willen and Evens (1972); King (1981); Ramakrishnaiah and Sarkar (1986); Shah (1988); Shyamsunder (1988) and Shastree et al. (1991). Seasonal fluctuations in nitrate nitrogen concentration were recorded by Munawar (1970a); Zutshi and Vass (1978); Shah (1988); Kumar (1996a, b); Bath (1996); Bath and Kaur (1997); Sarwar and Majid (1997) and Sajeev (1999).

Ellis et al. (1946) and Jhingran (1991b) opined that natural waters get nitrate nitrogen through weathering of rocks. Harold (1934); Harvey (1940); Vashisht and Sharma (1975); Ajmal et al. (1985); Trivedy and Goel (1986); Adoni and Joshi (1987); Das (1989); Bandopadhyay and Gopal (1991) and Bath (1996) recorded nitrate concentration in water bodies rich in macrophytic vegetation and phytoplanktons. Zutshi et al. (1980) compared nitrate nitrogen values in nine lakes of Kashmir during limnological studies of these lakes. Sarwar and Zutshi (1989) and Sarwar (1991a, b) recorded nitrate nitrogen in water harbouring rich growth of macrophytes.

The impact of sewage on nitrate nitrogen was observed by Zutshi and Vass (1971); Seenayya (1971); King (1981); Ramakrishnaiah and Sarkar (1986); Trivedy and Goel (1986); Das (1989); Sarwar (1991a, b); Bath (1996) and Savarna Somashekar (2000) and Kaur et al. (2002).

James et al. (1990) recorded nitrate nitrogen concentration in Uyyakondan channel, Tamilnadu where water received hospital waste. Alikunhi (1957) and Zutshi et al. (1980) described productivity and nutritive status of water on the basis of nitrate nitrogen values.

Total Phosphate Phosphorus

Phosphorus is one of the major nutrients responsible for biological productivity as has been demonstrated by Hutchinson (1957). Michael (1969); Kaul (1977); Zutshi and Vass (1978); Harshey et al. (1982); Bath (1996); Bhatt et al. (1999) and Kumar et al. (2003) recorded seasonal variations in phosphate phosphorus values. Wanganeo et al. (1996) recorded concentration phosphate phosphorus across the water column in a limnological study on a dimictic Himalayan lake.

Welch (1952); Ruttner (1953); Zutshi and Vass (1978); Trivedy and Goel (1986); Mishra (1988); Sarwar and Zutshi (1989); Das (1989); Kannan (1991); Sarwar (1991a, b); Bath (1996); Bath and Kaur (1998); Khabade et al. (2003) and Prasannakumari et al. (2003) studied the impact of domestic sewage on phosphate phosphorus concentration of water. Heron (1961); Singh and Roy (1990); Bandopadhyay and Gopal (1991); Sarwar (1991a, b); Thomas and Azis (1996); Bhatt et al. (1999) and Khabade et al. (2003) observed the impact of agricultural effluent on the concentration of phosphorus in various water bodies.

Sawyer (1947), US department of interior division of technical support (1969) and Vollenweider (1972) have prescribed maximum permissible limits of phosphate phosphorus in water.

Silicate

Pearsall (1932); Welch (1952); Stockner and Benson (1971) and Carter (1856b) reported the importance of silicon element in the development of diatom frustules.

Sarwar (1991a, b, 1999) described silicate to be an essential nutrient for phytoplankton. Jayangoudar (1980) and Wanganeo et al. (1996) recorded silicate concentration in surface and bottom waters.

Seasonal variations in silicate values of Kashmir lakes were recorded by Sarwar (1991a, b, 1999) while Kumar et al. (2003) recorded seasonal trends in tropical water bodies of Bihar.

Impact of sewage on silicate contents of water was investigated by Trivedy and Goel (1986); Sarwar and Zutshi (1989) and Sarwar and Majid (1997).

Total Dissolved Solids

Contribution by inorganic substance, organic matter, suspended particles silt, clay and plankton towards total dissolved solids in water was studied by Ganapati (1956). Trivedy and Goel (1986) suggested that total dissolved solids do not give a clear picture of the kind of pollution.

Conductivity

Conductivity, a property of water caused by the presence of various ions, was seen to be increased in water receiving abundance of nutrients by Trivedy and Goel (1986), Sarwar and Zutshi (1989); Sarwar et al. (1996); Bhatt et al. (1999); Sajeev (1999); Sarwar (1999); Sabu Thomas et al. (2001); Ara et al. (2003); Das et al. (2003); Khatri and Dhankhar (2003) and Prasannakumari et al. (2003). Mishra and

Saksena (1993); Jameel (1998) and Bhatt et al. (1999) suggested conductivity to be a rapid measure of total salt concentration. Wanganeo et al. (1996) recorded conductivity values of water at different depth. Oslen (1950); Wilcox (1955); Dunn (1962) and Trivedy and Goel (1986) classified the nutritive status of water on the basis of conductivity values.

Phytoplankton

The important role played by phytoplankton in transformation of energy from one trophic level to the next level was discussed by Sewell (1934). The studies related to distribution of seasonal variations and relative dominance of different phytoplankton groups in the Kashmir valley lakes were carried out by Kant and Kachroo (1974, 1977); Mir and Kachroo (1982); Zutshi and Vass (1982); Sarwar (1985); Sarwar and Zutshi (1988); Mir (1995); Sarwar et al. (1996) and Yousuf et al. (2002). Studies on similar lines were carried out in other water bodies by Gonzalves and Joshi (1946); Rao (1955); Shetty et al. (1961); Skuja (1964); Lund (1965); Zafar (1967); Richardson (1968); Venkateshwarlu (1969); Munawar (1970a, b); Crayton and Sommerfeld (1978); Vass et al. (1978); Manikya (1984); Bhowmick and Singh (1985); Venkateshwarlu and Reddy (1985); Sabater and Isabel (1990); Singh and Srivastava (1991); Gujarathi and Kanhere (1998); Varma (1998); Mallick and Padhi (1999); Bhatt et al. (1999); Biswas and Konar (2000); Chakraborty and Konar (2000); Katti et al. (2002); Nandan and Aher (2002); Pandey et al. (2002); Sedamkar and Angadi (2003) and Khanna and Bhutiani (2003).

Lowe (1974) labeled some eutrophic species as bioindicators during a study of diatom population dynamics in a central lowa drainage ditch. Similar studies have been done by Kolkwitz and Marsson (1908); Shetty et al. (1961); Singh (1965); David and Ray (1966); George et al. (1966); Jolly and Chapman (1966); Bulusu et al. (1967); Richardson (1968); Singh et al. (1969); Verma and Shukla (1968, 1969, 1970); Williams (1969); Munawar (1970a); Stockner and Benson (1971); Dickman (1975); Weilgolaski (1975); Conner (1977); Siver (1977); Verma et al. (1978); Cassie (1979); Govindan and Sundarsan (1979); Blinn et al. (1980); David et al. (1981); Gunale and Balakrishan (1981); Paramasivam and Sreenivasan (1981); Valecha and Bhatnagar (1988); Das (1989); Venkateshwarlu et al. (1990); Reddy et al. (1991); Ahmad (1996); Syal (1996); Gujarathi and Kanhere (1998); Varma (1998); Bhatt et al. (1999); Suvarna and Somashekar (2000); Chakraborty and Konar (2000) and Sedamkar and Angadi (2003).

Sarwar (1991a, b); Zutshi et al. (1980); Rishi and Kachroo (1981); Kawoosa (1985) and Khan and Bhat (2000) also identified some indicator species of phytoplankton of Kashmir lakes.

The abundance of phytoplankton in relation to the nutrients present in water was studied by Hentschel (1928). Gunther (1936) and Pearsall and Lind (1942) investigated the distribution of phytoplankton in nutrient rich Peru coastal current and some North-West Irish Loughs respectively.

Jarnefelt (1952); Shetty et al. (1961); Zafar (1964); David and Ray (1966); George et al. (1966); Bulusu et al. (1967); Richardson (1968); Singh et al. (1969); Verma and Shukla (1968, 1969, 1970); Munawar (1970a, b); Stockner and Benson (1971); Dickman (1975); Weilgolaski (1975); Conner (1977); Kumar (1997a); Siver (1977); Verma et al. (1978); Cassie (1979); Blinn et al. (1980); David et al. (1981); Gunale and Balakrishan (1981); Paramasivam and Sreenivasan (1981); Kar et al. (1987); Das (1989); Venkateshwarlu et al. (1990a); Ahmad (1996); Singh (1996); Syal (1996); Khan et al. (1998); Gujarathi and Kanhere (1998); Varma (1998); Bhatt et al. (1999); Biswas and Konar (2000); Chakraborty and Konar (2000); Katti et al. (2001) and Sedamkar and Angadi (2003) also studied the distribution and relative abundance of phytoplankton, in water receiving organic wastes.

Carter (1856b) and Siver (1977) studied effects of sewage effluence on distribution of phytoplankton. Bulusu et al. (1967) made observations on self purification of Khan river and its effects on Kshipra river with respect to phytoplankton. Stockner and Benson (1967); Vollenweider (1972) and Khan and Bhat (2000) carried out studies on distribution of phytoplankton in lakes harbouring aquatic weeds. Tilzer (1973) studied the diurnal periodicity of the phytoplankton assemblage in a mountain lake. Sarwar (1991a, b) recorded phytoplankton population of marshy lands around Srinagar. Syal (1996) studied the distribution pattern of phytoplankton in Satluj river under the impact of fertilizer wastes, sewage and several other industrial wastes. Vareethiah and Haniffa (1998) described phytoplankton community organization and species succession in Thengappattanam estuary in the south west coast of India making the record on contribution of Dinophyceae group towards phytoplankton.

Distribution of phytoplankton in relation to levels of nitrogen and phosphorus was described by Rodhe (1948) and MC Murry and Olive (1975) while Hedge and Bharati (1984) described a relationship between pH and the growth of phytoplankton.

Macrophytes

Hooker (1875) was the first to list aquatic plants of Kashmir. Arber (1920) presented a monograph of water plants. Mukherjee (1921, 1925, 1926, 1932) published reports on the aquatic and marshland vegetation of Dal lake Kashmir. Raunkier (1934) provided the life form classification of hydrophytes. Penfound (1952) made distinction between wetland plants and aquatic plants and divided them into emergents, submerged and floating type. Spence (1964) predicted that species composition of macrophytes is governed by flooding and water level fluctuations and not by predation. Rajmankova (1975); Hejney and Husak (1978) and Pandit (1984) documented some hydrophytes as pollution indicator during studies of macrophytes. Zutshi et al. (1980), Kaul and Handoo (1980) and Pandit (2002b) made extensive studies on various forms of aquatic plants in Kashmir lakes and found profuse growth of emergents to be indicative of productive nature of the lake water. Kaul (1970) correlated some weed complexes with eutrophication while Goulder and Boatman (1971) correlated some submergeds with nutrient enrichment of water. Westlake (1965); Kaul et al. (1978) and Pandit (2002b) used importance value index to describe productive status of the lake water.

Zooplankton

While studying a number of Swedish lakes of different trophic levels, Pejler (1957) made observations on zooplanktons inhabiting mesotrophic to eutrophic waters. Study of zooplankton fauna of various water bodies was done by Alikunhi (1957); George et al. (1966); Michael (1968); Allan (1976); Singh and Sahai (1978); Winner et al. (1979); Ferguson (1982); Singh et al. (1982); Carmonze et al. (1983); Serruya and Pollingher (1983); Hecky (1984); Sharma and Patnaik (1985); Drenner et al. (1987); Payne (1987); Sharma and Michael (1987); Mengaston and Fernando (1991); Shukla (1995); Kumar (1995a, b); Kaur et al. (1996a, b and 1999); Syal (1996); Unni and Naik (1997), and Biswas and Konar (2000) who observed dominance of zooplankton in different aquatic systems. Studies related to abundance and diversity of zooplanktons in Kashmir lakes were carried out by Khan and Zutshi (1979); Zutshi et al. (1980); Yousuf and Qadri (1985); Yousuf et al. (1986); Yousuf (1988) and Sarwar and Parveen (1996).

Seasonal dynamics of zooplankton in different waterbodies were discussed in the studies of Dutta et al. (1954); George (1966); Battish (1968); Patalas (1972); Patnaik (1973); Gabriel et al. (1975); Nassar (1975); Vasisht and Sharma (1975); O' Brein (1979); Moore (1980); Vasisht and Monga (1981); Kohli et al. (1982), Ramakrishnaiah and Sarkar (1982); Nogrady (1983); Orcutt and Pace (1984); Sukkumaran et al. (1984); Agarwal and Gautum (1985); Saha and Pandit (1988); Shukla (1995); Suguman (1995); Bath (1996); Syal (1996); Kaur et al. (1996a, b and 1997); Thomas and Azis (1996); Sajeev (1999), Biswas and Konar (2000); Sharma et al. (2000); Baruah and Das (2001) and Pulle and Khan (2003).

Yousuf and Qadri (1985) conducted studies on lake Manasbal of Kashmir and discussed seasonal trends in distribution pattern of zooplankton. Similar studies were done by Sarwar and Parveen (1996) on Khusalsar and Gilsar lakes of Kashmir and by Zuthsi et al. (1980) on nine lakes of Kashmir.

Zooplanktons were found to occur in a variety of aquatic habitats by Zafar (1959a); Phillipose (1960); Venkateshwarlu (1969) and Munawar (1970b) and Syal (1996). Vasisht and Sra (1979); Balkhi and Yousuf (1991); Pandit and Kaul (1981); Saxsena and Mishra (1990); Kaur et al. (1996a, b and 2003); Bath and Kaur (1997) and Sajeev (1999) recorded abundance of zooplanktons in waters harbouring rich vegetation.

Dexter (1959); Holden and Green (1960); Macan (1963); Brooks and Dodson (1965); Jolly and Chapman (1966); Krishnamoorthi and Visvesvara (1966); Pahwa and Mehrotra (1966); Ray et al. (1966); Gideiri (1969); Hynes (1970); Kochsiek et al. (1971); Hall et al. (1976); Threlkeld (1979); O'Brien (1979); Tonapi (1980);

Scullion et al. (1982); Singh (1982); Haberman (1983); Spencer et al. (1983); Orcutt and Pace (1984); Petts and Greenwood (1985); Saha and Pandit (1988); Bath (1996); Kaur et al. (1996a, b); Kumar and Singh (1996); Kumar (1997); Bath and Kaur (1997); Sajeev (1999) and Kaur et al. (2003) recorded and discussed the zooplankton fauna of water bodies receiving rich organic wastes.

During the study of zooplankton in various water bodies, Berzins (1949); Pejlar (1957); Arora (1966); David and Ray (1966); George et al. (1966); Jolly and Chapman (1966); Krishnamoorthi and Visvesvara (1966); Algarsamy et al. (1967); Venkateswarlu and Jayanti (1968); Verma and Shukla (1968); De Smet and Evens (1972); Klimowicz (1973); Verma and Dalela (1975); Vasisht and Sharma (1975); Radwan (1976); Hakkari (1978); Das and Pandey (1978) and Bilgrami and Datta Munshi (1979) designated some zooplanktons to be indicators of clean and eutrophicated waters. Jyoti and Sehgal (1979) while working on the ecology of Surinsar lake in Jammu and Kashmir with reference to zooplankton, also labeled some bioindicator species in this region. Some bioindicator species of zooplanktons were also identified by Khare et al. (1979); Vasisht and Sra (1979); Jayangouder (1980), Vasisht and Kapoor (1981); Maemets (1983); Sladecek (1983); Spencer et al. (1983); Suduzuki et al. (1983); Hutchinson (1941); Sukumaran et al. (1984); Orcutt and Pace (1984); Ponji et al. (1985); Sinha (1987); Sinha and Prasad (1988); Das (1989); Sharma et al. (1989); Goel and Chavan (1991); Jindal and Ghezta (1991); Bath (1996); Kaur et al. (1996a, b and 2003); Syal (1996) and Sajeev (1999). Zutshi et al. (1980); Yousuf and Qadri (1985) and Balkhi and Yousuf (1991) described certain zooplanktons characteristic of hard water in Kashmir lakes.

Zoobenthos

Odum (1971) described common inhabitants of sewage water with particular reference to oligochaetes. Learner et al. (1971) examined benthos assemblage above and below a point source of sewage and found upstream to be quite diverse while downstream to be left with only chironomids and oligochaetes. Dance and Hynes (1980) and Ajao and Fagado (1990) observed the distribution of macrozoobenthos in waters receiving complex mixtures of domestic waste.

Seasonal dynamics of macrozoobenthic organisms in different waterbodies were discussed by Munawar (1970a, b); Mandal and Moitra (1975); Das (1979); Rai and Datta Munshi (1979); Chowdhary (1984); Sharma (1986); Kaushik et al. (1991); Dhillon et al. (1993a); Syal (1996); Singh (1982) and Yousuf et al. (2002).

Rich vegetation provides food and shelter to the growing zoobenthos was suggested by Needham and Llyod (1916); Krecker (1939); Andrews and Hasler (1943); Tonapi (1980); Sharma (1988); Kaushik et al. (1991); Kumar (1996b); Bath and Kaur (1998) and Sajeev (1999).

Cordery (1976) studied the impact of stress imposed by motor boats on the aquatic insects. The impact of silt on the aquatic organisms was reported by Kaul et al. (1978) and Pennak (1978). In contaminated sections of the water body, chironomids and tubificid worms were the only taxa to survive (Winner et al. 1980). Chironomid larvae were found to be the most common component of benthos while toxicity of pesticides to benthic insect communities was discussed by William and Feltmate (1992). The stress of various environmental pollutants on the aquatic organisms was discussed by Kumar (1996a, b).

The distribution of macrozoobenthos designated as an indicator of clean and eutrophicated water was described by Gaufin and Tarzwell (1956); Curry (1962); Schneider (1962); Shrivastava (1962); Hussainy (1965); Verma and Shukla (1969); Serruya (1978); Pennak (1978); Allanson (1979); Das and Bisht (1979); Pandit (1980); King (1981); Roy and Sharma (1983); Sharma (1986); Sinha and Prasad (1988); Das (1989); Rao et al. (1991); Dhillon et al. (1993a) and Kumar (1996b).

Correlation

The understanding of abiotic and biotic factors in freshwaters and their correlations were studied by Pearsall (1932); Gonzalves and Joshi (1946); Welch (1952); Hutchinson (1957); Oswald (1960); Reid (1961); Banerjee and Roychowdhuri (1966); Pahwa and Mehrotra (1966); Hussainy (1967); Munawar (1970a, b); Agrawal et al. (1976); Zutshi and Vass (1978); Bagde and Verma (1985b); Patil et al. (1985); Yadava et al. (1987); Shastree et al. (1991); Bath (1996); Kaur et al. (1997) and Adak and Purohit (2001).

Arora (1966); Yadava et al. (1987); Abbasi et al. (1996); Aravinda et al. (1998); Bhavani et al. (2000); Dahiya et al. (2000) and Sedamkar and Angadi (2003) studied the inter-relationship between physico-chemical parameters.

The effect of organic enrichment on the physico-chemical parameters, biota and their correlations were studied by Macan (1963); Michael (1964); Hussainy (1967); Marshall and Orr (1972); Khalof and McDonald (1975); King (1981); Patra and Nayak (1982); Saha and Chowdhary (1985); Sharma and Shrivastava (1986); Gupta and Sexena (1997); Jain and Sharma (1997); Gujarathi and Kanhere (1998); Pillai et al. (1999); Ansari and Prakash (2000) and Mariappan and Vasudevan (2002).

The impact of sewage and industrial effluents on the physico-chemical and biological parameters and the inter-relations were studied by Elliot (1976); Harshey et al. (1982); Balkhi et al. (1984); Dudani et al. (1986); Chauhan (1988); Datta Munshi and Singh (1991); Bath and Kaur (1997); Dhembara and Pondhe (1997) and Garg et al. (1998).

Material and Methods

Selection of Sites

Dal Lake The lake is situated at an altitude of 1584 m a.s.l. in the northeast of Srinagar at a mean latitude of 34.0° 7′ North and longitude of 74.0° 52′ East (Fig. 1). The lake is surrounded on the east by Mahadev mountain range and on the south by kohi-Sulaiman. On its western and northern banks is the Srinagar city. Telbal Nallah bringing water from high altitude mountains, feeds the lake on its north. Towards the south-western side, an outflow channel of the Dal lake called Psunt kol discharges the lake water into river Jhelum. The lake is not a continuous sheet of water but is divided into five distinct but interconnected basins, viz, the Hazratbal, the Bod Dal, the Gagribal, the Branambal and the Nagin basin.

At the turn of the twentieth century, its water spread was 25 km^2 (Drew 1875; Stein 1896; Ennex 1978). At present it is bedeviled by human incursions both in the periphery as well as within the lake itself. The encroachments like conversion of the open water areas into floating vegetable gardens and land masses, construction of residential houses, hotels and houseboats have shrunk the lake from 25 km^2 to 11.3 km^2 (LAWDA 2004). These human activities have altered the immediate environment of the lake which is now fed with continous flow of sewage from hotels and houseboats and agricultural run-off from floating gardens. Keeping in consideration these environmental features, four collection sites were selected from this lake which are as follows.

M. Jeelani, *Lake Ecology in Kashmir, India*,
DOI 10.1007/978-3-319-40880-4_3

Emergents 52.20%
Rooted floating type 9.20%
Submergeds 38.60%

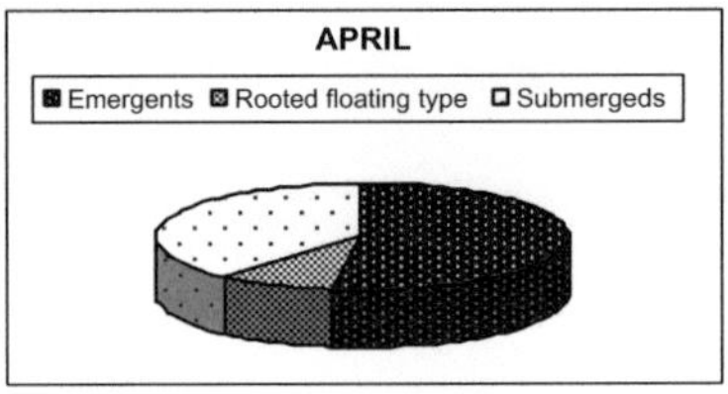

Emergents 51.50%
Rooted floating type 16.40%
Submergeds 32.10%

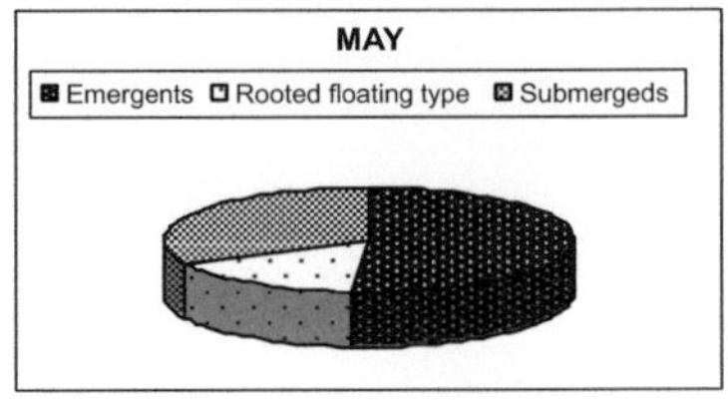

Emergents 49.80%
Rooted floating type 23.90%
Free floating 1.70%
Submergeds 24.60%

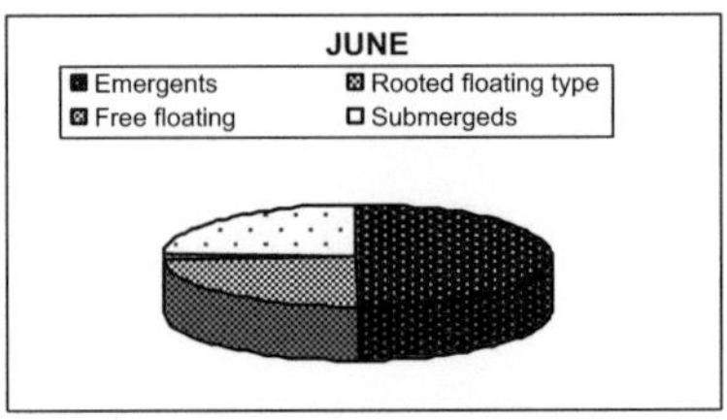

Emergents 51.00%
Rooted floating type 26.80%
Free floating 6.40%
Submergeds 15.80%

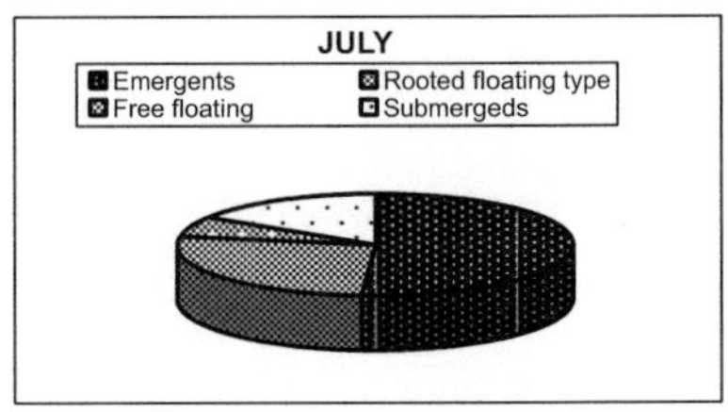

Emergents 49.30%
Rooted floating type 29.00%
Free floating 8.80%
Submergeds 12.90%

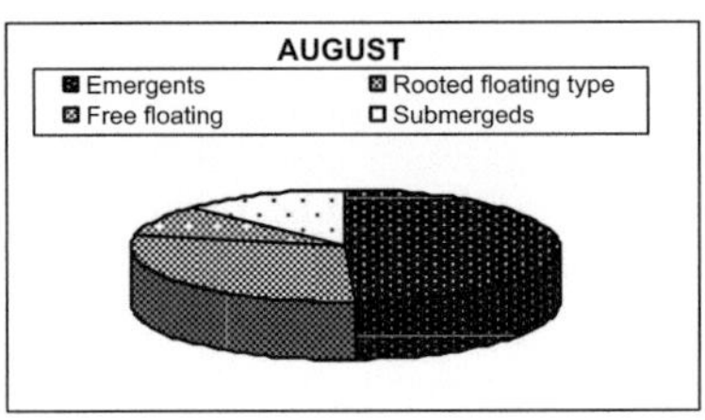

Emergents 38.20%
Rooted floating type 30.30%
Free floating 16.00%
Submergeds 15.50%

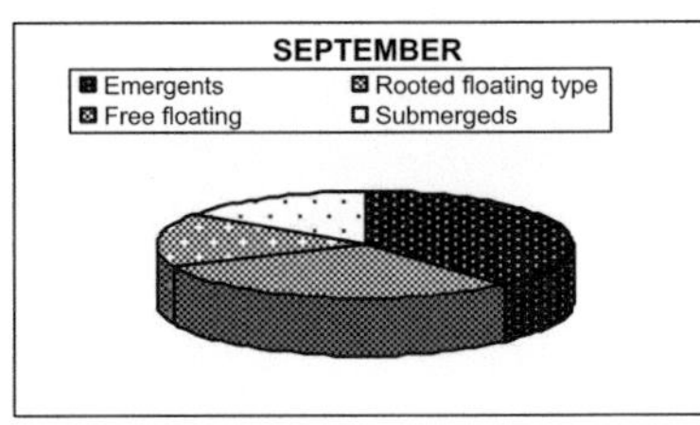

Emergents 41.00%
Rooted floating type 33.50%
Free floating 12.20%
Submergeds 13.30%

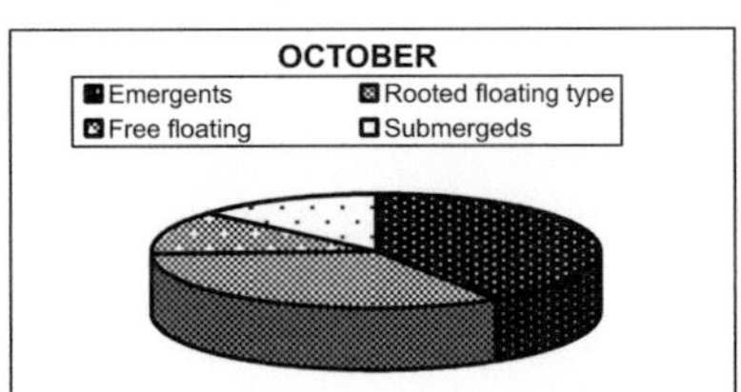

Fig. 1 Monthly percentage contribution of various life form classes in Dal lake

Emergents	63.50%
Rooted floating type	19.00%
Submergeds	17.50%

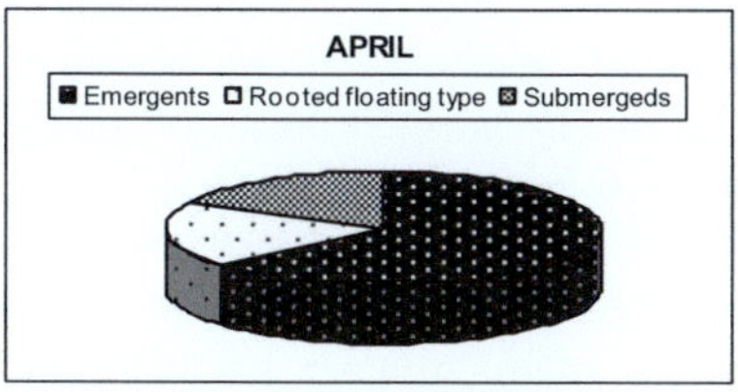

Emergents	65.20%
Rooted floating type	18.60%
Submergeds	16.20%

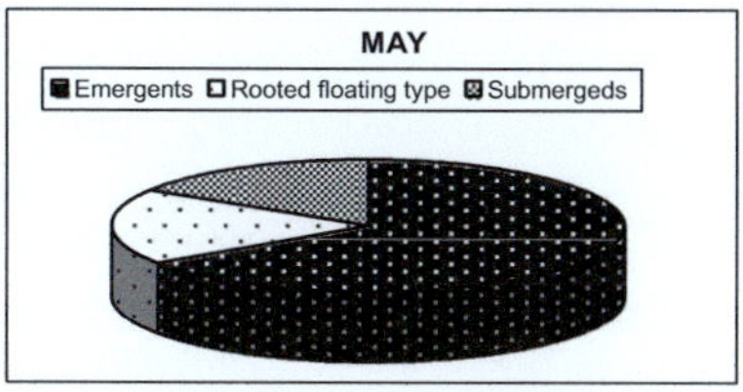

Emergents	62.30%
Rooted floating type	15.70%
Free floating	8.80%
Submergeds	13.20%

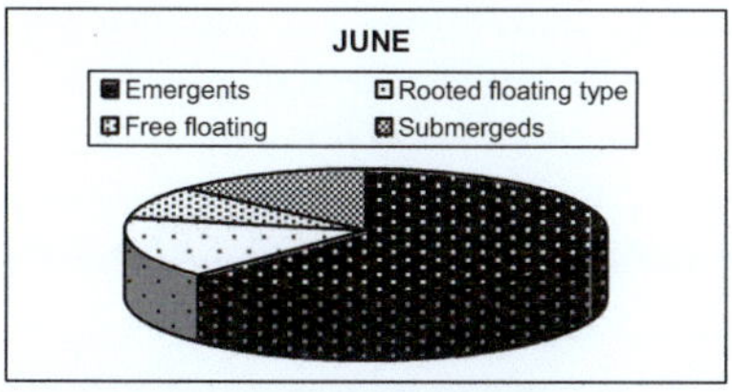

Emergents	55.00%
Rooted floating type	15.70%
Free floating	14.50%
Submergeds	14.80%

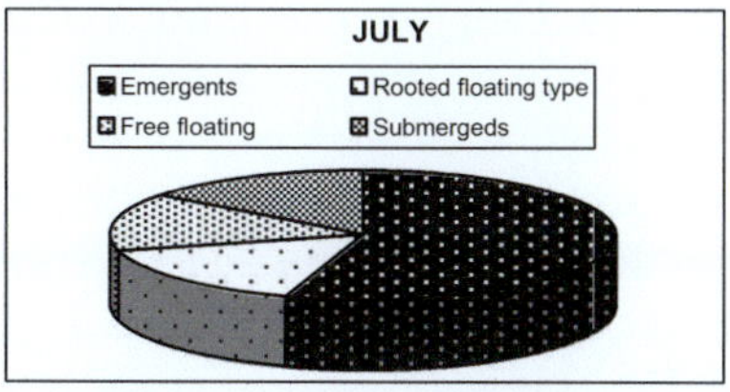

Emergents	59.50%
Rooted floating type	14.90%
Free floating	14.60%
Submergeds	11.20%

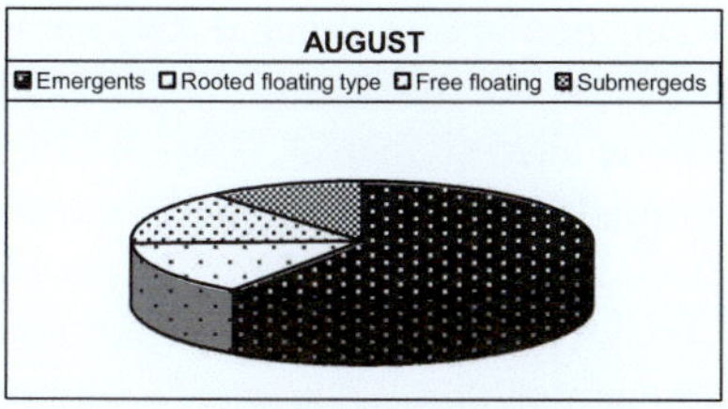

Emergents	63.40%
Rooted floating type	14.40%
Free floating	8.80%
Submergeds	13.40%

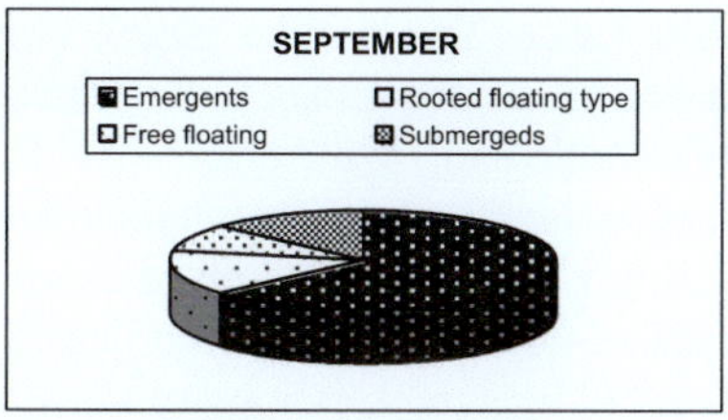

Emergents	68.50%
Rooted floating type	15.10%
Free floating	4.10%
Submergeds	12.50%

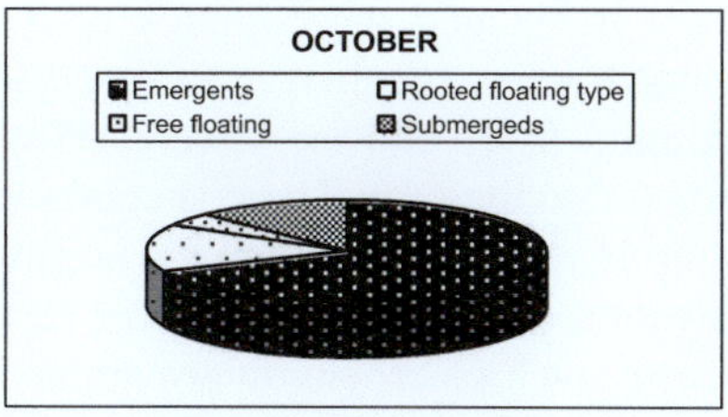

Fig. 2 Monthly percentage contributions of various life form classes in the Anchar lake

SITE-I: This sampling site is located in Lokut Dal basin which is demarcated by a tongue shaped projection from the Gagribal and a bund from the Bod Dal. This site is located in the open water area of the lake. Crystal clear water of the lake at this site is mainly dominated by *Potamogeton lucens*, *P.natans, Myriophyllum spicatum* and *Ceratophyllum demersum.* This site as such does not have direct source of pollution and has been taken as the reference site (Plate-I; 1 & 2).

SITE-II: This sampling site is located where the lake water is encroached upon by floating vegetable gardens (locally called Radh). The construction of floating gardens at this site has produced stretches of stagnant water and has promoted the invasion of obnoxious plant growth of *Lemna* sp. and *Salvinia* sp. Increased agricultural activity on the floating gardens is continuously adding soil particles into water at this site (Plate-II; 3 & 4).

SITE-III: This site is located in the area between Nehru park and the outlet of the lake near Drugjan. This portion of the lake is mainly used for anchoring house-boats, shikaras and hotels, and is subjected to intense human pressure due to tourist influx. On the banks of the area are located a large number of hotels with high bed capacity. This part of the lake shows the predominance of aquatic weeds like *Myriophyllum spicatum, Hydrilla verticellata, Potamogeton crispus, P. natans* and *Nymphoides peltatum* (Plate-III; 5 & 6).

SITE-IV: This site is located in the area of the lake where a large drain which is the recipient of several house hold wastes, hotel and commercial effluents, discharges its contents directly into the lake. Besides, a number of other sewerage pipes carrying wastes from hotels also directly open into the lake (Plate-IV; 7 & 8).

Anchar Lake The Anchar lake is shallow basined and is situated 12 kms to the northwest of Srinagar city within the geographical coordinates of 34° 20′ and 34° 26′ N latitude and 74°82′ and 74°85′ E longitude at 1584 m a.s.l. (Fig. 2). The lake is monobasined with its main catchment comprising Srinagar city and a number of bordering villages. A network of channels from the river Sind enters the lake on its western shore and serves as the main source of water. The littorals of the lake are surrounded by a thick canopy of trees, willows and poplar trees, providing the base material for the manufacture of baskets, cricket bats, wood carvings, wicker work, etc. According to Lawrence (1895), the area of the Anchar lake during 1893–1894 was 19.54 km^2 and since then there has been a considerable decrease in the surface area of the lake. Presently the area of the lake is 6.5 km^2 (LAWDA 2004). Towards the north east of the lake is situated the complex of S.K. Institute of Medical Sciences (SKIMS) hospital draining much of its effluents into the lake. The run-off from surrounding agricultural fields and the sewerage from the bordering human settlements are also drained into the lake. There is continuous and unabated silting due to silt loaded inflow from the river Sind. On the northwest side of the lake, the catchment is under cultivation and is used mainly for raising multiple crops. The lake is heavily infested with thick macrophytic growth and the littorals of the lake are dominated by *Phragmites* sp., *Typha* sp., *Sparganium* sp. and *Nelumbium nucifera* (Lotus).

Plate I Site-I: open water area of the Dal lake (*above*) and the periphery of the Dal Lake (*below*)

Plate II Site-II: floating gardens area of the Dal lake showing agricultural activities

Three sampling sites were selected from the Anchar lake established at different ecological regions under different environmental conditions.

Plate III Site-III: area of the Dal lake showing houseboats. Rich growth of Nymphoides

Plate IV Site-IV: area of the Dal lake showing hotels

SITE-I: The river Sind opens into the Anchar lake on its western side via a network of channels. This site lies at the entrance of the feeding channel from the Sind river into the Anchar lake. The site is characterized by flowing water. This perennial water source brings in lot of sediments along with. The site is devoid of any macrophytic vegetation (Plate-V; 9).

Plate V Site-I: Anchar lake showing silt deposits being carried from the river sind
Site-II: central site of the lake showing dense mats of free floating type of macrophytes

SITE-II: It is the deepest portion of the lake with standing water colonized by thick strands of macrophytes. Here water from site-I and site-III mixes up and its flow is mild (Plate-V; 10).

SITE-III: It represents the portion of the lake lying in front of the SK Institute of Medical Sciences (SKIMS) hospital. A large drain from the SKIMS directly discharges its effluents into the lake. This site is characterised by dark coloured water with foul odour. Solid waste dumps are seen scattered at the littorals. Dense growth of macrophytes is seen at this site (Plate-VI; 11 & 12).

11

12

Plate VI Site-III: Anchar lake showing SKIMS hospital at the bank (*above*)
Site-III: showing opening of the SKIMS drainage pipe into the Anchar lake (*below*)

Sample Collection

During the present study, water samples were collected from four sites of the Dal lake and three sites of the Anchar lake for a period of 2 years extending from September, 2000 to August, 2002. For physico-chemical analysis of water, sampling was done once in a month and samples were collected from both surface and bottom of the Dal lake while only surface samples were taken from the Anchar lake. For the study of phytoplankton, zooplankton and macrophytes also, sampling was carried out on monthly basis. The phytoplankton study was carried out for the Dal lake only. For Zoobenthic fauna, however, the sampling was done after every 3 months.

Physico-Chemical Analysis

For the physico-chemical analysis, standard methods as suggested by Welch (1948); Murphy and Riley (1962); Mackereth (1963); Golterman and Clymo (1969); Trivedy and Goel (1986) and APHA (1989) were followed. The physico-chemical parameters analysed for the present studies include the estimation of water temperature, depth, transparency, water turbidity, pH, dissolved oxygen, total alkalinity, total hardness, calcium, magnesium, iron, sodium, potassium, chloride, ammonical nitrogen, nitrate nitrogen, total phosphorus, silicates, total dissolved solids and conductivity. The water samples were collected in polyethylene bottles of one litre capacity. In the Dal lake in addition to surface water, bottom water samples were collected from pre-selected sites with the help of Ruttner type water sampler (Ruttner 1968). The sampling in both the lakes was done between 10a.m to 12 noon.

1. **Water Temperature:** The temperature of surface water was recorded on the spot by using graduated Celsius Thermometer. The bulb of the thermometer was dipped in water for at least 2 min. for obtaining the water temperature. The results were expressed in °C.
2. **Depth:** The water depth at each sampling site was recorded by sounding the lake bottom with a standard lead weight of 1 Kg attached to the marked rope. The results were expressed in meter (m).
3. **Transparency:** The transparency of water at different sampling sites was determined by using a standard secchi disc (diameter 20 cm). The mean of the depths at which the secchi disc disappeared and reappeared was taken as water transparency. The results were expressed in meter (m).
4. **Turbidity:** The turbidity was measured by using nephelo turbidity meter (N.T.U). One gram of hydrazine sulphate was dissolved in 100 ml of distilled water to prepare stock solution 'A'. The solution 'B' was prepared by dissolving 10 g of hexamethylene tetramine dissolved in 100 ml of distilled water. To prepare standard solution, 5 ml of each of the solutions 'A' and 'B' were mixed and final volume was raised to 100 ml by adding distilled water. This solution

has a value of 400 NTU. 10 ml of this solution was diluted to 100 ml. Instrument was set at 100 with 40 NTU standard suspension.

$$\text{One division on the scale} = 0.4\,\text{NTU turbidity}$$

$$\text{Turbidity, NTU} = \text{Nephelometer reading} \times 0.4 \times \text{dilution factor}$$

5. **pH:** The pH of water samples was determined by means of a composite digital device (EL Digital pH, conductivity, cum TDS meter, Model 181E). Before measuring the pH of water samples, the pH meter was standardized with known buffer solutions of pH 9.2 and 4.0.
6. **Dissolved Oxygen:** The dissolved oxygen content was determined by Winkler's method (Winkler 1888). Water sample was collected in 300 ml dissolved oxygen bottles and fixed on the spot with 1 mL each of alkali iodide (75 g of KOH and 50 g of KI in 200 mL distilled water) and manganous sulphate (100 g of $MnSO_4.2H_2O$ in 250 mL distilled water). The bottle was tightly stoppered and inverted a few times to mix the reagents thoroughly and was brought to the laboratory. The precipitates formed were dissolved by adding 1 ml of concentrated sulphuric acid to the sample. Then 50 ml of the sample was taken and titrated against 0.01 N sodium thiosulphate (2.482 g $Na_2S_2O_3.5H_2O$ in distilled water, 1.5 ml of 6 N NaOH and the final volume made to 1000 mL) using starch as indicator. The results were expressed in mg/L by using the following formula

$$\text{D.O.}\,(\text{mg/L}) = \frac{V_1 \times N \times e \times 1000}{V_2}$$

V_1 = Volume of 0.01 N sodium thiosulphate used
V_2 = Volume of the sample taken for titration
N = Normality of sodium thiosulphate
e = Equivalent weight of oxygen

7. **Total Alkalinity:** 100 ml of water sample was titrated against 0.02 N H_2SO_4 (0.55 mL of concentrated H_2SO_4 in 1000 mL distilled water) using phenolphthalein (0.25 g of pure phenolphthalein powder in 100 mL of 60 % ethanol) and methyl orange (0.5 g dry methyl orange powder in 100 mL of 95 % ethanol) as indicators for carbonate and bicarbonate alkalinity respectively. The same sample tested for carbonate alkalinity was used and titrated for the estimation of bicarbonate alkalinity using methyl orange as indicator. The carbonate alkalinity present, if any, was added to the bicarbonate alkalinity. The results were derived using the following formula and were expressed in mg/L.

$$\text{Total alkalinity}\,(\text{mg/L}) = \frac{V_1 - V_2 \times N \times e \times 1000}{V_3}$$

V_1 = Volume of titrant used for sample
V_2 = Volume of titrant used for blank
V_3 = Volume of the sample taken for titration
N = Normality of the titrant H_2SO_4
e = Equivalent weight of carbonate or bicarbonate

8. **Hardness:** It was determined by computing the results of calcium and magnesium by using following formula.
   Hardness mg equivalent CaCO3/L = 2.497 [Ca (mg/L)] + 4.118 [Mg (mg/L)]
9. **Calcium:** For estimation of calcium, 4 ml of 1 N sodium hydroxide (4 g of NaOH in 100 ml distilled water) and a pinch of solochrome dark blue (SDB) were added to 25 mL of sample which was then titrated against 0.01 N EDTA (3.723 g of disodium salt of EDTA in 1000 mL distilled water) till blue colour appeared. The concentration (mg/L) was determined by using the following formula

$$\text{Calcium (mg / L)} = \frac{V_1 - V_2 \times N \times e \times 1000}{V_3}$$

V_1 = Volume of EDTA used for sample
V_2 = Volume of EDTA used for blank
V_3 = Volume of the sample taken for titration
N = Normality of EDTA
e = Equivalent weight of calcium

10. **Magnesium:** For the estimation of magnesium, a pinch of Eriochrome black-T was added to 25 mL of water sample and then heated gently up to 70 °C in a water bath. After this, 4 mL of ammonia buffer (67.5 g of ammonium chloride dissolved in 570 mL of concentrated ammonia and diluted to 1000 mL) was added to the sample and titrated against 0.01 N EDTA (3.723 g of disodium salt of EDTA in 1000 ml distilled water) till blue colour appeared. As this titration would give total hardness due to calcium and magnesium together, so to get the magnesium hardness only, the volume of titrant used for calcium hardness was subtracted from that used for total hardness (i.e., Calcium and Magnesium together). The results were expressed in mg/L by the following formula

$$\text{Magnesium (mg / L)} = \frac{[(V_1 - V_2) - V_3] \times N \times e \times 1000}{V_4}$$

V_1 = Volume of EDTA used for sample
V_2 = Volume of EDTA used for blank
V_3 = Volume of the EDTA used for the calcium hardness
V_4 = Volume of the sample taken for titration
N = Normality of EDTA
e = Equivalent weight of magnesium

11. **Iron:** The concentration of iron was determined by taking 25 mL of water sample to which 0.5 ml of 1:1 HCl, and a drop of concentrated HNO_3 and 0.5 ml of 8 N KSCN (200 g of potassium thiocyanate dissolved in 150 mL distilled water with gentle heating and stirring) were added. After 20 min, the intensity of wine red colour developed was measured spectrophotometrically (Model Elico SL 150 UV/Visible) at 480 nm. Ferrous ammonium sulphate (slowly add 20 mL concentrated sulphuric acid to 50 mL water and dissolve 1.404 g ferrous ammonium sulphate and add 0.1 N potassium permanganate dropwise until a faint pink colour persists) was used as standard for the preparation of various dilutions for making standard curve. The results were expressed in μg/L.
12. **Sodium:** Sodium was estimated by using flame photometer (Systronics). Stock sodium solution was prepared by dissolving 2.542 g of NaCl dried at 140 °C in one litre distilled water. Ten dilutions of the stock solution were prepared and sodium was estimated using the respective sodium filter to draw the standard curve from which sodium contents of the sample were determined. The results were expressed as mg/L
13. **Potassium:** Potassium was also estimated by using flame photometer (Systronics). Standard potassium solution was prepared by dissolving 1.907 g KCl dried at 110 °C in one litre distilled water. The stock solution was diluted to prepare ten intermediate dilutions and potassium was estimated using respective filter. Standard curve was drawn from the intermediate dilutions of the stock solution from which potassium contents of the sample was determined. The results were expressed in mg/L.
14. **Chloride:** 100 mL of sample was titrated against 0.028 N silver nitrate solution (4.7 g of $AgNO_3$ in 100 mL distilled water) using potassium chromate (5 g of K_2CrO_4 in 100 mL distilled water) as indicator till brick red colour appeared at the end point. Concentration in mg/L was determined by using the following formula

$$\text{Chloride}\,(\text{mg}/\text{L}) = \frac{V_1 \quad V_2\,\text{x}\,N \times e \times 1000}{V_3}$$

V_1 = Volume of silver nitrate used for sample
V_2 = Volume of silver nitrate used for blank
V_3 = Volume of the sample taken for titration
N = Normality of silver nitrate solution
e = Equivalent weight of chlorine

15. **Ammonical-nitrogen:** The ammonical nitrogen was estimated by Nesselerization method. To 25 mL of water sample, 1.5 mL of Nessler's reagent (100 g of mercuric iodide and 70 g of potassium iodide dissolved in distilled water and mixed to a solution of 180 g NaOH in 500 mL distilled water and diluted to 1000 mL) and 0.5 mL of Signette's salt (20 g of potassium antimony tartarate tetrahydrate in 200 mL distilled water and traces of mercuric chloride for preservation) were added. After 20 min, intensity of yellow colour devel-

oped was measured at 420 nm with the help of spectrophotometer (Model Elico SL 150 UV/Visible). Anhydrous ammonium chloride (3.819 g in 1000 mL distilled water) standard was used for making various dilutions. The results were expressed in µg/L.

16. **Nitrate-nitrogen:** The diphenylamine sulphonic acid method was used for the estimation of nitrate-nitrogen. To 1 mL of filtered water sample, a drop of saturated NaCl solution and 4 ml of nitrate reagent [2 mL of stock solution (2.2 g of diphenylamine in a mixture of 150 mL H_2SO_4 and 50 mL distilled water in 380 mL distilled water and 100 mL H_2SO_4} were added. After 70 min, the intensity of blue colour developed was measured at 690 nm spectrophotometrically (Model Elico SL 150 UV/Visible). The standards were prepared form potassium nitrate (0.7218 g in 1000 mL of distilled water) and the results were expressed in µg/L.
17. **Total Phosphate-Phosphorus (TPP):** For the estimation of total phosphate phosphorus 25 ml of water sample was digested to dryness. After this, 1 ml of 70 % perchlorus acid was added and again digested. Then after cooling, added 10 ml distilled water, 1 mL of acid molybdate (25 g of ammonium molybdate dissolved in distilled water and mixed with a cool solution of 250 mL of concentrated H_2SO_4 and 400 mL distilled water and raising final volume to 1000 mL) and 1–3 drops of stannous chloride (2.5 g stannous chloride dissolved in 100 mL glycerol with heating and stirring) were added. The intensity of the blue colour developed was measured at 690 nm spectrophotometrically (Model Elico SL 150 UV/Visible) after 20 min. The various standards were made from potassium hydrogen phosphate (4.393 g in 1000 mL of distilled water) and the results were expressed in µg/L.
18. **Silicates:** For the estimation of silicates, 25 mL of water sample was treated with 2 mL of ammonium molybdate (10 g ammonium molybdate in 90 mL distilled water and 100 mL H_2SO_4) and 0.5 mL oxalic acid (5 g oxalic acid in 50 ml distilled water). Various intermediate dilutions of standard sodium silicate (4.73 g in 1000 mL of distilled water) solution were prepared and the intensity of colour developed was measured after 20 min spectrophotometrically at 690 nm (Model Elico SL 150 UV/Visible). The results were expressed in mg/L.
19. **Total dissolved solids:** Total dissolved solids (TDS) were estimated using (EL Digital pH, conductivity cum TDS meter, Model 181E). The results are expressed in mg/L.
20. **Conductivity:** The conductivity of water samples was measured at 25 °C with the help of a composite digital device (EL Digital pH, conductivity cum TDS meter, Model 181E) which was calibrated before use with standard potassium chloride solution (0.01 M). The results were expressed in µS/cm at 25 °C.

Biological Analysis

The sampling and quantitative estimation of plankton and zoobenthos was done in accordance with Welch (1948) and APHA (1989).

Phytoplankton The study of phytoplankton community was divided into two parts, viz., qualitative and quantitative analysis.

Qualitative Analysis The standard planktonic net (64 nm pore diameter) was hauled through vertical and horizontal planes at the selected sites of the Dal lake once every month. The plankters collected in the 50 ml polyethylene bottle connected at the lower end of the net were preserved in lugols solution. The whole sample was studied under the phase contrast inverted microscope (Nikon). Identification was done with the help of keys given in the books of Heurk (1896); Smith (1950); Desikachary (1959); Randhawa (1959); Ward and Whipple (1959); Philipose (1967); Round (1973); Prescott (1978) and Palmer (1980).

Quantitative Analysis To obtain representative samples of the phytoplankton, two litres of water was filtered through a standard plankton net (64 nm pore diameter). The water was allowed to filter through the net and the plankters were concentrated in the 50 mL polyethylene bottle connected at the lower end of the net. The sample was preserved in Lugols solution and then further concentrated by centrifugation to 5 mL. The entire sample was studied under the phase contrast inverted microscope (Nikon) by taking 1 ml of concentrated preserved sample at a time in a Sedgwick Rafter Chamber and number of individuals of each phytoplankton species were counted. The results are expressed as individuals per mL.

Zooplankton The study of zooplankton was divided into two parts, viz., qualitative and quantative analysis.

Qualitative Analysis The standard planktonic net (64 nm pore diameter) was hauled through vertical and horizontal planes of the lake (Dal and Anchar lake) at selected sites once every month. The plankters collected in the 50 ml polyethylene bottle connected at the lower end of the net were preserved in 5 % formalin. Then 1 mL of this sample was taken at a time in a Sedgwick Rafter chamber and studied under the phase contrast inverted microscope (Nikon). The identification was done with the help of keys given by Pennak (1978); Ward and Whipple (1959); Mellanby (1963); Tonapi (1980); Koste and Shiel (1980) and Sharma (1980).

Quantitative Analysis To collect sample for quantitative analysis of zooplankton population, 10 l of water was filtered through the plankton net and the water was allowed to filter through the net, the plankters concentrated in the 50 mL polyethylene bottle connected at the lower end of the net. The sample thus obtained was preserved in 5 % formalin and further reduced in volume to 5 ml by centrifugation. 1 ml of concentrated preserved sample was taken at a time in a Sedgwick Rafter Chamber (Whipple et al. 1927) and counting was done for each zooplankton taxon.

The entire 5 ml of the concentrated sample was studied under phase contrast inverted microscope (Nikon). The results are expressed as individuals per litre.

Macrozoobenthos For the collection of Macrozoobenthos, sampling from the selected sites of the Dal and the Anchar lake was carried out after every 3 months by Ekman's grab. The mud sample collected was passed through a series of sieves (0.5,0.6,0.2 mm) with addition of water on the spot. The samples were brought to the laboratory, where the animals were picked up with forceps and pipettes and also by hand picking and preserved in 70 % alcohol. The density of organisms was calculated per square meter of bottom area by the formula given below. The identification was done with the help of the keys given by Pennak (1978); Ward and Whipple (1959); Mellanby (1963); Tonapi (1980) and Williams and Feltmate (1992).

$$N = \frac{C}{E \times S} \times 10,000$$

N = Number of organism/m^2
C = Number of organism counted
E = Area of Ekman's grab (240.25/cm^2).
S = Number of samples taken at each station (3 replicates)

Macrophytes The community features of macrophytes were worked out on monthly basis from April to October during both the years. Quadrat method was followed for the macrophytic study (Curtis 1959). Quadrats of definite size (1 m^2) were laid randomly at and around the selected sites covering almost all the water body. For the submerged plants, Ekman Dredge was used to collect the plants falling in the quadrat. The macrophytes occurring in each quadrat/sampling unit were listed species wise and the number of individuals of each species was counted for various community characteristics of macrophytes.

The frequency, relative frequency, density, relative density, abundance, relative abundance and Importance Value Index (IVI) were determined as follows

$$\text{Frequency (\%)} = \frac{\text{No. of quadrats in which the species occurred}}{\text{Total no. of quadrats studied}} \times 100$$

$$\text{Relative Frequency} = \frac{\text{Frequency of one species}}{\text{Total Frequencies of all species}} \times 100$$

$$\text{Density } (m^{-2}) = \frac{\text{No. of individuals of a species}}{\text{Total no. of quadrats studied}}$$

$$\text{Relative Density } (m^{-2}) = \frac{\text{Density of one species}}{\text{Density of all the species}} \times 100$$

$$\text{Abundance} = \frac{\text{Total no. of individualas of a species}}{\text{No. of quadrats in which species occur}}$$

$$\text{Relative Abundance} = \frac{\text{Abundance of one species}}{\text{Abundance of all the species}} \times 100$$

$$\text{Importance Value Index (IVI)} = \text{Relative Frequency} + \text{Relative density} + \text{Relative Abundance}$$

Statistical Analysis

Average value was obtained by summing the values and then dividing the results by their number. The standard deviation was obtained by calculating the square root of the variance. It represents the amount of variation in the data. Average values and standard deviation were derived in the **Windows Microsoft Excel**.

For the study of correlation between any two variables corresponding to physico-chemical parameters, phytoplanktons and zooplanktons, the quantitative data for each month was subjected to correlation analysis by calculating the index termed as correlation coefficient ("r"). The value was calculated by a **computer software "Primer" by Stanton and Glantz (1996).**

$$r = \frac{\sum (x - \overline{x})(y - \overline{y})}{\sqrt{\sum (x - \overline{x})^2 \sum (y - \overline{y})^2}}$$

The measured coefficient with its value may or may not coincide with the true value of coefficient of the entire population. Therefore, a significant test ('t') for 'r' was calculated, assuming null hyposis (Ho) as the true coefficient. The value of "t" was also calculated by a computer software "primer".

$$t = \frac{\sqrt{n-2}}{\sqrt{1-\gamma^2}}$$

The calculated value of 't' was then compared with critical value of 't' obtained upto n–2° of freedom at 5 % level while referring to student's 't' distribution table. For the present studies, the critical value of 't' against 24–2° of freedom came to be 2.07. These values give the indication of significance for the purpose of calculating relationship.

Results and Discussion

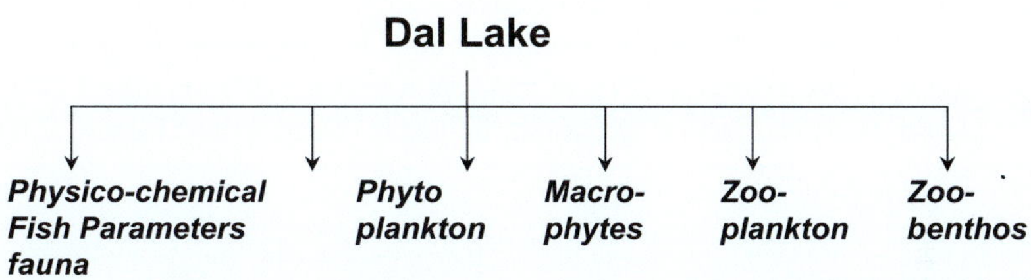

Physico-Chemical Parameters

Climatology of the Study Area

The valley of Kashmir occupies northernmost geographical position in India. It is a Lacustrine basin of Pleistocene origin formed as a result of tectonic activity (Zutshi and Khan 1978). The basin exists between the lesser and the Great Himalaya and is characterised by numerous aquatic ecosystems (Khan 1984). Due to diverse physical features and location between the weak monsoon zone of Punjab and cold dry belt of Ladakh and Tibet, the climate of Kashmir is highly variable. Bagnoulus and Homji (1959) has classified the climate of Kashmir as submediterranian type showing four seasonal phases which are as follows:

Season	Months	Duration	Mean temp (°C)
Spring	March–May	3	8.4–17.9
Summer	June–August	3	21.7–23.9
Autumn	September–November	3	20.5–7.7
Winter	December–February	3	1.05–3.5

M. Jeelani, *Lake Ecology in Kashmir, India*,
DOI 10.1007/978-3-319-40880-4_4

According to Koul (1979), the climate of Kashmir is highly variable and does not conform to any definite type. He suggested the following scheme.

Season	Months	Duration	Mean temp (°C)
Cool season or (Winter)	Nov–March	5	0–10
Hot season or (Summer) Trasnsition season	June–Sept.	4	20–30
(a) warm (spring)	Apr–May	2	10–20
(b) 2nd warm	October	1	10–20

In the present investigation, the seasonal pattern suggested by Bagnoulus and Homji (1959) has been used where each season is of 3 months duration. Zutshi et al. (1980) also followed the same pattern and suggested that lakes of Kashmir valley experience a climate with features common with submediterranian conditions. Macan (1963), Patra and Nayak (1982) and Yousuf et al. (1986) also followed similar seasonal scheme for Kashmir water bodies.

The monthly data pertaining to minimum and maximum air temperature and precipitation over the period of 2 years from September, 2000 to August 2002 was collected from meteorological department, Srinagar (Table 1, and Figs. 1, 2 and 3).

Water Temperature (°C)

Temperature plays a key role in determination of various other parameters such as pH, conductivity, saturation level of gases and various forms of alkalinity (Trivedy and Goel 1986; Esmaeili and Johal 2005).

In the Dal lake, water temperature showed fluctuations in accordance with atmospheric temperature. The water temperature ranged from 4 to 27 °C in the lake water over the period of 2 years (Table 2 and Fig. 4).

Maximum temperature was recorded in August, thereafter it started declining and recorded its minimum in January followed by gradual increase in the following months to reach the maximum in August. This trend was observed at all the selected sites. In overall terms, the temperature do follow set pattern dictated by seasonal cycle. Similar seasonal variations in temperature has been observed in Kashmir water bodies by Qadri and Yousuf (1978), Sarwar (1987) and Shah (1988). In other water bodies also, a similar trend has been observed by Welch (1952), Rao (1955), Zafar (1955), Munawar (1970a, b), Hannan and Young (1974), Swarup and Singh (1979), Goel et al. (1980), Harshey et al. (1982), Patra and Nayak (1983), Bagde and Varma (1985a, b), Bist (1986), Palharya and Malviya (1988), Shyamsunder (1988), Adholia et al. (1991), Prakash (1990), Jindal and Kumar (1993), Kumar (1995a, b), Bath (1996), Kumar and Singh (1996), Syal (1996), Bath and Kaur (1998), Kaur et al. (1999), Roy (2000) and Khabade et al. (2003).

Four selected sites of the lake show little difference in water temperature showing that discharges from houseboats, hotels and floating gardens fail to affect the temperature of water. On the other hand, Jolly and Chapman (1966), Venkateshwarlu

Table 1 Showing the record of climatic data of Srinagar city

	Sep-00	Oct-00	Nov-00	Dec-00	Jan-01	Feb-01	Mar-01	Apr-01	May-01	Jun-01	Jul-01	Aug-01	Sep-01	Oct-01	Nov-01	Dec-01	Jan-02	Feb-02	Mar-02	Apr-02	May-02	Jun-02	Jul-02	Aug-02
Temp°C (Minimum)	11.6	6.6	0.7	−0.8	−2.8	−0.5	4.5	8.6	11.3	15.2	17.1	18.2	6.6	1.3	−3.6	−5.6	−6.7	−4.8	−1.2	5.2	9.8	12.8	16	13.6
(Maximum)	25.1	23.3	18.5	10	9.3	9.5	16.8	20.6	26.5	28.9	30.5	29.8	32.3	20.5	24.4	15	11.6	17.4	22.9	28.3	34.2	34.6	34.7	33.5
Precipitation (mm^3)	54.5	8.2	0	8.9	35.5	105.8	105	77.7	47.3	82.2	16.8	75.3	22.6	0	12.3	15.5	21.3	46.1	56	57	37	52.8	32.3	35.9
Relative humidity % at 8.30 h.	73	76	75	64	88	84	73	72	63	68	66	77	75	68	77	8.6	83	76	68	71	64	70	75	77
Relative humidity % at 17.30 h.	49	53	53	65	60	60	47	51	44	48	51	60	48	44	55	62	47	45	42	47	43	52	55	53
No. of rainy days.	5	2	4	5	7	7	4	4	5	7	9	6	6	0	2	4	2	7	8	11	7	6	9	6

Precipitation = ppt of 1.0 mm or more is taken as a rainy day
Source = Office of the Director Meteorological centre Rambagh; Srinagar – 190015

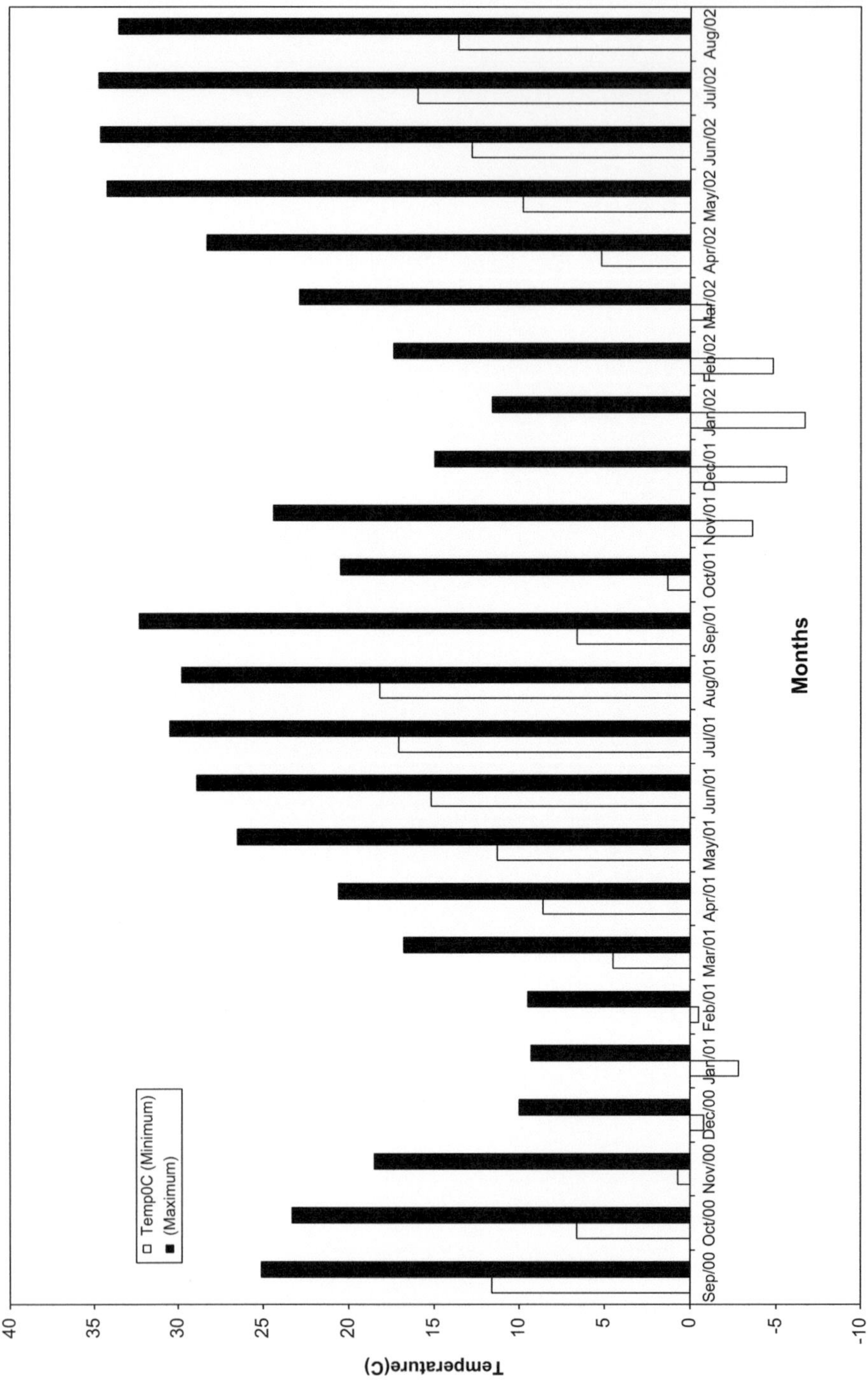

Fig. 1 Atmospheric temperature variations during the year 2000–2002 of the Srinagar city of Kashmir valley

Fig. 2 The pattern of total precipitation during the year 2000–2002

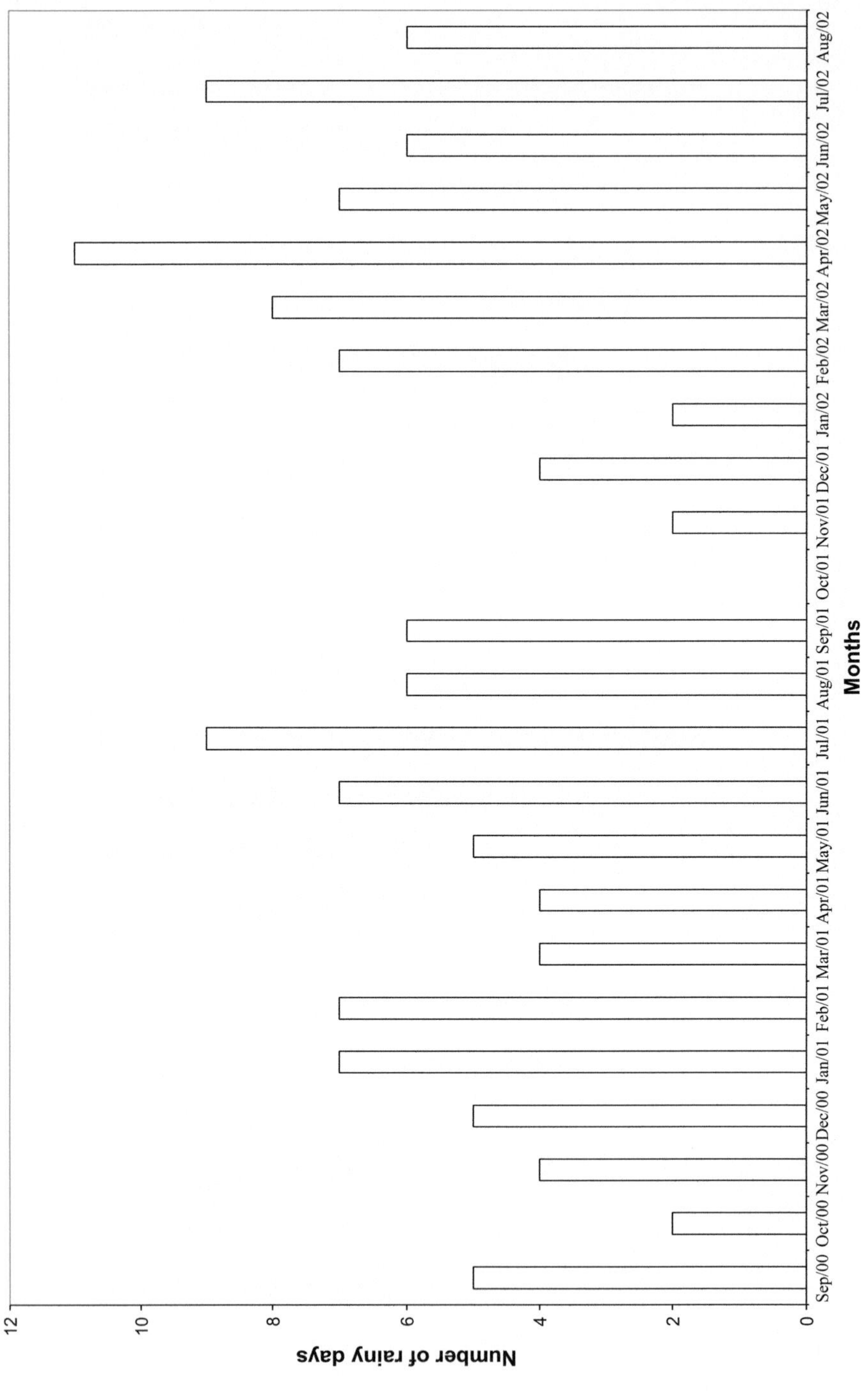

Fig. 3 Showing the number of rainy days during the year 2000–2002

Table 2 Monthly fluctuations in water temperature (°C) at the investigated sites of the Dal lake

	Months																			
2000–2001	**Sep**	**Oct**	**Nov**	**Avg**	**±S.D.**	**Dec**	**Jan**	**Feb**	**Avg**	**±S.D.**	**Mar**	**April**	**May**	**Avg**	**±S.D.**	**June**	**July**	**Aug**	**Avg**	**±S.D.**
SITE-I	24.0	20.0	15.0	19.7	3.7	10.0	5.0	6.0	7.0	2.2	10.0	15.0	19.0	14.7	3.7	18.0	24.0	23.0	21.7	2.6
SITE-II	24.0	20.0	15.0	19.7	3.7	10.0	5.0	6.0	7.0	2.2	10.0	15.0	19.0	14.7	3.7	18.0	24.0	23.0	21.7	2.6
SITE-III	24.0	20.0	15.0	19.7	3.7	10.0	5.0	6.0	7.0	2.2	10.0	15.0	19.0	14.7	3.7	18.0	24.0	25.0	22.3	3.1
SITE-IV	24.0	20.0	15.0	19.7	3.7	10.0	5.0	6.0	7.0	2.2	10.0	15.0	19.0	14.7	3.7	18.0	24.0	24.0	22.0	2.8
2001–2002	**Sep**	**Oct**	**Nov**	**Avg**	**±S.D.**	**Dec**	**Jan**	**Feb**	**Avg**	**±S.D.**	**Mar**	**April**	**May**	**Avg**	**±S.D.**	**June**	**July**	**Aug**	**Avg**	**±S.D.**
SITE-I	22.0	20.0	19.0	20.3	1.2	14.0	5.0	9.9	9.6	3.7	11.0	15.0	20.0	15.3	3.7	20.0	25.0	27.0	24.0	2.9
SITE-II	22.0	20.0	19.0	20.3	1.2	14.0	4.0	7.0	8.3	4.2	11.0	15.0	20.0	15.3	3.7	20.0	25.0	27.0	24.0	2.9
SITE-III	22.0	20.0	19.0	20.3	1.2	14.0	5.0	6.0	8.3	4.0	11.0	15.0	20.0	15.3	3.7	20.0	25.0	27.0	24.0	2.9
SITE-IV	22.0	20.0	19.0	20.3	1.2	14.0	5.0	9.0	9.3	3.7	11.0	15.0	20.0	15.3	3.7	20.0	25.0	27.0	24.0	2.9

Avg.: Average
S.D.: Standard deviation

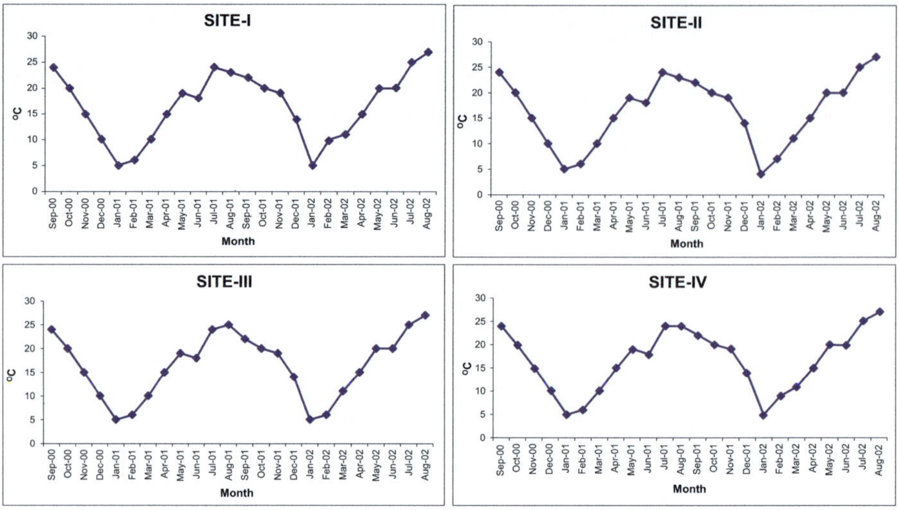

Fig. 4 Monthly fluctuations in water temperature (°C) at the investigated sites of the Dal lake

and Jayanti (1968), Munawar (1970a, b), Kumar and Sharma (1979), Sharma et al. (1981), Mahadevan and Krishnaswamy (1983), Palharya and Malviya (1988), Bath (1996), Syal (1996), Katti et al. (2002) and Kaur et al. (2003) recorded an increase in water temperature with addition of effluents.

The water temperature of the lake does not fall below 4 °C throughout the study period and can be classified as warm monomictic as suggested by Hutchinson (1957) and Reid and Wood (1976).

Depth (m)

In the Dal lake, water level is mainly maintained by snow melt water entering through Telbal nallah. Only a small contribution is made by local precipitation towards the water level as has also been observed in other lakes of Kashmir by Kaul (1979). In the Dal lake overall depth of water ranged from 0.8 to 2.80 m with marked variation at different study sites. No definite seasonal pattern was recorded in the depth.

The water level fluctuated between 1.56 and 2.80 m at site-I, between 0.8 and 1.9 m at site-II, between 0.8 and 1.6 m at site-III, and between 0.8 and 1.83 m at site-IV. Data on depth reveals that site-I represents the deepest zone of the lake whereas at other sites, the lake bed is raised (Table 3 and Fig. 5).

At Site-II, floating gardens which encroach upon year after year the lake area have decreased the depth. The mud is being carried out from other places, settled on the floating islands and ultimately these islands are converted to marshy lands. Death and decomposition of macrophytes further raise the lake bed.

Table 3 Monthly fluctuations in water depth (m) at the investigated sites of the Dal lake

	Months																		
2000–2001	**Sep**	**Oct**	**Nov**	**Avg**	**±S.D.**	**Dec**	**Jan**	**Feb**	**Avg**	**±S.D.**	**Mar**	**April**	**May**	**Avg**	**±S.D.**	**June**	**July**	**Aug**	**Avg**	**±S.D.**
SITE-I	2.0	1.9	2.1	2.0	0.1	2.3	2.3	1.9	2.2	0.2	2.2	2.2	2.4	2.2	0.1	2.3	2.1	2.3	2.2	0.1
SITE-II	1.8	1.8	1.7	1.8	0.1	1.5	1.5	1.5	1.5	0.0	1.5	1.5	1.6	1.5	0.1	1.4	0.9	1.4	1.2	0.2
SITE-III	1.2	1.2	1.2	1.2	0.0	1.2	1.3	1.1	1.2	0.1	1.2	1.2	1.4	1.2	0.1	1.2	1.2	1.2	1.2	0.0
SITE-IV	1.8	1.8	1.8	1.8	0.0	1.5	1.5	1.4	1.5	0.1	1.4	1.5	1.5	1.5	0.0	1.4	1.1	0.8	1.1	0.2
2001–2002	**Sep**	**Oct**	**Nov**	**Avg**	**±S.D.**	**Dec**	**Jan**	**Feb**	**Avg**	**±S.D.**	**Mar**	**April**	**May**	**Avg**	**±S.D.**	**June**	**July**	**Aug**	**Avg**	**±S.D.**
SITE-I	2.1	2.0	2.0	2.1	0.1	2.0	2.8	2.6	2.5	0.4	1.9	2.0	1.6	1.8	0.1	1.6	1.8	2.2	1.9	0.3
SITE-II	2.0	2.0	1.9	1.9	0.0	2.0	1.7	1.5	1.7	0.2	1.5	1.5	1.7	1.6	0.1	1.5	1.3	1.3	1.4	0.1
SITE-III	1.4	1.4	1.3	1.3	0.0	1.3	1.5	1.5	1.4	0.1	1.5	1.5	1.6	1.5	0.1	1.4	0.9	1.4	1.2	0.2
SITE-IV	0.8	0.8	0.8	0.8	0.0	0.8	1.5	1.5	1.3	0.3	1.5	1.4	1.5	1.4	0.0	1.5	1.4	1.1	1.3	0.2

Avg.: Average
S.D.: Standard deviation

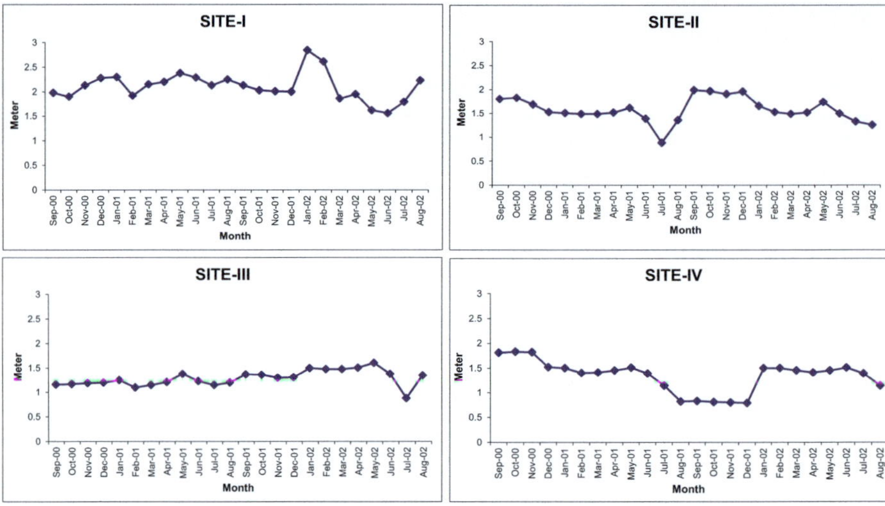

Fig. 5 Monthly fluctuations in depth (m) at the investigated sites of the Dal lake

At site-III, the lake is occupied by houseboats, from where degradable and non-degradable waste is dumped into the lake. This site is characterised by extensive growth of macrophytes which upon death and decay pile up and raise the level of the lake bed.

Site-IV is situated where the lake water flows into the river Jhelum and its flow is controlled by weir and lock system. The water from site-I, II and III collects at this site as a result of which the silt gets deposited at the bottom of the lake. Furthermore, extensive growth of macrophytes add to the sediment on death, decay and decomposition. The lake therefore, has now assumed a status of shallow water body (Pandit 2002a). Such lakes tend to become eutrophic as suggested by Naumann (1919) and Thienemann (1925) and Rawson (1955).

Earlier Zutshi et al. (1980) recorded a depth of 6 m in the Dal lake. In the present investigation which incorporates the work done during 2000–2002, about 24 years after the work of Zutshi, the maximum depth has been recorded to be 2.80 m.

Water Transparency (m)

Secchi disc transparency is basically a function of reflection of light from the surface of the disc and therefore, is affected by the absorption characteristics of the water and of dissolved and, particulate matter contained in the water (Wetzel and Likens 2000).

The transparency in the Dal lake fluctuated between 0.4 and 1.62 m during the period of study with the maximum recorded at site-I and minimum at site-IV. No seasonal trend was observed in the transparency values of the lake. Transparency values were seen to fluctuate from 1.1 to 1.6 m at site-I, from 0.5 to 1.6 m at site-II, from 0.5 to 1.0 m at site-III and from 0.4 to 1.4 m at site-IV (Table 4 and Fig. 6).

Table 4 Monthly fluctuations in water transparency (m) at the investigated sites of the Dal lake

	Months																		
2000–2001	**Sep**	**Oct**	**Nov**	**Avg**	**±S.D.**	**Dec**	**Jan**	**Feb**	**Avg**	**±S.D.**	**Mar**	**April**	**May**	**Avg**	**±S.D.**	**June**	**July**	**Aug**	**Avg**	**±S.D.**
SITE-I	1.4	1.3	1.2	1.3	0.1	1.5	1.5	1.5	1.5	0.0	1.4	1.1	1.3	1.3	0.1	1.4	1.1	1.1	1.2	0.1
SITE-II	1.6	1.7	1.0	1.4	0.3	1.1	1.0	0.8	1.0	0.1	0.8	0.7	0.8	0.8	0.0	0.6	0.5	0.8	0.6	0.1
SITE-III	0.8	0.9	1.0	0.9	0.1	0.8	0.9	0.9	0.9	0.0	0.7	0.7	0.7	0.7	0.0	0.9	0.9	1.1	0.9	0.1
SITE-IV	1.3	1.4	1.2	1.3	0.1	1.4	1.2	1.0	1.2	0.2	1.0	1.0	0.9	0.9	0.1	0.6	0.4	0.6	0.5	0.1
2001–2002	**Sep**	**Oct**	**Nov**	**Avg**	**±S.D.**	**Dec**	**Jan**	**Feb**	**Avg**	**±S.D.**	**Mar**	**April**	**May**	**Avg**	**±S.D.**	**June**	**July**	**Aug**	**Avg**	**±S.D.**
SITE-I	1.1	1.1	1.1	1.1	0.0	1.2	1.5	1.6	1.4	0.2	1.5	1.3	1.3	1.4	0.1	1.3	1.1	1.1	1.2	0.1
SITE-II	1.2	1.2	1.2	1.2	0.0	1.2	1.1	0.8	1.0	0.2	0.8	0.8	0.8	0.8	0.0	0.6	0.6	0.7	0.7	0.1
SITE-III	0.9	0.9	1.0	0.9	0.0	1.0	1.0	0.8	0.9	0.1	0.8	0.7	0.8	0.8	0.0	0.6	0.5	0.8	0.6	0.1
SITE-IV	0.6	0.7	0.7	0.7	0.0	0.7	1.3	1.2	1.0	0.3	1.0	1.0	0.9	1.0	0.0	0.7	0.6	0.6	0.6	0.0

Avg.: Average
S.D.: Standard deviation

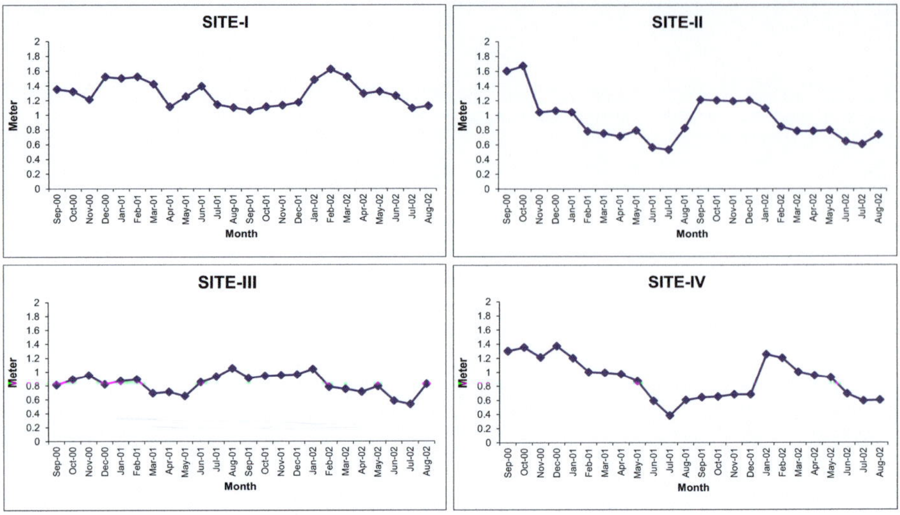

Fig. 6 Monthly fluctuations in water transparency (m) at the investigated sites of the Dal lake

Site-I represents the zone where maximum transparency values were recorded. This sampling site is least interfered by human habitation and therefore, does not receive domestic sewage or any other type of waste. The water here is crystal clear. Furthermore, the water at this site remains undisturbed with suspended particles settled down to the bottom.

Site-II is extensively used for vegetable cultivation on floating gardens. The low values in transparency might be due to soil erosion from the adjoining floating gardens. In addition, decomposible material of macrophytes is also add to the water thereby lowering its transparency. Frequent plying of boats stirs the water and disturbs the sedimentation of solids. Wanganeo et al. (1996) observed fall in transparency due to addition of silt from the catchment area. David (1956), Sengar and Singh (1986), Thorat and Sultana (2000), Das et al. (2003) and Verma (2003) also made similar findings in different water bodies.

At site-III, lake is occupied by houseboats which discharge sewage and other degradable and non-degradable material into the lake water and the site harbours extensive growth of macrophytes. The sewage and organic matter added upon death and decay of macrophytes decrease transparency of water. Kumar and Sharma (1999), Maruthi et al. (2000), Maruthi et al. (2000) also recorded low transparency in waters receiving sewage.

At site-IV transparency was found to be minimum. Water from other sites of the lake ultimately pool here and its flow from here into the river Jhelum. Apart from the solid matter added at previous sites water here receives fresh discharges from hotels and houseboats. Moreover, water is frequently disturbed by movement of shikaras at this site which also adds suspended materials to the water by stirring. All these factors together lower the transparency of water at this site. Kumar and Sharma (1999), Maruthi et al. (2000) and Esmaeili and Johal (2005) also reported that addition of sewage lowers the transparency of water.

Earlier Zutshi et al. (1980) recorded 0.4–5.0 m transparency in this lake. The present studies show that the transparecy has decreased during these years due to death and decay of macrophytes and clay and silt particles discharged from the above mentioned human perturbations.

Turbidity (NTU)

Any substance having more than 10^{-9}m size produces turbidity which is actually the expression of optical property and is caused by clay, silt, organic matter, phytoplankton and other microscopic organisms (Trivedy and Goel 1986).

In the Dal lake, the turbidity values ranged from 7 to 21 NTU at four sites. No seasonal trend is indicated in the turbidity values of the Dal lake in the present investigation.

At site-I, the turbidity values varied from 7.0 to 15 NTU, at site-II from 8 to 21 NTU, at site-III from 11 to 20 NTU and at site-IV from 12 to 21 NTU (Table 5 and Fig. 7).

At Site-I, turbidity is found to be lowest. This site is located in the open water area of the lake where water remains undisturbed so that suspended particles are able to settle down.

At site-II turbidity is higher than the site-I. This site is close to the floating gardens where organic matter is being regularly added to enrich the soil bed for better yield. Moreover, the rich growth of submerged and emergent plants add suspended organic matter on their decomposition. The rate of sedimentation of organic and inorganic matter is slow due to frequent human disturbances. All these factors result in increase in the turbidity of water at this site. Pagioro et al. (1997) and Thorat and Sultana (2000) observed increase in turbidity in shallow water bodies due to frequent disturbances leading to resuspension of solids. Khan and Zutshi (1980) and Wanganeo et al. (1996) attributed high turbidity to silt from the catchment area while Das et al. (2003) and Sharma and Verma (2003) attributed it to high amount of suspended solids.

At site-III water receives sewage and other domestic discharges from houseboats whereas at site-IV discharges from hotels are released. Both these sites harbour extensive growth of macrophytes which upon death and decay add suspended organic matter. Sewage discharges and suspended organic matter increase the turbidity of water at site-III and site-IV. Maruthi et al. (2000) and Das et al. (2003) also observed an increase in turbidity in water bodies receiving sewage. David (1956), Sengar and Singh (1986), Pagioro et al. (1997), Kumar and Sharma (1999), Maruthi et al.(2000), Thorat and Sultana (2000) and Sharma and Verma (2003) also recorded increase in turbidity due to organic matter.

There are no earlier records of turbidity in this lake.

ISI (1986) have suggested the maximum turbidity value for drinking water as 10 NTU. Based on this, the turbidity values which range from 7 to 21 NTU at the investigated sites almost exceed the limits.

Table 5 Monthly fluctuations in water turbidity (N.T.U.) at the investigated sites of the Dal lake

	Months																		
2000–2001	**Sep**	**Oct**	**Nov**	**Avg**	**±S.D.**	**Dec**	**Jan**	**Feb**	**Avg**	**±S.D.**	**Mar**	**April**	**May**	**Avg**	**±S.D.**	**June**	**July**	**Aug**	**Avg**	**±S.D.**
SITE-I	8.0	12.0	9.0	9.7	1.7	8.0	10.0	11.0	9.7	1.2	15.0	14.0	12.0	13.7	1.2	10.0	11.0	14.0	11.7	1.7
SITE-II	16.0	15.0	16.0	15.7	0.5	12.0	14.0	15.0	13.7	1.2	20.0	15.0	14.0	16.3	2.6	15.0	16.0	15.0	15.3	0.5
SITE-III	16.0	15.0	15.0	15.3	0.5	11.0	12.0	16.0	13.0	2.2	18.0	15.0	15.0	16.0	1.4	14.0	14.0	15.0	14.3	0.5
SITE-IV	15.0	14.0	16.0	15.0	0.8	14.0	16.0	12.0	14.0	1.6	15.0	19.0	16.0	16.7	1.7	14.0	16.0	17.0	15.7	1.2
2000–2001	**Sep**	**Oct**	**Nov**	**Avg**	**±S.D.**	**Dec**	**Jan**	**Feb**	**Avg**	**±S.D.**	**Mar**	**April**	**May**	**Avg**	**±S.D.**	**June**	**July**	**Aug**	**Avg**	**±S.D.**
SITE-I	11.0	8.0	7.0	8.7	1.7	7.0	8.0	10.0	8.3	1.2	11.0	12.0	13.0	12.0	0.8	11.0	12.0	11.0	11.3	0.5
SITE-II	19.0	20.0	21.0	20.0	0.8	18.0	8.0	10.0	12.0	4.3	8.0	9.0	9.0	8.7	0.5	9.0	11.0	11.0	10.3	0.9
SITE-III	15.0	14.0	12.0	13.7	1.2	12.0	14.0	15.0	13.7	1.2	20.0	15.0	14.0	16.3	2.6	15.0	16.0	15.0	15.3	0.5
SITE-IV	21.0	19.0	17.0	19.0	1.6	17.0	14.0	16.0	15.7	1.2	15.0	13.0	14.0	14.0	0.8	16.0	16.0	17.0	16.3	0.5

Avg.: Average
S.D.: Standard deviation

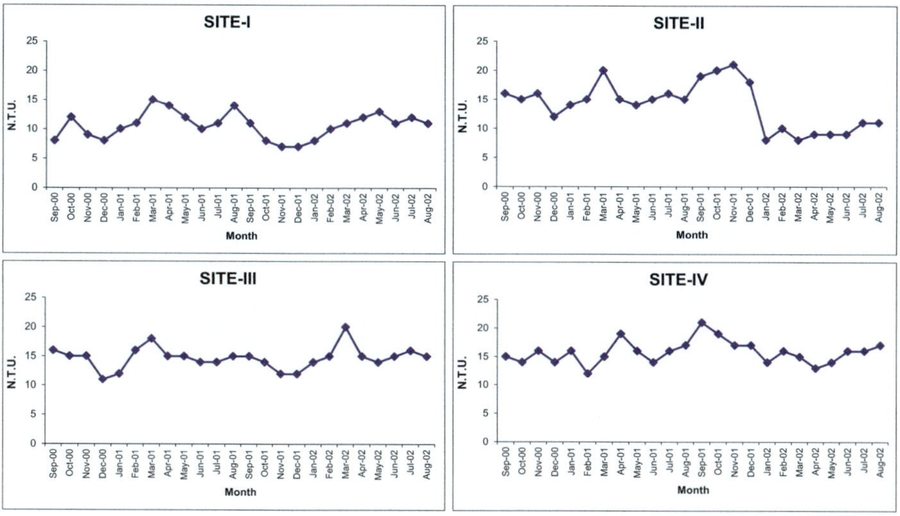

Fig. 7 Monthly fluctuations in water turbidity (N.T.U.) at the investigated sites of the Dal lake

pH

The pH a measure of concentration of hydrogen ions in water, is an important factor which influences the metabolism of organisms inhabiting it (Trivedy and Goel 1986 and Esmaeili and Johal 2005).

The pH values of water in the Dal lake ranged from 7.0 to 9.3 for both surface and bottom waters. No difference in pH values between surface and bottom waters was recorded at site-I and site-III whereas a difference of only 0.8 units and 0.1 units between surface and bottom waters was recorded at site-I and site-IV respectively. The site-II receive rich amounts of fertilizers and pesticides through agricultural run-off from the adjoining floating gardens which on sedimentation might have increased the pH of bottom water. Wanganeo et al. (1996) observed a difference of 1.7 in surface and bottom water while Singh et al. (1982) found a difference of 0.9 and both these authors regarded this much difference to be non-appreciable.

A look at the 2 year data on pH values at all the investigated sites reveals no definite seasonal pattern. Sarwar (1999) for the Anchar lake of Kashmir and Syal (1996) and Valarmathi et al. (2002) for other lakes also failed to record any seasonal trend in pH values. However, Zutshi and Vass (1978), Khan and Zutshi (1980) and Wanganeo (1984) for Kashmir lakes and Mohanty (1975), Singh et al. (1982), Sharma and Dhaneshwar (1986), Mishra (1988), Jindal and Kumar (1993), Khalique and Afser (1995), Bath (1996), Islam and Islam (1996) and Narain and Chauhan (2000) for other lakes recorded higher pH values during summer.

At site-I, pH ranged from 7.4 to 9.3 for both surface and bottom waters, at site-II it ranged from 7.0 to 7.9 for surface waters and from 7.1 to 8.7 for bottom waters, at site-III it ranged from 7.0 to 7.8 for surface as well as bottom waters and at site-IV it fluctuated from 7.1 to 8.9 for surface water and from 7.3 to 9.0 for bottom waters (Table 6 and Fig. 8).

Table 6 Monthly fluctuations in pH at the investigated sites of the Dal lake

	Months																		
2000–2001	**Sep**	**Oct**	**Nov**	**Avg**	**±S.D.**	**Dec**	**Jan**	**Feb**	**Avg**	**±S.D.**	**Mar**	**April**	**May**	**Avg**	**±S.D.**	**June**	**July**	**Aug**	**Avg**	**±S.D.**
SITE-I (S)	7.4	7.4	8.6	7.8	0.6	8.2	8.6	8.3	8.4	0.2	8.4	8.8	8.7	8.6	0.2	8.6	9.0	9.1	8.9	0.2
SITE-I (B)	7.4	7.4	8.8	7.9	0.7	8.0	8.4	8.1	8.2	0.2	8.4	8.6	8.6	8.5	0.1	8.5	9.2	9.2	9.0	0.3
SITE-II (S)	7.3	7.6	7.4	7.4	0.1	7.1	7.5	7.4	7.3	0.2	7.5	7.3	7.2	7.3	0.1	7.3	7.2	7.5	7.3	0.1
SITE-II (B)	8.7	8.5	8.0	8.4	0.3	7.8	7.7	7.5	7.7	0.1	7.5	7.4	7.3	7.4	0.1	7.4	7.2	7.4	7.3	0.1
SITE-III (S)	7.7	7.8	7.6	7.7	0.1	7.2	7.2	7.2	7.2	0.0	7.5	7.4	7.3	7.4	0.1	7.5	7.9	7.4	7.6	0.2
SITE-III (B)	7.6	7.6	7.7	7.6	0.0	7.4	7.2	7.1	7.2	0.1	7.6	7.5	7.4	7.5	0.1	7.8	7.8	7.3	7.6	0.2
SITE-IV (S)	8.0	8.0	7.9	8.0	0.0	7.6	7.6	7.6	7.6	0.0	7.2	7.2	7.5	7.3	0.1	7.5	7.3	7.4	7.4	0.1
SITE-IV (B)	8.1	8.1	8.0	8.1	0.0	7.8	7.8	7.8	7.8	0.0	7.3	7.4	7.6	7.4	0.1	7.6	7.4	7.5	7.5	0.1
2001–2002	**Sep**	**Oct**	**Nov**	**Avg**	**±S.D.**	**Dec**	**Jan**	**Feb**	**Avg**	**±S.D.**	**Mar**	**April**	**May**	**Avg**	**±S.D.**	**June**	**July**	**Aug**	**Avg**	**±S.D.**
SITE-I (S)	9.3	8.5	8.7	8.8	0.3	8.4	8.0	8.2	8.2	0.2	8.8	8.9	8.8	8.8	0.0	8.1	8.8	7.7	8.2	0.5
SITE-I (B)	9.3	8.5	8.7	8.8	0.3	8.4	7.9	8.3	8.2	0.2	8.6	8.6	8.4	8.5	0.1	7.9	7.4	7.7	7.7	0.2
SITE-II (S)	7.1	7.3	7.3	7.2	0.1	7.4	7.4	7.5	7.4	0.0	7.3	7.3	7.3	7.3	0.0	7.4	7.5	7.9	7.6	0.2
SITE-II (B)	7.1	7.3	7.3	7.2	0.1	7.4	7.4	7.5	7.4	0.0	7.8	7.8	7.6	7.7	0.1	7.4	7.5	7.5	7.5	0.0
SITE-III (S)	7.0	7.5	7.4	7.3	0.2	7.3	7.6	7.6	7.5	0.1	7.2	7.2	7.5	7.3	0.1	7.5	7.5	7.3	7.4	0.1
SITE-III (B)	7.0	7.5	7.4	7.3	0.2	7.3	7.8	7.8	7.6	0.2	7.3	7.4	7.6	7.4	0.1	7.6	7.4	7.4	7.5	0.1
SITE-IV (S)	7.4	7.6	7.8	7.6	0.2	7.5	7.1	8.9	7.8	0.8	8.2	8.0	8.2	8.1	0.1	8.2	8.9	8.9	8.7	0.3
SITE-IV (B)	7.4	7.6	7.8	7.6	0.2	7.5	8.5	8.6	8.2	0.5	9.0	7.9	8.1	8.3	0.5	8.1	8.1	8.4	8.2	0.1

Avg.: Average (S): Surface
S.D.: Standard deviation (B): Bottom

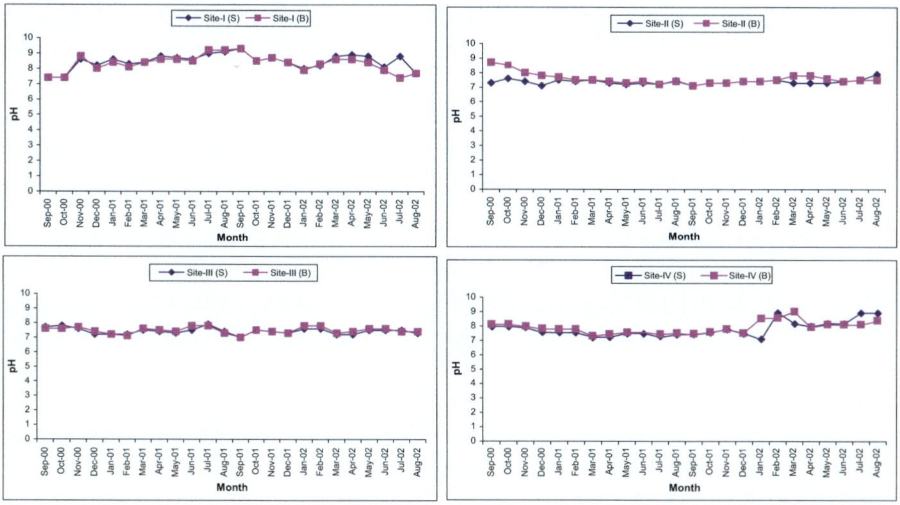

S = Surface; B = Bottom

Fig. 8 Monthly fluctuations in pH at the investigated sites of the Dal Lake

The pH fall on alkaline side for both surface and bottom water which is attributed to rich and intense macrophytic vegetation as has also been suggested by Qadri and Yousuf (1978), Khan (1979), Sarwar (1987, 1991a, b) and Wanganeo et al. (1996) in various Kashmir lakes and Kumar (1998) in fish ponds of Santal Pargana. On the other hand Zutshi et al. (1980) and Sarwar and Majid (1997) attributed the alkaline nature of the Kashmir lakes to the presence of calcium rich rocks in the catchment areas while Sarwar (1999) attributed it to higher silicate content in the rocky beds and sediments due to the production of silicic acid by hydrolysis of the silicate. The alkaline nature of other water bodies have also been reported by Bist (1986), Prakash (1990), Bath (1996), Maruthi et al. (2000), Khabade et al. (2003) and Esmaeili and Johal (2005).

The data on pH values reveals no marked difference at the study sites showing little effect of hotels, house boats and floating gardens on the pH of water. This might be due to buffering capacity of water as it absorbs acid or alkaline inputs without undergoing much change in its pH values. Singh (1960), Zutshi and Vass (1978), Sharma and Dhaneshwar (1986), Abel (1989), Kundangar and Ahmad (1991) and Jameson and Rana (1996) also observed similarly.

Earlier Zutshi et al. (1980) recorded the pH of this lake with wide range towards alkaline side. Sarwar and Zutshi (1989) recorded the average pH to be 8.4 in this lake.

Dissolved Oxygen (mg/L)

Dissolved oxygen is one of the important factor which measure the extent of organic as well as biological pollution load in the aquatic systems. The dissolved oxygen in the aquatic systems is also a function of biological processes such as photosynthesis and respiration (Trivedy and Goel 1986; Esmaeili and Johal 2005).

Dissolved oxygen values registered for Dal lake fluctuated from 0.8 to 12.8 mg/L. The data on dissolved oxygen at different sites reveals dissolved oxygen to be lower during summer and higher during winter at sites I, III and IV. At high temperature during summer, oxygen holding capacity of water gets decreased and thus dissolved oxygen values decline where as at lower temperature, oxygen remains dissolved in water increasing its dissolved oxygen contents. Hutchinson (1957), Chakraboarty et al. (1959), Reid (1961), Pahwa and Mehrotra (1966), Ray et al. (1966), Vasisht and Sharma (1975), Badola and Singh (1981), Bhowmick and Singh (1985), Bandopadhyay and Gopal (1991), Datta Munshi and Singh (1991), Shastree et al. (1991) and Esmaeili and Johal (2005) have also given similar explanations. At site-II, definite seasonal trend is not indicated. Here water is shallow and is frequently disturbed by plying of shikaras resulting in the regular aeration of water.

Dissolved oxygen values over 2 years study ranged from 6.0 to 11.4 mg/L at site-I, from 8.3 to 12.8 at site-II, from 2.0 to 10.6 mg/L at site-III and from 0.8 to 7.8 mg/L at site-IV (Table 7 and Fig. 9).

The overall pattern of dissolved oxygen is as follows

$$\text{Site II} > \text{Site I} > \text{site III} > \text{site IV}$$

At Site-I water holds good amount of dissolved oxygen. This sampling site is located in the open water area of the lake and is least interfered by human habitation and, therefore, does not receive domestic sewage or any other type of waste. The water here is crystal clear and harbours rich mats of submerged macrophytes but lacks floating type of macrophytic vegetation. Submerged macrophytes liberate oxygen into water during photosynthetic activity. Sculthrope (1967) and Sarwar (1987) also recorded higher content of dissolved oxygen in water due to active photosynthesis by harbouring luxuriant growth of macrophytes. Furthermore, water is free from surface vegetation and thus is well aerated as is also suggested by Roy (2000) and Khatri and Dhankhar (2003).

At site-II dissolved oxygen was found to be maximum. Here water area has been extensively used for formation of floating gardens. The site harbours rich growth of macrophytes. High contents of dissolved oxygen in water at this place might be due to rich amount of oxygen produced during photosynthesis by macrophytes. Sculthrope (1967) and Sarwar (1987) recorded higher content of dissolved oxygen due to active photosynthesis by luxuriant growth of macrophytes. Moreover, frequent plying of shikara in this region causes ripples in the water which also add atmospheric oxygen into water. Roy (2000) also reported high contents of dissolved oxygen in water due to turbulence.

The dissolved oxygen contents of water were found to be lower at site-III and the lowest at site-IV. Both these sites harbour rich growth of submerged and floating type of macrophytes, the surface being covered by *Salvinia* sp. Reduction in the oxygen content of Kashmir lake waters due to free floating macrophyte *Salvinia* sp. cover has also been reported by Zutshi and Vass (1971). Secondly, these two sites receive ample amount of domestic sewage from hotels and houseboats due to intense washing and bathing. The decomposition of organic matter contained in sewage consume dissolved oxygen available in water and thus lowers the dissolved oxygen

Table 7 Monthly fluctuations in dissolved oxygen (mg/L) at the investigated sites of the Dal lake

	Months																		
2000–2001	**Sep**	**Oct**	**Nov**	**Avg**	**±S.D.**	**Dec**	**Jan**	**Feb**	**Avg**	**±S.D.**	**Mar**	**April**	**May**	**Avg**	**±S.D.**	**June**	**July**	**Aug**	**Avg**	**±S.D.**
SITE-I	6.8	8.8	9.6	8.4	1.2	10.0	10.2	9.5	9.9	0.3	9.0	9.3	8.4	8.9	0.4	6.0	6.1	9.6	7.2	1.7
SITE-II	8.8	9.0	10.2	9.3	0.6	9.6	10.4	10.0	10.0	0.3	9.2	10.2	9.2	9.5	0.5	12.8	12.8	11.0	12.2	0.9
SITE-III	8.4	5.8	6.8	7.0	1.1	8.0	7.6	9.4	8.3	0.8	9.4	10.6	8.0	9.3	1.1	5.6	5.4	6.5	5.8	0.5
SITE-IV	1.2	2.0	1.2	1.5	0.4	4.0	6.8	7.6	6.1	1.5	4.8	4.8	3.2	4.3	0.8	2.4	2.0	0.8	1.7	0.7
2000–2001	**Sep**	**Oct**	**Nov**	**Avg**	**±S.D.**	**Dec**	**Jan**	**Feb**	**Avg**	**±S.D.**	**Mar**	**April**	**May**	**Avg**	**±S.D.**	**June**	**July**	**Aug**	**Avg**	**±S.D.**
SITE-I	8.0	8.5	9.5	8.7	0.6	9.3	7.2	7.2	7.9	1.0	11.0	11.4	10.0	10.8	0.6	8.0	7.5	7.8	7.8	0.2
SITE-II	10.8	11.2	11.5	11.2	0.3	11.0	9.0	10.0	10.0	0.8	10.0	8.3	10.0	9.4	0.8	12.0	12.0	10.4	11.5	0.8
SITE-III	7.5	7.9	8.5	8.0	0.4	8.9	6.8	7.6	7.8	0.9	4.8	4.8	3.2	4.3	0.8	2.4	2.0	2.0	2.1	0.2
SITE-IV	2.5	3.5	3.5	3.2	0.5	4.0	7.6	6.8	6.1	1.5	6.5	7.6	7.8	7.3	0.6	8.0	5.6	5.4	6.3	1.2

Avg.: Average
S.D.: Standard deviation

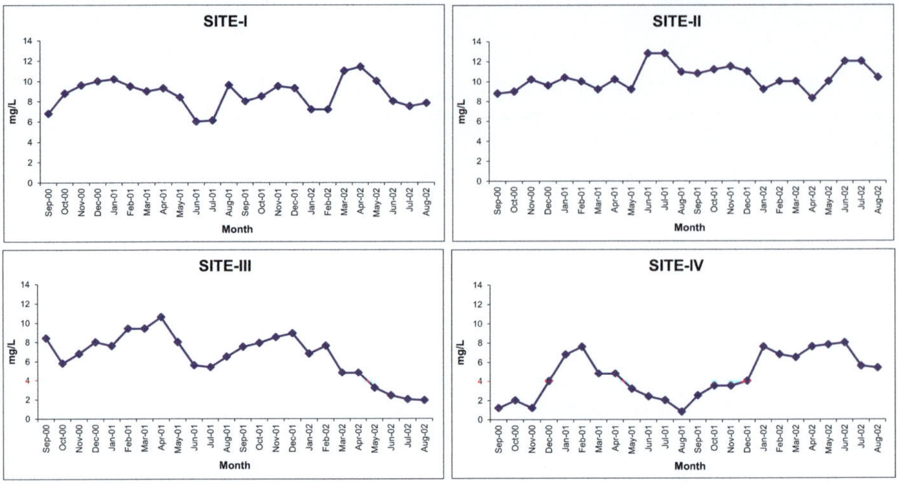

Fig. 9 Monthly fluctuations in dissolved oxygen (mg/L) at the investigated sites of the Dal lake

content. This has also been suggested by Gonzalves and Joshi (1946), Jolly and Chapman (1966), Venkateshwarlu and Jayanti (1968), De Smet and Evens (1972), Cairns et al. (1975), Vasisht and Sharma (1975), King (1981), Woodword (1984), Palharya and Malviya (1988), Rana and Palria (1988), Goel and Chavan (1991), Jhingran (1991a, b), Tripathi et al. (1991), Mathuthu et al. (1993), Bath (1996), Syal (1996), Kaur et al. (1996a, b), Jameel (1998), Kumar (1998), Valarmathi et al. (2002), Kumar et al. (2003), Kaur et al. (2003) and Prasannakumari et al. (2003).

Dissolved oxygen was found to be very low at site-IV where water was found to be highly turbid. Pahwa and Mehrotra (1966), Badola and Singh (1981) and Bhatt et al. (1988) while interrelating the turbidity, photosynthetic activity and dissolved oxygen suggested that in turbid water, photosynthetic activity is retarded leading to decline in the release of oxygen into water.

Earlier Sarwar and Zutshi (1989) recorded average dissolved oxygen to be 11.1 mg/L in the Dal lake. A comparison with present studies shows that dissolved oxygen has decreased in the lake specially at site -III and IV.

Bennet (1971) and ICMR (report no. 44) has prescribed minimum permissible limit of dissolved oxygen in water to be 15 mg/L, whereas, CBCWP (1982) prescribed this limit as 6 mg/L.On the basis of Bennet and ICMR limits, the dissolved oxygen in this lake is too low. However, as per the limits given by CBCWP the limits are within the range.

Total Alkalinity (mg/L)

Alkalinity of the water is its capacity to neutralise a strong acid and is characterised by the presence of all hydroxyl ions capable of combining with the hydrogen ion. Alkalinity is primarily due to presence of carbonates and bicarbonates which come from weathering of rocks (Trivedy and Goel 1986).

Total alkalinity values varied from 38 to 460 mg/L in surface water and from 78 to 480 mg/L for bottom waters showing higher values in bottom waters. Wanganeo et al. (1996) in Nilnag found higher values of total alkalinity for bottom water and recorded an increase in total alkalinity with increase in the depth across the column.

Total alkalinity was observed to be lower during summer. The water harbours rich growth of macrophytes at all the sites which is maximum during summer. The low values of total alkalinity during this period might be due to absorption of nutrients by the growing macrophytes. Kaul and Handoo (1980) also recorded lower values of total alkalinity from May to September and attributed this to the consumption of dissolved carbonates as a carbon source during rapid carbon assimilation by autotrophs.

Total alkalinity ranged from 42 to 112 mg/L in surface water and from 38 to 160 mg/L in bottom water at site-I, from 38 to 94 mg/L in surface water and from 70 to 162 mg/L in bottom waters at site-II, from 73 and 195 mg/L in surface water and 71 and 210 mg/L in bottom water and at site-III, from 82 to 460 mg/L in surface waters and from 102 to 480 mg/L in bottom water at site-IV (Table 8 and Fig. 10).

The overall site wise trend is

$$\text{Site IV} > \text{Site III} > \text{Site I} > \text{Site II}.$$

At site-I, with least human interferences, the amount of total alkalinity present is mainly contributed by carbonaceous salts present in rocks as has been suggested by Gibbs (1970), Sreenivasan (1970), Zutshi et al. (1980), Patra and Nayak (1982), Weiler (1988), Chapman and Kimstach (1992), Meybeek et al. (1992) and Syal (1996).

At site-II, total alkalinity was found to be minimum. This site is located where the lake is encroached by floating gardens and is extensively harbored by macrophytes. The low values of total alkalinity at this site might be due to absorption of carbonates and bicarbonates by the luxuriantly growing macrophytes as has also been suggested by Kaul and Handoo (1980).

The Site-III and IV are located where water receives sewage from the house boats and hotels respectively. Alkalinity was found to be very high at both of these sites, the maximum being at site-IV. High alkalinity values at these two sites might be due to addition of large quantities of sewage from hotels and house boats. Water (1957), Sarwar and Zutshi (1987a, b), Sarwar (1991a, b), Sarwar et al. (1996), Valarmathi et al. (2002), Kumar et al. (2003) and Prasannakumari et al. (2003) also observed increase in total alkalinity due to addition of sewage. Maximum values of total alkalinity at site-IV may be due to accumulation of water from all other sites bringing in and concentrating the nutrients at this site.

Freiser and Fernando (1966) and Qadri and Yousuf (1980a, b) while interpreting a relationship between pH and total alkalinity revealed that the carbonic acid predominates in water with pH between 6.0 or below 6 and bicarbonates predominates when pH is between 6 and 10 and carbonates predominate when pH is above 10. In the Dal lake, pH fluctuated between 7.0 and 9.3 suggesting that the alkalinity at all the sites of the lake was primarily due to bicarbonates. Zutshi et al.(1980) made similar observations for Kashmir lakes.

Table 8 Monthly fluctuations in total alkalinity (mg/L) at the investigated sites of the Dal lake

	Months																		
2000–2001	**Sep**	**Oct**	**Nov**	**Avg**	**±S.D.**	**Dec**	**Jan**	**Feb**	**Avg**	**±S.D.**	**Mar**	**April**	**May**	**Avg**	**±S.D.**	**June**	**July**	**Aug**	**Avg**	**±S.D.**
SITE-I (S)	49.0	48.0	45.0	47.3	1.7	79.0	50.0	49.0	59.3	13.9	55.0	59.0	49.0	54.3	4.1	49.0	49.0	42.0	46.7	3.3
SITE-I (B)	130.0	105.0	89.0	108.0	16.9	160.0	115.0	105.0	126.7	23.9	109.0	110.0	89.0	102.7	9.7	115.0	95.0	52.0	87.3	26.3
SITE-II (S)	49.0	45.0	41.0	45.0	3.3	75.0	55.0	57.0	62.3	9.0	49.0	43.0	46.0	46.0	2.4	40.0	49.0	64.0	51.0	9.9
SITE-II (B)	115.0	89.0	95.0	99.7	11.1	138.0	119.0	106.0	121.0	13.1	99.0	97.0	89.0	95.0	4.3	79.0	70.0	71.0	73.3	4.0
SITE-III (S)	85.0	80.0	90.0	85.0	4.1	125.0	115.0	110.0	116.7	6.2	98.0	85.0	109.0	97.3	9.8	75.0	115.0	73.0	87.7	19.3
SITE-III (B)	95.0	81.0	82.0	86.0	6.4	149.0	129.0	115.0	131.0	14.0	105.0	89.0	101.0	98.3	6.8	85.0	115.0	71.0	90.3	18.4
SITE-IV (S)	260.0	208.0	315.0	261.0	43.7	265.0	460.0	195.0	306.7	112.1	180.0	125.0	115.0	140.0	28.6	115.0	120.0	82.0	105.7	16.9
SITE-IV (B)	240.0	215.0	315.0	256.7	42.5	280.0	480.0	210.0	323.3	114.4	195.0	160.0	131.0	162.0	26.2	120.0	170.0	187.0	159.0	28.4
2001–2002	**Sep**	**Oct**	**Nov**	**Avg**	**±S.D.**	**Dec**	**Jan**	**Feb**	**Avg**	**±S.D.**	**Mar**	**April**	**May**	**Avg**	**±S.D.**	**June**	**July**	**Aug**	**Avg**	**±S.D.**
SITE-I (S)	112.0	79.0	84.0	91.7	14.5	62.0	128.0	86.0	92.0	27.3	100.0	65.0	97.0	87.3	15.8	65.0	74.0	64.0	67.7	4.5
SITE-I (B)	38.0	80.0	86.0	68.0	21.4	62.0	125.0	100.0	95.7	25.9	70.0	59.0	117.0	82.0	25.2	56.0	86.0	125.0	89.0	28.2
SITE-II (S)	94.0	90.0	90.0	91.3	1.9	92.0	45.0	77.0	71.3	19.6	74.0	38.0	46.0	52.7	15.4	38.0	43.0	42.0	41.0	2.2
SITE-II (B)	88.0	87.0	88.0	87.7	0.5	89.0	134.0	112.0	111.7	18.4	162.0	118.0	96.0	125.3	27.4	86.0	78.0	70.0	78.0	6.5
SITE-III (S)	89.0	85.0	85.0	86.3	1.9	80.0	160.0	195.0	145.0	48.1	180.0	125.0	115.0	140.0	28.6	115.0	120.0	182.0	139.0	30.5
SITE-III (B)	96.0	95.0	90.0	93.7	2.6	91.0	180.0	210.0	160.3	50.5	195.0	160.0	131.0	162.0	26.2	120.0	170.0	105.0	131.7	27.8
SITE-IV (S)	170.0	172.0	172.0	171.3	0.9	170.0	110.0	166.0	148.7	27.4	136.0	188.0	106.0	143.3	33.9	170.0	100.0	172.0	147.3	33.5
SITE-IV (B)	196.0	180.0	185.0	187.0	6.7	189.0	120.0	162.0	157.0	28.4	144.0	184.0	196.0	174.7	22.2	180.0	110.0	102.0	130.7	35.0

Avg.: Average (S): Surface
S.D.: Standard deviation (B): Bottom

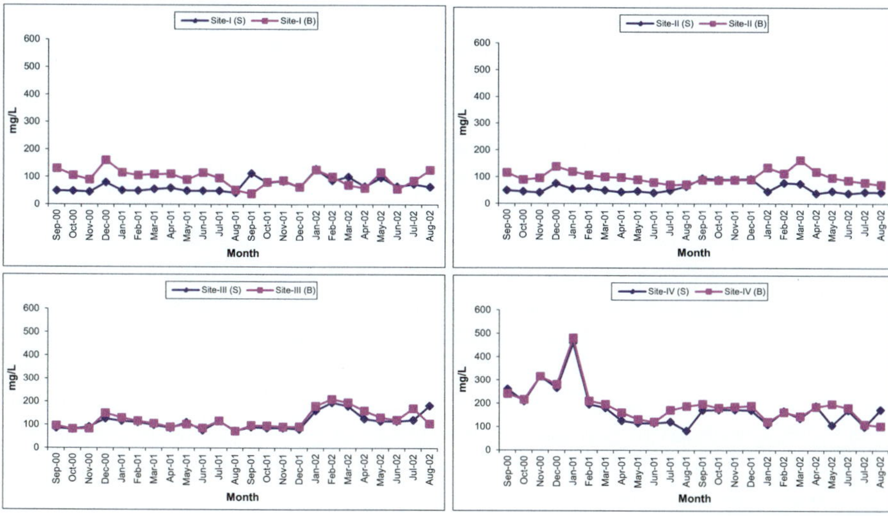

S = Surface; B = Bottom

Fig. 10 Monthly fluctuations in total alkalinity (mg/L) at the investigated sites of the Dal lake

Moyle (1946) while classifying lake waters on the basis of the total alkalinity, labelled waters having alkalinity upto 40 mg/L as soft, with 40–90 mg/L as medium and above 90 mg/L as hard. On the basis of this categorization, the Dal lake can be considered as medium to hard water lake. Sarwar and Rifat (1991), on the other hand, suggested that total alkalinity level of 40 mg/L provides a dividing line between soft and hard water. Philipose (1960) and Spence (1964) classified the lake as nutrient rich if the bicarbonate alkalinity is more than 60 mg/L. Based on this classification, the Dal lake water can be considered nutrient rich and productive.

Earlier Sarwar and Zutshi (1989) have recorded an average value of total alkalinity to be 71.9 mg/L. The present studies show an intensive increase in total alkalinity in the Dal lake over the years.

ISI (1986) recommended the maximum permissible limit of total alkalinity in water as 300 mg/L. During the present study, total alkalinity contents of water were found to be below 300 mg/L at site-I, II and III but above the limit at site-IV. ICMR (report no. 44) recommended the maximum permissible limits of total alkalinity in water as 120 mg/L. On the basis of these recommendations it can be concluded that the water at all the sites exceed the limits.

Total Hardness (mg/L)

Cations that form insoluble compounds with soap contribute to what is termed as "hardness". The two most common bivalent cations of lake water, calcium and magnesium, usually account for most of the hardness (Cole 1975; Esmaeili and Johal 2005).

The total hardness values in the Dal lake ranged from 73 to 464 mg/L for surface waters and from 92 to 462 mg/L for bottom waters without any marked difference in surface and bottom waters. The perusal of the data on the hardness concentration does not reveal any definite seasonal pattern.

The total hardness values ranged from 84 to 226 mg/L in surface water and 92 to 246 mg/L in bottom waters at site-I, from 85 to 329 mg/L in surface water and 111 to 266 mg/L in bottom waters at site-II, from 73 to 280 mg/L in surface water and from 119 and 285 mg/L in bottom waters at site-III and from 176 to 464 mg/L in surface water and 171 to 462 mg/L in bottom waters at site-IV (Table 9 and Fig. 11).

At site-I, hardness values for surface waters were found to be the lowest among the studied sites. Here the only source of hardness are the rocks from which bivalent salts like calcium and magnesium which contribute towards hardness dissolve into water as any other pollutant source is absent. Zutshi (1968), Vass (1973), Zutshi et al. (1980), Kundangar and Zutshi (1985), Sarwar (1987, 1991a, b), Sarwar and Wazir (1988), Sarwar and Rifat (1991) and Sarwar et al. (1996) who worked on the Dal and various other lakes of Kashmir also suggested calcium, one of the component of hardness, to be contributed from the rocks. Thomson (1952), Bhuyan (1970), Upadhyaya and Roy (1982), Trivedy and Goel (1986), Chapman and Kimstach (1992) and Meybeek et al. (1992) also suggested that calcium get dissolved in water from the rocks along its course.

At site-II, site-III and site-IV, the concentration of hardness was recorded to be high as compared to site-I. In addition to natural concentration of bivalents in water, these sites receive agricultural fertilizer run-off from floating gardens, sewage and domestic wastes from the houseboats and hotels. Singh et al. (1969), Bagde and Verma (1985a, b), Trivedy and Goel (1986), Sinha (1987), Sinha and Prasad (1988), Gopal (1990), Sinha et al. (1991), Kumar and Singh (1996), Syal (1996) and Kaur et al. (2003) also attributed high bivalent level in water containing sewage.

There are no earlier reports of the studies of total hardness in this lake.

Naturally occurring waters are usually classified into hard and soft waters. Kannan (1991) has provided the following chart depicting the nature of water on the basis of total hardness value.

Hardness (PPM)	Water quality
0–60	Soft
61–120	Moderately hard
121–180	Hard
>180	Very hard

Based on these values, the water of the Dal lake can be categorized as moderately hard to very hard.

Table 9 Monthly fluctuations in total hardness (mg/L) at the investigated sites of the Dal lake

	Months																			
2000–2001	**Sep**	**Oct**	**Nov**	**Avg**	**±S.D.**	**Dec**	**Jan**	**Feb**	**Avg**	**±S.D.**	**Mar**	**April**	**May**	**Avg**	**±S.D.**	**June**	**July**	**Aug**	**Avg**	**±S.D.**
SITE-I (S)	141.0	148.0	84.0	124.3	28.7	101.0	98.0	171.0	123.3	33.7	144.0	184.0	127.0	151.7	23.9	92.0	113.0	172.0	125.7	33.9
SITE-I (B)	167.0	103.0	94.0	121.3	32.5	122.0	127.0	92.0	113.7	15.5	238.0	229.0	109.0	192.0	58.8	103.0	106.0	159.0	122.7	25.7
SITE-II (S)	164.0	245.0	165.0	191.3	38.0	296.0	329.0	281.0	302.0	20.0	276.0	235.0	191.0	234.0	34.7	181.0	215.0	238.0	211.3	23.4
SITE-II (B)	193.0	241.0	266.0	233.3	30.3	228.0	218.0	254.0	233.3	15.2	182.0	158.0	192.0	177.3	14.3	169.0	126.0	186.0	160.3	25.2
SITE-III (S)	137.0	138.0	196.0	157.0	27.6	205.0	183.0	280.0	222.7	41.5	141.0	152.0	97.0	130.0	23.8	73.0	134.0	159.0	122.0	36.1
SITE-III (B)	120.0	204.0	204.0	176.0	39.6	285.0	273.0	284.0	280.7	5.4	230.0	183.0	187.0	200.0	21.3	183.0	235.0	247.0	221.7	27.8
SITE-IV (S)	264.0	374.0	439.0	359.0	72.2	366.0	464.0	396.0	408.7	41.0	322.0	225.0	220.0	255.7	46.9	213.0	241.0	258.0	237.3	18.6
SITE-IV (B)	256.0	401.0	462.0	373.0	86.4	415.0	415.0	403.0	411.0	5.7	293.0	268.0	225.0	262.0	28.1	223.0	186.0	174.0	194.3	20.9
2001–2002	**Sep**	**Oct**	**Nov**	**Avg**	**±S.D.**	**Dec**	**Jan**	**Feb**	**Avg**	**±S.D.**	**Mar**	**April**	**May**	**Avg**	**±S.D.**	**June**	**July**	**Aug**	**Avg**	**±S.D.**
SITE-I (S)	158.0	171.0	183.0	170.7	10.2	158.0	226.0	141.0	175.0	36.7	139.0	89.0	113.0	113.7	20.4	190.0	197.0	222.0	203.0	13.7
SITE-I (B)	141.0	198.0	191.0	176.7	25.4	170.0	246.0	201.0	205.7	31.2	183.0	163.0	116.0	154.0	28.1	117.0	83.0	173.0	124.3	37.1
SITE-II (S)	169.0	189.0	239.0	199.0	29.4	226.0	306.0	236.0	256.0	35.6	287.0	167.0	88.0	180.7	81.8	85.0	112.0	110.0	102.3	12.3
SITE-II (B)	192.0	212.0	241.0	215.0	20.1	232.0	114.0	178.0	174.7	48.2	215.0	200.0	192.0	202.3	9.5	169.0	125.0	111.0	135.0	24.7
SITE-III (S)	139.0	151.0	165.0	151.7	10.6	174.0	209.0	244.0	209.0	28.6	272.0	200.0	220.0	230.7	30.3	213.0	241.0	233.0	229.0	11.8
SITE-III (B)	119.0	131.0	163.0	137.7	18.6	190.0	262.0	250.0	234.0	31.5	243.0	243.0	200.0	228.7	20.3	223.0	161.0	149.0	177.7	32.4
SITE-IV (S)	261.0	259.0	241.0	253.7	9.0	257.0	347.0	297.0	300.3	36.8	275.0	181.0	199.0	218.3	40.7	176.0	235.0	190.0	200.3	25.2
SITE-IV (B)	257.0	265.0	242.0	254.7	9.5	271.0	310.0	346.0	309.0	30.6	344.0	250.0	253.0	282.3	43.6	171.0	195.0	237.0	201.0	27.3

Avg.: Average (S): Surface
S.D.: Standard deviation (B) : Bottom

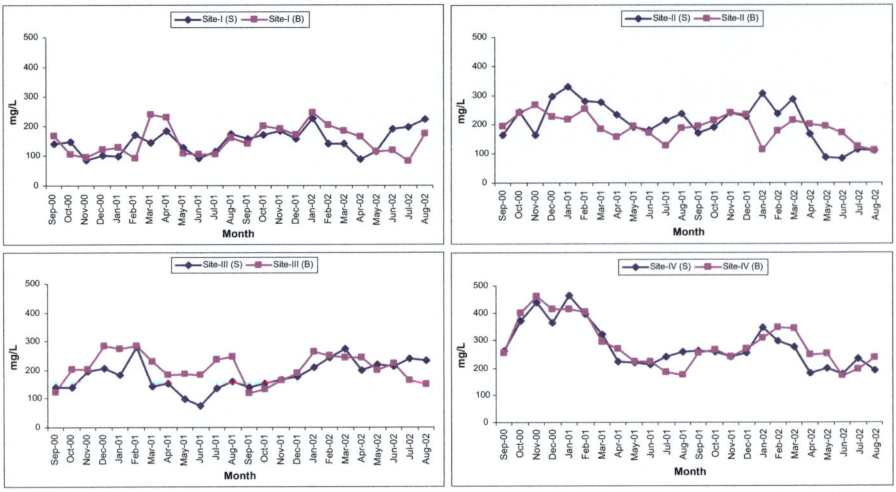

S = Surface; B = Bottom

Fig. 11 Monthly fluctuations in total hardness (mg/L) at the investigated sites of the Dal Lake

Calcium (mg/L)

Calcium is one of the most abundant substances of the natural waters. Being present in high quantities in the rocks, it is leached from these to contaminate the water. The quantities in natural waters generally vary from 10 to 100 mg/L depending upon the type of the rocks. Disposal of sewage and industrial wastes are also important sources of calcium. It has got a high affinity to get adsorbed on the soil particles and presence of other cations greatly influences its concentration in waters (Trivedy and Goel 1986).

The calcium values in the Dal lake ranged from 15.2 to 53.3 mg/L for surface waters and from 15.6 to 56.1 mg/L for bottom waters with slightly higher values in bottom waters.

The perusal of the data on the calcium concentration does not reveal any definite seasonal pattern. Zutshi and Vass (1978), Qadri and Yousuf (1980a, b), also failed to observe any seasonal trend in Kashmir lakes. Similarly in other lakes also, seasonal trend was not recorded by Pearsall (1932), Ganapati (1940), Zafar (1964), Husainy (1965), Boznaik and Kennedy (1968), Mac Crimmon and Kelso (1970), Munawar (1970a, b), Daborn (1976), Swarup and Singh (1979), Bagde and Verma (1985a, b), Sinha and Prasad (1988), Chapman and Kimstach (1992), Saradhamani and Siwakumar (1995) and Syal (1996). However, Shyamsunder (1988) observed rise of calcium ions during autumn and winter months.

The calcium values ranged from 18.6 to 28.0 mg/L in surface water and 15.6 to 39.8 mg/L in bottom waters at site-I, from 19.2 to 36.8 mg/L in surface water and 18.3 to 35.2 mg/L in bottom waters at site-II, from 16.0 to 29.5 mg/L in surface water and 17.3 and 29.6 mg/L in bottom waters at site-III, from 15.2 to 55.3 mg/L in surface water and 17.6 to 56.1 mg/L in bottom waters at site-IV. The data reveals calcium values to be lower in the surface waters than the bottom waters. Nutrients tend to settle at the bottom and are absorbed by the soil. The higher calcium values in the bottom waters might be due to solubility of nutrients from the soil into the deeper layer of water (Table 10 and Fig. 12).

At site-I, calcium values for surface waters were found to be the lowest among the studied sites. Here the only source of calcium are the rocks from which calcium salts dissolve into water as any other pollutant source is absent. This has also been suggested by Zutshi (1968), Vass (1973), Zutshi et al. (1980), Kundangar and Zutshi (1985), Sarwar (1987, 1991a, b), Sarwar and Wazir (1988), Sarwar and Rifat (1991) and Sarwar et al. (1996) who worked on the Dal and various other lakes of Kashmir. Thomson (1952), Bhuyan (1970), Upadhyaya and Roy (1982), Trivedy and Goel (1986), Chapman and Kimstach (1992) and Meybeek et al. (1992) also suggested that calcium get dissolved in water from the rocks along its course.

At site-II, site-III and site-IV, the concentration of calcium was recorded higher as compared to site-I. In addition to natural concentration of calcium in water, these sites receive wastes from one or the other polluting source. The increase in calcium concentration at site-II is due to addition of agricultural fertilizer wastes from the adjoining floating gardens. Sarwar (1999) also observed high calcium contents in waters receiving fertilizer waste from adjoining fields. At site-III and site-IV, the calcium enrichment of water is due to addition of sewage and domestic wastes from the house boats and hotels. At site-IV, concentration of calcium was found to be very high which is due to accumulation of water from previous sites. Singh et al. (1969), Bagde and Verma (1985a, b), Trivedy and Goel (1986), Sinha (1987), Sinha and Prasad (1988), Gopal (1990), Sinha et al. (1991), Syal (1996) and Kaur et al. (2003) also attributed high calcium level in water to sewage.

Ohle (1934) while laying stress on the importance of calcium in determining nutrient status of water, classified lakes with less than 10 mg/L of calcium as poor, with 10–25 mg/L as medium and with more than 25 mg/L as nutrient rich. As per this classification, the Dal lake water can be considered as ranging from medium to nutrient rich in terms of calcium concentration.

Earlier Sarwar and Zutshi (1989) recorded an average concentration of calcium to be 19.4 in this lake. The present studies show that calcium concentration has increased during these years due to multifarious human activities.

ISI (1986) have suggested the maximum permissible limit for calcium in drinking water to be 75 mg/L. Based on this, the calcium values which range from 15.2 to 53.3 mg/L for surface water at the investigated sites does not exceed the limit.

Magnesium (mg/L)

Magnesium occurs with calcium in all kinds of natural waters, but its concentration remains generally lower than calcium. The principal sources in the natural waters are various kinds of rocks. Sewage and industrial wastes are also important contributors of magnesium (Trivedy and Goel 1986).

Table 10 Monthly fluctuations in calcium (mg/L) at the investigated sites of the Dal lake

	Months																		
2000–2001	**Sep**	**Oct**	**Nov**	**Avg**	**±S.D.**	**Dec**	**Jan**	**Feb**	**Avg**	**±S.D.**	**Mar**	**April**	**May**	**Avg**	**±S.D.**	**June**	**July**	**Aug**	**Avg**	**±S.D.**
SITE-I (S)	20.0	20.5	24.0	21.5	1.8	28.0	24.8	25.4	26.1	1.4	21.0	27.0	27.0	25.0	2.8	25.0	18.6	22.5	22.0	2.6
SITE-I (B)	32.0	20.5	23.4	25.3	4.9	31.2	28.0	27.0	28.7	1.8	39.8	29.6	20.8	30.1	7.8	27.5	20.5	25.4	24.5	2.9
SITE-II (S)	20.4	23.4	24.4	22.7	1.7	28.5	30.6	28.5	29.2	1.0	26.5	26.4	26.5	26.5	0.0	22.4	19.3	24.3	22.0	2.1
SITE-II (B)	23.4	24.4	26.5	24.8	1.3	32.8	28.5	30.4	30.6	1.8	28.5	28.5	24.0	27.0	2.1	20.8	18.3	20.5	19.9	1.1
SITE-III (S)	16.3	20.4	24.8	20.5	3.5	24.1	20.5	28.0	24.2	3.1	26.0	20.4	23.4	23.3	2.3	17.5	21.4	25.7	21.5	3.3
SITE-III (B)	17.3	20.3	20.3	19.3	1.4	25.6	25.5	29.6	26.9	1.9	24.0	20.5	24.4	23.0	1.8	19.5	21.6	25.4	22.2	2.4
SITE-IV (S)	40.5	39.5	55.3	45.1	7.2	41.4	60.5	43.5	48.5	8.6	44.4	30.6	29.5	34.8	6.8	24.4	27.5	31.5	27.8	2.9
SITE-IV (B)	40.8	34.6	56.1	43.8	9.0	44.8	44.8	43.8	44.5	0.5	41.5	35.7	33.5	36.9	3.4	24.4	30.6	30.7	28.6	2.9
2001–2002	**Sep**	**Oct**	**Nov**	**Avg**	**±S.D.**	**Dec**	**Jan**	**Feb**	**Avg**	**±S.D.**	**Mar**	**April**	**May**	**Avg**	**±S.D.**	**June**	**July**	**Aug**	**Avg**	**±S.D.**
SITE-I (S)	20.0	25.5	27.5	24.3	3.2	24.4	26.3	24.8	25.2	0.8	25.4	21.3	27.6	24.8	2.6	27.0	25.0	23.0	25.0	1.6
SITE-I (B)	32.8	27.0	27.8	29.2	2.6	25.0	25.8	31.1	27.3	2.7	28.3	42.4	23.6	31.4	8.0	16.5	15.6	23.4	18.5	3.5
SITE-II (S)	21.6	24.5	23.9	23.3	1.2	25.0	28.3	40.8	31.4	6.8	36.8	27.2	26.4	30.1	4.7	22.4	19.2	20.2	20.6	1.3
SITE-II (B)	23.2	25.4	24.7	24.4	0.9	25.0	28.0	32.0	28.3	2.9	35.2	26.4	24.0	28.5	4.8	20.8	18.4	21.6	20.3	1.4
SITE-III (S)	16.0	17.5	18.5	17.3	1.0	18.5	30.1	23.5	24.0	4.8	24.4	20.6	29.5	24.8	3.6	24.4	27.5	21.5	24.5	2.4
SITE-III (B)	17.6	17.8	18.0	17.8	0.2	18.9	24.8	23.8	22.5	2.6	21.5	25.7	23.5	23.6	1.7	24.4	20.6	20.7	21.9	1.8
SITE-IV (S)	40.8	38.2	35.1	38.0	2.3	35.0	55.3	40.8	43.7	8.5	34.4	24.8	23.2	27.5	4.9	17.6	21.6	15.2	18.1	2.6
SITE-IV (B)	40.8	38.2	35.5	38.2	2.2	35.2	25.6	47.2	36.0	8.8	36.0	27.2	24.0	29.1	5.1	17.6	21.6	22.4	20.5	2.1

Avg.: Average (S): Surface
S.D.: Standard deviation (B): Bottom

S = Surface; B = Bottom

Fig. 12 Monthly fluctuations in calcium (mg/L) at the investigated sites of the Dal lake

The magnesium values in the Dal lake ranged from 2.2 to 30.5 mg/L for surface water and from 3.4 to 31.4 mg/L for bottom water showing slightly higher concentration in the bottom waters. No definite seasonal trend was observed. However, at site-II, Site-III and Site-IV, magnesium ion concentration was found to be higher in winter. This may be because of less macrophytic growth and therefore, lower uptake of magnesium by plants. Zafar (1964), Zutshi and Vass (1978), Zutshi et al. (1980), Trivedy and Goel (1986), Sarwar and Rifat (1991) and Saradhamani and Siwakumar (1995) also failed to observe any seasonal change in magnesium concentration intheir studies.

The magnesium ranged from 2.4 to 16.1 mg/L for surface waters and from 2.4 to 17.7 mg/L for bottom waters at site-I, from 2.2 to 24.6 mg/L for surface water and from 4.3 to 19.5 mg/L for bottom waters at site-II, from 2.9 to 20.6 mg/L for surface water and 7.3 to 21.4 mg/L for bottom water at site-III and from 11.6 to 30.5 mg/L for surface and 9.5 to 31.4 mg/L for bottom waters at site-IV (Table 11 and Fig. 13).

The overall trend at various sites is

$$\text{Site IV} > \text{Site II} > \text{Site III} > \text{Site I}$$

At site-I, magnesium contents were found to be lowest for surface as well for bottom waters. The only source of magnesium here seems to be the rocks from which magnesium salts dissolve in water while flowing over them. This has also been suggested by Zutshi et al. (1980) for Kashmir lakes and Bhuyan (1970), Trivedy and Goel (1986), Chapman and Kimstach (1992) and Meybeek et al. (1992) for other water bodies.

At Site-II, located in the area where floating gardens cover a large area of the lake, magnesium concentration was higher than at site-I. This site receive agricultural run-off from the floating gardens and dead organic matter by death and decay of macrophytes. These factors appear to increase the magnesium content of water. Site-III and site-IV are located at the places which receive effluence from house-

Table 11 Monthly fluctuations in magnesium (mg/L) at the investigated sites of the Dal lake

	Months																		
2000–2001	**Sep**	**Oct**	**Nov**	**Avg**	**±S.D.**	**Dec**	**Jan**	**Feb**	**Avg**	**±S.D.**	**Mar**	**April**	**May**	**Avg**	**±S.D.**	**June**	**July**	**Aug**	**Avg**	**±S.D.**
SITE-I (S)	8.9	9.5	2.4	6.9	3.2	3.1	3.6	10.5	5.7	3.4	9.0	11.4	5.8	8.7	2.3	2.9	6.5	11.3	6.9	3.4
SITE-I (B)	8.5	5.1	3.5	5.7	2.1	4.3	5.6	2.4	4.1	1.3	13.5	15.1	5.6	11.4	4.2	3.4	5.4	9.3	6.0	2.4
SITE-II (S)	11.0	18.2	10.2	13.1	3.6	21.9	24.6	20.5	22.3	1.7	20.6	16.5	12.2	16.4	3.4	12.2	16.3	17.3	15.3	2.2
SITE-II (B)	13.1	17.6	19.5	16.7	2.7	14.3	14.3	17.4	15.3	1.5	10.8	8.5	12.9	10.7	1.8	11.4	7.9	13.2	10.8	2.2
SITE-III (S)	9.4	8.5	13.1	10.3	2.0	14.1	12.9	20.5	15.8	3.3	7.4	9.9	3.8	7.0	2.5	2.9	7.9	9.3	6.7	2.7
SITE-III (B)	7.5	15.0	15.0	12.5	3.5	21.5	20.4	20.5	20.8	0.5	16.6	12.9	12.3	13.9	1.9	13.1	17.7	17.9	16.2	2.2
SITE-IV (S)	15.9	26.8	29.3	24.0	5.8	25.6	30.5	28.0	28.0	2.0	20.6	14.5	14.3	16.5	2.9	14.8	16.8	17.5	16.4	1.1
SITE-IV (B)	15.0	30.6	31.4	25.7	7.5	29.5	29.5	28.6	29.2	0.4	18.5	17.4	13.8	16.6	2.0	15.8	10.7	9.5	12.0	2.7
2001–2002	**Sep**	**Oct**	**Nov**	**Avg**	**±S.D.**	**Dec**	**Jan**	**Feb**	**Avg**	**±S.D.**	**Mar**	**April**	**May**	**Avg**	**±S.D.**	**June**	**July**	**Aug**	**Avg**	**±S.D.**
SITE-I (S)	8.2	10.5	11.2	10.0	1.3	9.5	15.6	7.7	10.9	3.4	7.4	3.5	4.3	5.1	1.7	12.0	13.1	16.1	13.7	1.7
SITE-I (B)	5.8	12.7	11.9	10.1	3.1	10.5	17.7	12.0	13.4	3.1	11.0	5.6	5.6	7.4	2.5	7.4	4.3	11.2	7.6	2.8
SITE-II (S)	11.2	12.5	17.5	13.7	2.7	15.5	22.9	13.1	17.2	4.2	19.0	9.7	2.2	10.3	6.9	2.9	6.3	5.8	5.0	1.5
SITE-II (B)	13.1	14.5	17.5	15.0	1.8	16.5	4.3	9.6	10.1	5.0	12.4	13.1	12.9	12.8	0.3	11.4	7.7	5.6	8.2	2.4
SITE-III (S)	9.7	10.5	11.6	10.6	0.8	12.5	13.1	18.1	14.6	2.5	20.6	14.5	14.3	16.5	2.9	14.8	16.8	17.5	16.4	1.1
SITE-III (B)	7.3	8.5	11.5	9.1	1.8	13.9	19.5	18.6	17.3	2.5	18.5	17.4	13.8	16.6	2.0	15.8	10.7	9.5	12.0	2.7
SITE-IV (S)	15.6	16.0	15.0	15.5	0.4	16.5	20.4	19.0	18.6	1.6	18.5	11.6	13.8	14.6	2.9	12.9	17.7	14.8	15.1	2.0
SITE-IV (B)	15.1	16.5	15.0	15.5	0.7	17.9	24.0	22.2	21.4	2.6	24.8	17.8	18.8	20.5	3.1	12.4	13.8	17.7	14.6	2.2

Avg.: Average (S): Surface
S.D.: Standard deviation (B): Bottom

Fig. 13 Monthly fluctuations in magnesium (mg/L) at the investigated sites of the Dal lake

boats and hotels. The increasing trend in the concentration of magnesium at these sites seems to be due to contribution of magnesium by sewage as has also been reported by Singh et al. (1969), Munawar (1970a, b), Bagde and Verma (1985a, b), Sinha et al. (1991), Syal (1996) and Kaur et al. (2003). Maximum level of magnesium was observed at Site-IV where water from all the sites accumulates and thus magnesium contents get concentrated.

At all the study sites, the concentration of magnesium was lower as compared to calcium. Trivedy and Goel (1986) and Sarwar (1999) also observed similarly and opined that magnesium ions are essential component of Chlorophyll-Magnesium-Porphyrine and are readily absorbed by plants.

Earlier Sarwar and Zutshi (1989) recorded average magnesium concentration to be 4.1 mg/L, much lower than the level recorded in the present study. It is therefore, evident that magnesium concentration in the Dal lake has increased over the years due to various anthropogenic activities.

ISI (1986) have suggested the maximum permissible limit for magnesium in drinking water to be 30 mg/L. Based on this, the magnesium values which range from 2.2 to 30.5 mg/L for surface at the investigated sites do not exceed the limits.

Iron (μg/L)

Iron is one of the most abundant elements of the rocks and soil. All kinds of waters have appreciable quantities of iron. Iron occurs in two valence forms viz., iron (II), ferrous and iron (III), ferric. Reduced iron is generally more soluble than oxidised iron (Trivedy and Goel 1986)

The iron concentration in the Dal lake ranged from 120 to 380 μg/L for surface water and 120 to 370 μg/L for bottom waters. No appreciable difference in the iron concentration between surface and bottom waters was recorded. Wanganeo et al. (1996) found slightly higher concentration of iron in the bottom water.

No definite seasonal pattern with respect to iron concentration of water was recorded at all the studied sites of the lake. Sajeev (1999) also failed to observe any seasonal trend in iron concentration in Ropar reservoir of Punjab. Hasan and Pandey (1985) observed lower values in both winter and summer, while Wanganeo et al. (1996) and Suvarna and Somashekar (2000) recorded maximum values in summer.

The iron concentration ranged from 120 to 320 in surface and 120 to 320 in bottom water at site-I, 210 to 380 in surface water and 220 to 370 in bottom water at site-II, 200 to 349 in surface water and 280 to 345 in bottom water at site-III and 140 to 349 in surface water and 229 to 350 in bottom water at site-IV (Table 12 and Fig. 14).

At site-I, concentration of iron was found to be lower than other studied sites which may be attributed to absence of disturbances caused by human habitation.

Site-II is located within the floating garden area of the lake. The concentration of iron at this place was found to be maximum among all the studied sites. Here, water receives nutrient loaded run-off from floating gardens, which increase iron concentration in water. Sarwar and Wazir (1988) and Sarwar (1991a, b) also attributed high content of iron in water to nutrient loaded run-off from adjoining vegetable fields and leaching from the catchment area.

Site-III and site-IV are located in the lake where water receives domestic sewage discharged from house boats and hotels at these sites. High values of iron contents in water suggest that sewage contributed good amount of iron to water. Trivedy and Goel (1986), Israili (1991), and Sajeev (1999) also suggested that iron is contribute in water by sewage and organic wastes.

Earlier Sarwar and Zutshi (1989) recorded average value of iron to be 144 μg/L which is much lower than the values recorded during the present study, thereby showing iron concentration to be on the rise due to sewage and agricultural run-off.

ISI (1986) has prescribed a maximum permissible limit of iron for drinking water as 300 μg/L. The concentration of the investigated lake ranges between 120 and 420 μg/L suggesting thereby that it exceeds the limit prescribed for drinking water.

Sodium (mg/L)

The monovalent cation sodium is one of the most abundant elements in the lithosphere. This alkali metal is very reactive and soluble when leached from the rocks and its compounds tend to remain in solution. For this reason it is at least the third most abundant metal in lakes (Cole 1975). Sodium, one of the natural cation is added in waters by weathering of various rocks while domestic sewage is also rich source of sodium (Gonzalves and Joshi 1946; Singh 1965 and Trivedy and Goel 1986).

Table 12 Monthly fluctuations in iron (μg/L) at the investigated sites of the Dal lake

	Months																		
2000–2001	**Sep**	**Oct**	**Nov**	**Avg**	**±S.D.**	**Dec**	**Jan**	**Feb**	**Avg**	**±S.D.**	**Mar**	**April**	**May**	**Avg**	**±S.D.**	**June**	**July**	**Aug**	**Avg**	**±S.D.**
SITE-I (S)	291.0	285.0	299.0	291.7	5.7	281.0	291.0	305.0	292.3	9.8	301.0	295.0	261.0	285.7	17.6	259.0	241.0	251.0	250.3	7.4
SITE-I (B)	281.0	285.0	296.0	287.3	6.3	280.0	270.0	301.0	283.7	12.9	285.0	261.0	245.0	263.7	16.4	230.0	235.0	240.0	235.0	4.1
SITE-II (S)	305.0	319.0	310.0	311.3	5.8	350.0	349.0	305.0	334.7	21.0	315.0	321.0	351.0	329.0	15.7	305.0	316.0	270.0	297.0	19.6
SITE-II (B)	300.0	311.0	305.0	305.3	4.5	335.0	340.0	310.0	328.3	13.1	310.0	305.0	305.0	306.7	2.4	310.0	329.0	370.0	336.3	25.0
SITE-III (S)	319.0	321.0	321.0	320.3	0.9	305.0	340.0	349.0	331.3	19.0	341.0	343.0	325.0	336.3	8.1	315.0	310.0	205.0	276.7	50.7
SITE-III (B)	320.0	320.0	325.0	321.7	2.4	310.0	345.0	345.0	333.3	16.5	335.0	340.0	317.0	330.7	9.9	319.0	308.0	296.0	307.7	9.4
SITE-IV (S)	319.0	316.0	321.0	318.7	2.1	301.0	343.0	351.0	331.7	21.9	349.0	340.0	310.0	333.0	16.7	305.0	315.0	200.0	273.3	52.0
SITE-IV (B)	325.0	320.0	321.0	322.0	2.2	315.0	351.0	350.0	338.7	16.7	330.0	335.0	314.0	326.3	9.0	309.0	300.0	230.0	279.7	35.3
2001–2002	**Sep**	**Oct**	**Nov**	**Avg**	**±S.D.**	**Dec**	**Jan**	**Feb**	**Avg**	**±S.D.**	**Mar**	**April**	**May**	**Avg**	**±S.D.**	**June**	**July**	**Aug**	**Avg**	**±S.D.**
SITE-I (S)	279.0	285.0	284.0	282.7	2.6	280.0	240.0	120.0	213.3	68.0	300.0	210.0	150.0	240.0	88.3	320.0	320.0	280.0	360.0	58.9
SITE-I (B)	280.0	285.0	285.0	283.3	2.4	281.0	220.0	160.0	220.3	49.4	320.0	120.0	220.0	220.0	81.6	320.0	280.0	320.0	306.7	18.9
SITE-II (S)	315.0	309.0	317.0	313.7	3.4	320.0	240.0	220.0	260.0	43.2	260.0	210.0	250.0	240.0	21.6	220.0	380.0	280.0	293.3	66.0
SITE-II (B)	314.0	310.0	315.0	313.0	2.2	309.0	245.0	235.0	263.0	32.8	250.0	235.0	245.0	243.3	6.2	220.0	295.0	280.0	265.0	32.4
SITE-III (S)	321.0	325.0	325.0	323.7	1.9	329.0	343.0	351.0	341.0	9.1	349.0	340.0	310.0	333.0	16.7	305.0	315.0	200.0	273.3	52.0
SITE-III (B)	315.0	316.0	319.0	316.7	1.7	325.0	340.0	345.0	336.7	8.5	330.0	335.0	315.0	326.7	8.5	320.0	310.0	280.0	303.3	17.0
SITE-IV (S)	301.0	305.0	305.0	303.7	1.9	310.0	180.0	180.0	223.3	61.3	200.0	140.0	268.0	202.7	52.3	180.0	185.0	200.0	188.3	8.5
SITE-IV (B)	301.0	307.0	308.0	305.3	3.1	312.0	340.0	320.0	324.0	11.8	300.0	240.0	229.0	256.3	31.2	280.0	300.0	340.0	306.7	24.9

Avg.: Average (S): Surface
S.D.: Standard deviation (B): Bottom

Fig. 14 Monthly fluctuations in iron (µg/L) at the investigated sites of the Dal lake

The level of sodium ranged from 2 to 18 mg/L for surface water as well as bottom waters. No seasonal trend is revealed from the data.

Sodium concentration varied from 2 to 14 mg/L for surface water and 2 to 18 mg/L for bottom waters at site-I, from 2 to 13 mg/L for surface water and 5 to 14 mg/L for bottom waters at site-II, from 4 to 15 mg/L for surface water and 3 to 10 mg/L for bottom waters at site-III, and from 2 to 18 mg/L for surface water and 3 to 18 mg/L for bottom waters at site-IV (Table 13 and Fig. 15).

The site wise trend is as follows

$$\text{Site IV} > \text{Site III} > \text{Site I} > \text{Site II}$$

Site-I is located where water does not receive any effluent, here water is least interfered by human activities and whatever sodium is present is due to the weathering of rocks.

Site-II is located where water receives run-off from agricultural fields. Here also, concentration of sodium is lower showing little contribution of agricultural run-off towards sodium concentration.

At site–III, water receives domestic sewage from houseboats and at site–IV from the hotels. Increase in concentration of sodium at these two sites seems to be due to addition of sewage. According to Gonzalves and Joshi (1946), Singh (1960), Munawar (1970a) and Trivedy and Goel (1986), the domestic sewage is rich in sodium and increases its concentration in waters after disposal.

The concentration of sodium was recorded lower than calcium and magnesium. Zutshi et al. (1980) and Sarwar and Zutshi (1989) while observing over all dominance of divalent cations, found sodium to be lower than calcium in most of the lakes of Kashmir except in Trigam lake where sodium is dominant over calcium which is also true for the present study.

Table 13 Monthly fluctuations in sodium (mg/L) at the investigated sites of the Dal lake

	Months																			
2000–2001	**Sep**	**Oct**	**Nov**	**Avg**	**±S.D.**	**Dec**	**Jan**	**Feb**	**Avg**	**±S.D.**	**Mar**	**April**	**May**	**Avg**	**±S.D.**	**June**	**July**	**Aug**	**Avg**	**±S.D.**
SITE-I (S)	4.0	4.0	2.0	3.3	0.9	5.0	3.0	4.0	4.0	0.8	5.0	11.0	14.0	10.0	3.7	14.0	13.0	8.0	11.7	2.6
SITE-I (B)	4.0	4.0	2.0	3.3	0.9	5.0	3.0	4.0	4.0	0.8	5.0	10.0	14.0	9.7	3.7	18.0	13.0	8.0	13.0	4.1
SITE-II (S)	10.0	10.0	13.0	11.0	1.4	12.0	11.0	6.0	9.7	2.6	8.0	6.0	5.0	6.3	1.2	7.0	5.0	6.0	6.0	0.8
SITE-II (B)	11.0	11.0	14.0	12.0	1.4	12.0	11.0	7.0	10.0	2.2	8.0	6.0	5.0	6.3	1.2	7.0	6.0	8.0	7.0	0.8
SITE-III (S)	9.0	7.0	4.0	6.7	2.1	6.0	8.0	9.0	7.7	1.2	8.0	11.0	14.0	11.0	2.4	15.0	13.0	8.0	12.0	2.9
SITE-III (B)	10.0	9.0	3.0	7.3	3.1	6.0	5.0	4.0	5.0	0.8	4.0	7.0	9.0	6.7	2.1	5.0	7.0	10.0	7.3	2.1
SITE-IV (S)	8.0	8.0	3.0	6.3	2.4	6.0	6.0	5.0	5.7	0.5	5.0	14.0	18.0	12.3	5.4	18.0	15.0	9.0	14.0	3.7
SITE-IV (B)	9.0	8.0	3.0	6.7	2.6	6.0	6.0	6.0	6.0	0.0	5.0	18.0	18.0	13.7	6.1	18.0	16.0	10.0	14.7	3.4
2001–2002	**Sep**	**Oct**	**Nov**	**Avg**	**±S.D.**	**Dec**	**Jan**	**Feb**	**Avg**	**±S.D.**	**Mar**	**April**	**May**	**Avg**	**±S.D.**	**June**	**July**	**Aug**	**Avg**	**±S.D.**
SITE-I (S)	6.0	5.0	7.0	6.0	0.8	5.0	4.0	2.0	3.7	1.2	4.0	2.0	8.0	4.7	2.5	8.0	8.0	4.0	6.7	1.9
SITE-I (B)	6.0	5.0	7.0	6.0	0.8	4.0	4.0	4.0	4.0	0.0	8.0	2.0	6.0	5.3	2.5	4.0	4.0	2.0	3.3	0.9
SITE-II (S)	9.0	7.0	7.0	7.7	0.9	6.0	4.0	2.0	4.0	1.6	4.0	2.0	8.0	4.7	2.5	6.0	6.0	7.0	6.3	0.5
SITE-II (B)	9.0	7.0	7.0	7.7	0.9	6.0	5.0	6.0	5.7	0.5	7.0	8.0	7.0	7.3	0.5	7.0	8.0	7.0	7.3	0.5
SITE-III (S)	8.0	8.0	8.0	8.0	0.0	8.0	6.0	5.0	6.3	1.2	5.0	4.0	8.0	5.7	1.7	8.0	5.0	9.0	7.3	1.7
SITE-III (B)	8.0	8.0	8.0	8.0	0.0	8.0	6.0	6.0	6.7	0.9	5.0	8.0	8.0	7.0	1.4	6.0	7.0	8.0	7.0	0.8
SITE-IV (S)	8.0	8.0	8.0	8.0	0.0	8.0	6.0	10.0	8.0	1.6	8.0	6.0	7.0	7.0	0.8	4.0	2.0	2.0	2.7	0.9
SITE-IV (B)	8.0	8.0	8.0	8.0	0.0	8.0	6.0	4.0	6.0	1.6	5.0	6.0	11.0	7.3	2.6	3.0	6.0	3.0	4.0	1.4

Avg.: Average (S): Surface
S.D.: Standard deviation (B): Bottom

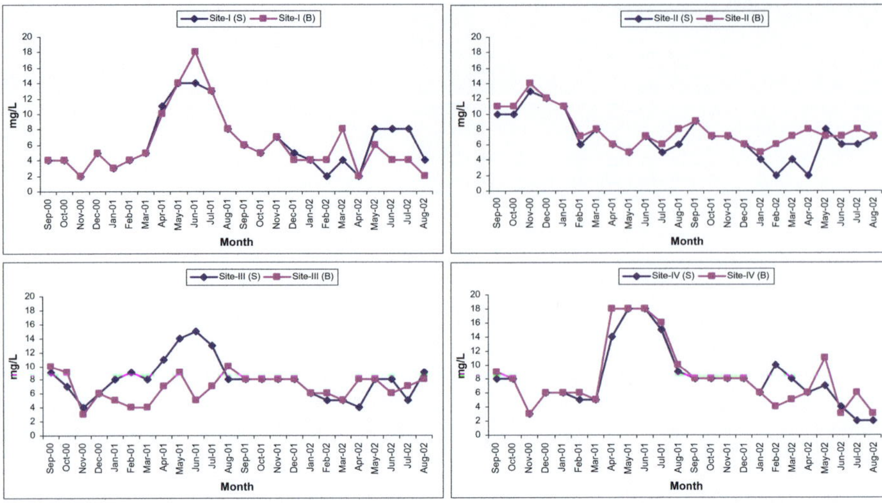

S = Surface; B = Bottom

Fig. 15 Monthly fluctuations in sodium (mg/L) at the investigated sites of the Dal lake

Earlier Sarwar and Zutshi (1989) recorded an average value of sodium in the Dal lake to be 3.3 mg/L. The present study shows the range in the Dal lake to be 2.0–18 mg/L, which is much higher indicating that sewage addition has increased over the years.

Potassium (mg/L)

Potassium, a close relative of sodium, is usually the fourth ranking cation in lake waters after calcium, magnesium and sodium. It is weathered from various feldspars but does not remain in solution so well as sodium. It combines with other products and is being removed from solution by adsorption on clay. Like sodium, potassium is also a naturally occurring element. (Cole 1975; Trivedy and Goel 1986).

The potassium concentration ranged from 1 to 10 mg/L for surface and 1 to 11 mg/L for bottom waters without any difference across the column. No seasonal trend is recorded from the 2 year study.

Potassium concentration varied from 2 to 8 mg/L for surface water and 1 to 8 mg/L for bottom water at site-I, from 1 to 9 mg/L for surface water and 3 to 10 mg/L bottom water at site-II, from 1 to 10 mg/L for surface water and 3 to 8 mg/L for bottom water at site-III and from 3 to 9 mg/L for surface water and 1 to 11 mg/L for bottom water at site-IV (Table 14 and Fig. 16).

The overall trend in the potassium values is as under

$$\text{Site IV} > \text{Site III} > \text{Site II} > \text{Site I}$$

Table 14 Monthly fluctuations in potassium (mg/L) at the investigated sites of the Dal lake

	Months																		
2000–2001	**Sep**	**Oct**	**Nov**	**Avg**	**±S.D.**	**Dec**	**Jan**	**Feb**	**Avg**	**±S.D.**	**Mar**	**April**	**May**	**Avg**	**±S.D.**	**June**	**July**	**Aug**	**Avg**	**±S.D.**
SITE-I (S)	3	3	2	2.67	0.47	4	3	2	3	0.82	3	6	8	5.67	2.05	7	6	5	6	0.82
SITE-I (B)	3	3	3	3	0	4	3	3	3.33	0.47	4	6	8	6	1.63	8	7	5	6.67	1.25
SITE-II (S)	8	7	8	7.67	0.47	4	5	9	6	2.16	4	6	4	4.67	0.94	3	4	7	4.67	1.7
SITE-II (B)	8	8	3	6.33	2.36	4	5	10	6.33	2.62	5	6	5	5.33	0.47	6	3	4	4.33	1.25
SITE-III (S)	4	4	3	3.67	0.47	4	6	4	4.67	0.94	3	5	6	4.67	1.25	10	10	7	9	1.41
SITE-III (B)	7	5	3	5	1.63	5	4	4	4.33	0.47	3	4	5	4	0.82	4	3	8	5	2.16
SITE-IV (S)	6	3	5	4.67	1.25	3	4	5	4	0.82	4	7	9	6.67	2.05	8	8	4	6.67	1.89
SITE-IV (B)	5	4	3	4	0.82	5	5	4	4.67	0.47	4	5	9	6	2.16	9	11	8	9.33	1.25
2001–2002	**Sep**	**Oct**	**Nov**	**Avg**	**±S.D.**	**Dec**	**Jan**	**Feb**	**Avg**	**±S.D.**	**Mar**	**April**	**May**	**Avg**	**±S.D.**	**June**	**July**	**Aug**	**Avg**	**±S.D.**
SITE-I (S)	4	2	2	2.67	0.94	2	1	4	2.33	1.25	6	5	3	4.67	1.25	2	1	2	1.67	0.47
SITE-I (B)	3	2	3	2.67	0.47	2	2	4	2.67	0.94	3	4	2	3	0.82	2	2	1	1.67	0.47
SITE-II (S)	3	3	3	3	0	3	3	1	2.33	0.94	3	1	4	2.67	1.25	3	3	3	3	0
SITE-II (B)	3	3	3	3	0	3	4	5	4	0.82	5	6	4	5	0.82	4	5	5	4.67	0.47
SITE-III (S)	1	2	2	1.67	0.47	2	2	6	3.33	1.89	4	6	2	4	1.63	2	1	1	1.33	0.47
SITE-III (B)	3	3	3	3	0	3	5	4	4	0.82	4	5	6	5	0.82	3	3	4	3.33	0.47
SITE-IV (S)	3	3	3	3	0	3	5	4	4	0.82	3	3	6	4	1.41	5	4	7	5.33	1.25
SITE-IV (B)	1	2	2	1.67	0.47	2	2	2	2	0	1	4	4	3	1.41	2	2	1	1.67	0.47

Avg.: Average (S): Surface
S.D.: Standard deviation (B): Bottom

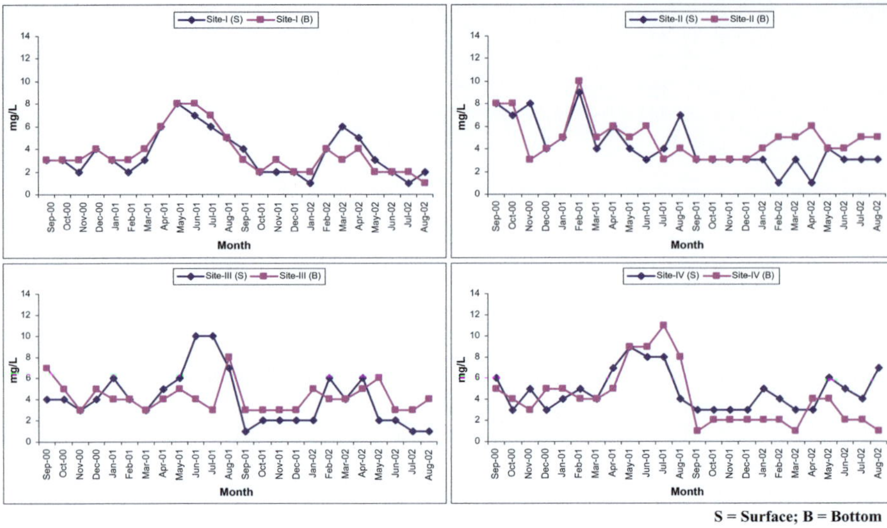

Fig. 16 Monthly fluctuations in potassium (mg/L) at the investigated sites of the Dal lake

Minimum values are recorded at site-I which is free of pollution. At site-II, III and IV, only a slight increase is observed by addition of agricultural run-off and sewage.

A comparison with the other cations shows the following trend

$$Ca > Mg > Na > K.$$

A similar trend was observed by Zutshi et al. (1980), Sarwar and Zutshi (1989) and Sarwar (1999) who recorded an overall dominance of bivalent cations over monovalent cations, and the potassium being the minimum of all the cations. Trivedy and Goel (1986) also observed that concentration of potassium remained lower than sodium, calcium and magnesium.

Earlier Sarwar and Zutshi (1989) reported average potassium concentration to be 2 mg/L. Their study also depicted potassium to be minimum of all the cations.

Chloride (mg/L)

Chloride occurs naturally in all types of waters. The most important source of chloride in waters is the domestic sewage. Man and other animals excrete very high quantities of chloride together with nitrogenous compounds. Sodium chloride which is the main source of chloride in waste water is a common article of diet and a part of it passes unchanged through the digestive system Therefore, the chloride concentration serves as an indicator of pollution by sewage. Chlorides are highly soluble with most of the naturally occurring cations and do not precipitate and cannot be removed biologically in treatment of the wastes. (Trivedy and Goel 1986 and APHA, 1989).

The chloride concentration fluctuated between 14 and 85 mg/L for surface water and between 10 and 92 mg/L for bottom waters. No appreciable difference in chloride contents was detected across the column of water.

From the 2 year study of the lake, definite seasonal pattern in chloride concentration could not be detected. Shastree et al. (1991), Bath (1996) and Sajeev (1999) also failed to observe any seasonal trend in their studies. Zutshi and Vass (1978) recorded lower chloride contents in autumn in the Dal lake. Ownbey and Kee (1967) and Ajmal et al. (1985) recorded higher chloride values during the winter and lower during summer.

The chloride concentration ranged from 14 to 37 mg/L in surface water and 10 to 35 mg/L in bottom water at site- I, from 8.5 to 75 mg/L in surface water and from 28 to 41 mg/L in bottom waters at site-II, from 15 to 39 mg/L in surface water and 12 to 32 mg/L in bottom waters at site-III, 15 to 85 mg/L in surface water and 20 to 92 mg/L in bottom waters at site-IV (Table 15 and Fig. 17).

The overall concentration of chloride at different sites show the pattern as

$$\text{Site IV} > \text{Site II} > \text{Site III} > \text{Site I}.$$

Site-I, which is located in the lake where water is least interfered by human habitation and does not receive any effluence, the concentration of chloride was found to be minimum among all the studied sites. The low concentration of chloride may be due to absence of addition of any type of domestic wastes. Prasannakumari et al. (2000) also observed low chloride contents in water devoid of sewage contamination.

Site-II is located where water is encroached upon by floating gardens which are used for agricultural activities. Here the soil is enriched with nutrients by extensive use of fertilizers. Chloride contents seem to be added to the water by agricultural run-off from the catchment floating garden.

Site-III is located where water receives faecal matter discharged from several lavatories of the house boats directly. At site-IV also, water receives sewage discharged from hotels. Water here collects from all the sites thereby further increasing the concentration of chloride in water. High concentration of chloride in water at sites receiving sewage indicates that chloride is added in good amount at these sites by the sewage. Drischel (1940), Thresh et al. (1944), Gonzalves and Joshi (1946), Blum (1957), Sawyer (1960), Zafar (1964),Singh (1965), Ownbey and Kee (1967), Venkateshwarlu (1969), Munawar (1970a, b), Cole (1975), Goel et al. (1980), Mehra (1986), Trivedy and Goel (1986), Das (1989), APHA (1989), Shastree et al. (1991), Bath (1996), Sarwar (1999), Prasannakumari et al. (2000) and Das et al. (2003) have also observed high contents of chloride in water due to addition of sewage.

Earlier Sarwar and Zutshi (1989) recorded average value of chloride to be 18.9 mg/L, which is much lower than the values recorded during the present study

Table 15 Monthly fluctuations in chloride (mg/L) at the investigated sites of the Dal lake

	Months																		
2000–2001	**Sep**	**Oct**	**Nov**	**Avg**	**±S.D.**	**Dec**	**Jan**	**Feb**	**Avg**	**±S.D.**	**Mar**	**April**	**May**	**Avg**	**±S.D.**	**June**	**July**	**Aug**	**Avg**	**±S.D.**
SITE-I (S)	17.0	21.0	18.0	18.7	1.7	38.0	23.0	21.0	27.3	7.6	25.0	14.0	19.0	19.3	4.5	18.0	17.0	18.0	17.7	0.5
SITE-I (B)	19.0	21.0	25.0	21.7	2.5	39.0	21.0	19.0	26.3	9.0	19.0	15.0	18.0	17.3	1.7	17.0	10.0	15.0	14.0	2.9
SITE-II (S)	19.0	25.0	19.0	21.0	2.8	39.0	21.0	19.0	26.3	9.0	19.0	16.0	19.0	18.0	1.4	8.5	9.9	28.0	15.5	8.9
SITE-II (B)	30.0	38.0	35.0	34.3	3.3	41.0	39.0	37.0	39.0	1.6	39.0	34.0	36.0	36.3	2.1	37.0	36.0	38.0	37.0	0.8
SITE-III (S)	18.0	25.0	26.0	23.0	3.6	35.0	25.0	20.0	26.7	6.2	20.0	19.0	15.0	18.0	2.2	17.0	18.0	16.0	17.0	0.8
SITE-III (B)	20.0	22.0	21.0	21.0	0.8	32.0	24.0	20.0	25.3	5.0	21.0	15.0	15.0	17.0	2.8	14.0	19.0	35.0	22.7	9.0
SITE-IV (S)	52.0	39.0	69.0	53.3	12.3	49.0	85.0	39.0	57.7	19.8	36.0	21.0	17.0	24.7	8.2	19.0	15.0	39.0	24.3	10.5
SITE-IV (B)	45.0	49.5	66.0	53.5	9.0	55.0	90.0	35.5	60.2	22.5	35.0	22.5	20.0	25.8	6.6	21.0	30.0	46.0	32.3	10.3
2001–2002	**Sep**	**Oct**	**Nov**	**Avg**	**±S.D.**	**Dec**	**Jan**	**Feb**	**Avg**	**±S.D.**	**Mar**	**April**	**May**	**Avg**	**±S.D.**	**June**	**July**	**Aug**	**Avg**	**±S.D.**
SITE-I (S)	26.0	25.0	22.0	24.3	1.7	20.0	25.0	31.0	25.3	4.5	28.0	37.0	27.0	30.7	4.5	19.0	51.0	23.0	31.0	14.2
SITE-I (B)	26.0	25.0	22.0	24.3	1.7	28.0	25.0	31.0	28.0	2.4	28.0	42.0	23.0	31.0	8.0	16.0	56.0	23.0	31.7	17.4
SITE-II (S)	68.0	75.0	60.0	67.7	6.1	60.0	42.5	49.8	50.8	7.2	48.0	39.8	34.2	40.7	5.7	34.2	38.5	49.2	40.6	6.3
SITE-II (B)	32.0	28.0	39.0	33.0	4.5	34.0	39.8	35.4	36.4	2.5	36.8	34.2	35.6	35.5	1.1	37.1	31.3	34.2	34.2	2.4
SITE-III (S)	28.0	27.0	25.0	26.7	1.2	16.0	25.0	39.0	26.7	9.5	36.0	21.0	17.0	24.7	8.2	19.0	15.0	29.0	21.0	5.9
SITE-III (B)	20.0	15.0	15.0	16.7	2.4	12.0	25.0	20.0	19.0	5.4	21.0	15.0	15.0	17.0	2.8	16.0	19.0	28.0	21.0	5.1
SITE-IV (S)	72.0	61.0	59.0	64.0	5.7	70.0	32.6	19.8	40.8	21.3	56.8	24.1	14.2	31.7	18.2	14.2	17.1	18.4	16.6	1.8
SITE-IV (B)	92.0	82.0	71.0	81.7	8.6	75.0	31.2	25.5	43.9	22.1	25.2	26.9	35.6	29.2	4.6	35.8	31.3	37.0	34.7	2.5

Avg.: Average (S): Surface
S.D.: Standard deviation (B): Bottom

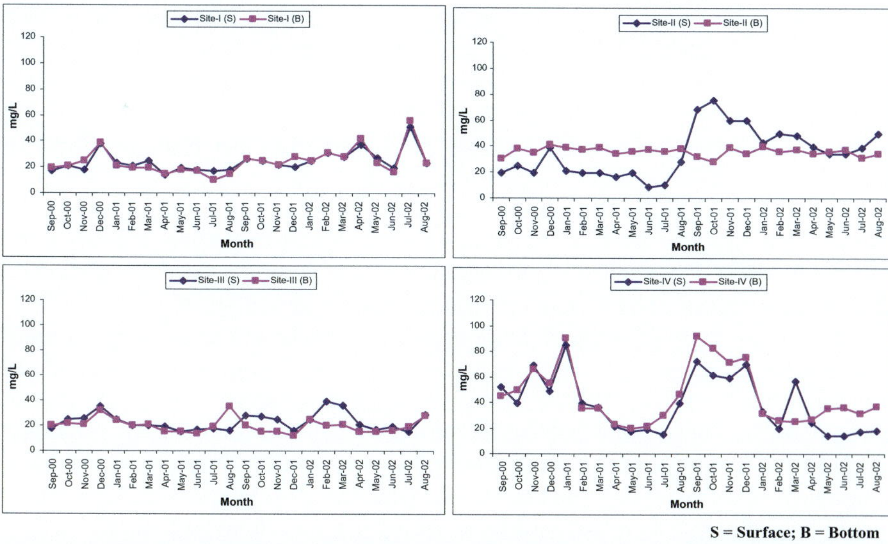

Fig. 17 Monthly fluctuations in chloride (mg/L) at the investigated sites of the Dal lake

(10–92 mg/L), thereby showing that sewage contamination has increased over the years because of encroachment of the lake by human habitation.

During the entire period of study, the chloride values were observed to remain within the maximum permissible limits which is 250 mg/L as suggested by I.S.I. (1986).

Ammonical Nitrogen (μg/L)

Most of the ammonia in fresh water exists in ionic form. Organic matter is the main source of ammonia in water. Ammonia of mineral origin is rare in natural waters. (Trivedy and Goel 1986). According to Bruce (1958), high value of free ammonia is a sign of organic pollution.

During the present investigation, ammonical nitrogen concentration ranged from 104 to 395 μg/L for surface waters and 110 to 490 μg/L for bottom waters thereby indicating higher amounts in the bottom waters. Higher values at the bottom might be due to decomposition of organic matter as suggested by Wanganeo et al. (1996) and Bandela et al. (1999).

No definite seasonal trend in the ammonical-nitrogen contents was revealed. Zutshi and Vass (1978) and Ramakrishnaiah and Sarkar (1986) also failed to record well marked and consistent seasonal pattern with respect to this form of nitrogen in the Dal lake and Konar reservoir (Bihar) respectively. On the other hand, Harold (1934), Adoni and Joshi (1987) and Bandopadhyay and Gopal (1991) observed lower values during summer.

Ammonical-nitrogen concentration varied from 104 to 280 µg/L for surface water and 110 to 290 µg/L for bottom water at site-I, from 120 to 209 µg/L for surface water and 189 to 290 µg/L for bottom waters at site-II, from 107 to 395 µg/L for surface water and 210 to 490 µg/L for bottom waters at site-IV and from 107 and 219 µg/L for surface water and 111 to 269 µg/L for bottom water at site-IV (Table 16 and Fig. 18).

Site-I is located in the lake where water does not receive any effluence from the human habitation like sewage etc. Ammonical nitrogen present here is likely to be due to microbial decomposition of macrophytes and excretory products of aquatic biota as also suggested by Wetzel (1975) and Bhatt et al. (1999).

Site-II is located within the floating gardens area. The site receives agricultural run-off from these gardens which is generally rich in nitrogenous compounds as also suggested by Sarwar (1991a, b). However, an expected increase in ammonical nitrogen was not observed which can be attributed to high level of dissolved oxygen at this site. In the presence of oxygen, ammonical nitrogen is readily converted into ammonical-nitrate (Trivedy and Goel 1986) which is also evident from high values of ammonical nitrate recorded at this site during the present investigation.

Site-III is located in the lake where water receives wastes from houseboats. This site is under intense pressure of human habitation. Waste disposal which brings in household effluents from houseboats and residues of detergents enrich the water with ammonical-nitrogen. This is evident by the high value of ammonical nitrogen recorded at this site. Thresh et al. (1944), Seenayya (1971), King (1981), Trivedy and Goel (1986), Das (1989), Sarwar (1991a, b), Bhatt et al. (1999) and Narain and Chauhan (2000) also suggested that direct discharge of sewage increases concentration of ammonical-nitrogen in water. According to Trivedy and Goel (1986), the presence of high amounts of ammonical-nitrogen indicates discharge of fresh faecal matter. Some of the ammonical-nitrogen is also added as a result of decomposition of macrophytes at this site.

At site-IV fresh sewage from hotels is added but ammonical nitrogen values have been found to be lower. This is likely to be due to conversion of ammonical nitrogen to ammonical nitrate by the time water reaches at the sample collection site. This is evident from the fact that ammonical nitrate concentration is very high at this site. Trivedy and Goel (1986) suggested that occurrence of ammonical nitrate along with ammonical nitrogen indicates that some time has lapsed since the sewage has been added.

Earlier Sarwar and Zutshi (1989) recorded average value of 10 µg/L which is much lower than the values recorded during the present study, thereby showing concentration of ammonical nitrogen on the rise due to sewage and agricultural run-off.

ISI. (1986) has recommended maximum permissible limit of this form of nitrogen to be 500 µg/L. The present data reveals that ammonical nitrogen is within the permissible limits in the Dal lake water.

Table 16 Monthly fluctuations in ammonical nitrogen (μg/L) at the investigated sites of the Dal lake

	Months																		
2000–2001	**Sep**	**Oct**	**Nov**	**Avg**	**±S.D.**	**Dec**	**Jan**	**Feb**	**Avg**	**±S.D.**	**Mar**	**April**	**May**	**Avg**	**±S.D.**	**June**	**July**	**Aug**	**Avg**	**±S.D.**
SITE-I (S)	104.0	155.0	280.0	179.7	73.9	249.0	220.0	150.0	206.3	41.6	105.0	108.0	280.0	164.3	81.8	150.0	109.0	182.0	147.0	29.9
SITE-I (B)	289.0	285.0	290.0	288.0	2.2	245.0	235.0	170.0	216.7	33.2	125.0	110.0	295.0	176.7	83.9	186.0	115.0	185.0	162.0	33.2
SITE-II (S)	150.0	150.0	141.0	147.0	4.2	141.0	150.0	151.0	147.3	4.5	188.0	188.0	136.0	170.7	24.5	150.0	126.0	120.0	132.0	13.0
SITE-II (B)	255.0	250.0	245.0	250.0	4.1	241.0	251.0	250.0	247.3	4.5	290.0	220.0	240.0	250.0	29.4	255.0	240.0	229.0	241.3	10.7
SITE-III (S)	140.0	140.0	180.0	153.3	18.9	280.0	310.0	395.0	328.3	48.7	380.0	380.0	250.0	336.7	61.3	285.0	268.0	161.0	238.0	54.9
SITE-III (B)	259.0	275.0	210.0	248.0	27.7	391.0	425.0	490.0	435.3	41.1	305.0	309.0	301.0	305.0	3.3	325.0	395.0	382.0	367.3	30.4
SITE-IV (S)	150.0	141.0	109.0	133.3	17.6	79.0	147.0	119.0	115.0	27.9	174.0	154.0	107.0	145.0	28.1	160.0	180.0	187.0	175.7	11.4
SITE-IV (B)	155.0	150.0	115.0	140.0	17.8	85.0	150.0	122.0	119.0	26.6	181.0	160.0	111.0	150.7	29.3	269.0	200.0	190.0	219.7	35.1
2001–2002	**Sep**	**Oct**	**Nov**	**Avg**	**±S.D.**	**Dec**	**Jan**	**Feb**	**Avg**	**±S.D.**	**Mar**	**April**	**May**	**Avg**	**±S.D.**	**June**	**July**	**Aug**	**Avg**	**±S.D.**
SITE-I (S)	169.0	160.0	190.0	173.0	12.6	184.0	209.0	190.0	194.3	10.7	189.0	196.0	186.0	190.3	4.2	208.0	209.0	215.0	210.7	3.1
SITE-I (B)	172.0	162.0	195.0	176.3	13.8	186.0	194.0	182.0	187.3	5.0	175.0	184.0	165.0	174.7	7.8	179.0	208.0	204.0	197.0	12.8
SITE-II (S)	150.0	155.0	149.0	151.3	2.6	149.0	209.0	190.0	182.7	25.0	189.0	196.0	186.0	190.3	4.2	208.0	208.0	205.0	207.0	1.4
SITE-II (B)	251.0	252.0	250.0	251.0	0.8	250.0	209.0	190.0	216.3	25.0	196.0	189.0	208.0	197.7	7.8	205.0	252.0	251.0	236.0	21.9
SITE-III (S)	145.0	149.0	149.0	147.7	1.9	135.0	147.0	119.0	133.7	11.5	174.0	154.0	107.0	145.0	28.1	160.0	180.0	187.0	175.7	11.4
SITE-III (B)	346.0	349.0	350.0	348.3	1.7	335.0	335.0	400.0	356.7	30.6	335.0	345.0	310.0	330.0	14.7	325.0	385.0	362.0	357.3	24.7
SITE-IV (S)	150.0	149.0	135.0	144.7	6.8	137.0	174.0	147.0	152.7	15.6	119.0	154.0	180.0	151.0	25.0	216.0	215.0	219.0	216.7	1.7
SITE-IV (B)	151.0	145.0	140.0	145.3	4.5	140.0	147.0	150.0	145.7	4.2	122.0	181.0	160.0	154.3	24.4	120.0	210.0	190.0	173.3	38.6

Avg.: Average (S): Surface
S.D.: Standard deviation (B): Bottom

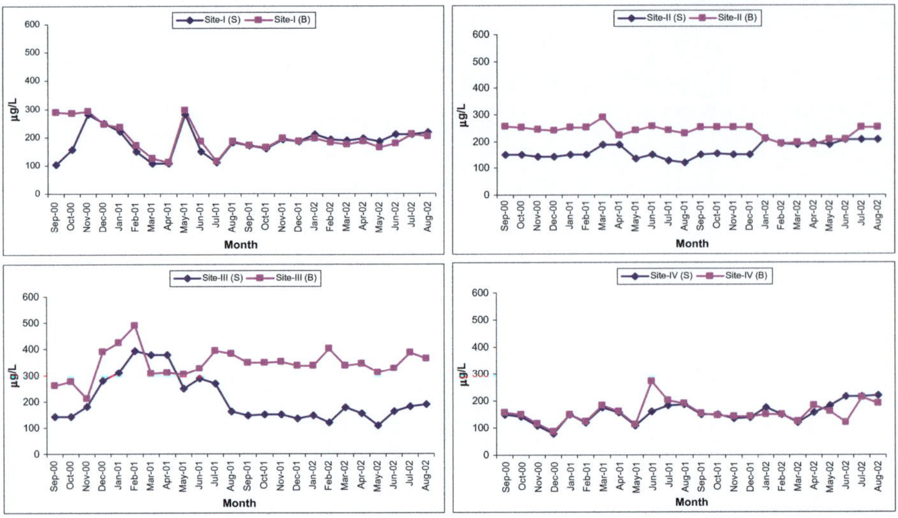

S = Surface; B = Bottom

Fig. 18 Monthly fluctuations in ammonical nitrogen (µg/L) at the investigated sites of the Dal lake

Nitrate-Nitrogen (µg/L)

The present investigation revealed nitrate-nitrogen concentration at different sites of the lake to vary from 113 to 910 µg/L for surface waters and 215 to 850 µg/L for bottom waters. In overall terms, nitrate-nitrogen was higher in the bottom water than the surface water. This may be due to addition of nitrate-nitrogen into bottom water as a result of death and decay of macrophytes. Zutshi et al. (1980) also reported similarly. However, Wanganeo et al. (1996) showed a decreasing trend across the column and has not given any explanation for this observation.

No definite seasonal trend was observed during the present study. Irregular high and low concentration was recorded during some months. Zutshi and Vass (1970), Ramakrishnaiah and Sarkar (1986), Sarwar and Majid (1997), Bath and Kaur (1998) and Sajeev (1999) also failed to depict any seasonal trend and showed irregular fluctuations throughout the study period. Wanganeo et al. (1996) recorded nitrate-nitrogen contents to be decreasing from June onwards. Harold (1934), Harvey (1940), Vashisht and Sharma (1975), Ajmal et al. (1985), Adoni and Joshi (1987) and Bandopadhyay and Gopal (1991) recorded lower values of nitrate nitrogen during the summer season while Willem et al. (1972), King (1981), Shyamsunder (1988), Shah (1988) and Shastree et al. (1991) observed maximum nitrate during the winter.

Nitrate-nitrogen ranged form 113 to 700 µg/L for surface and 215 to 761 µg/L for bottom water at site-I, from 152 to 710 µg/L for surface water and 405 to 850 µg/L for bottom water at site-II, from 208 to 570 µg/L for surface water and 235 to 580 µg/L for bottom water at site-III, and from 215 to 910 µg/L for surface water and 217 to 750 µg/L for bottom water at site-IV (Table 17 and Fig. 19).

Table 17 Monthly fluctuations in nitrate nitrogen (μg/L) at the investigated sites of the Dal lake

	Months																		
2000–2001	**Sep**	**Oct**	**Nov**	**Avg**	**±S.D.**	**Dec**	**Jan**	**Feb**	**Avg**	**±S.D.**	**Mar**	**April**	**May**	**Avg**	**±S.D.**	**June**	**July**	**Aug**	**Avg**	**±S.D.**
SITE-I (S)	570.0	350.0	600.0	506.7	111.5	113.0	420.0	420.0	317.7	144.7	360.0	440.0	700.0	500.0	145.1	760.0	800.0	387.0	649.0	186.0
SITE-I (B)	590.0	360.0	610.0	520.0	113.4	215.0	415.0	305.0	311.7	81.8	271.0	449.0	761.0	493.7	202.5	610.0	601.0	580.0	597.0	12.6
SITE-II (S)	460.0	460.0	620.0	513.3	75.4	870.0	710.0	610.0	730.0	107.1	305.0	210.0	160.0	225.0	60.1	170.0	187.0	152.0	169.7	14.3
SITE-II (B)	470.0	470.0	630.0	523.3	75.4	850.0	740.0	615.0	735.0	96.0	430.0	450.0	490.0	456.7	24.9	480.0	480.0	450.0	470.0	14.1
SITE-III (S)	570.0	310.0	310.0	396.7	122.6	208.0	260.0	295.0	254.3	35.7	390.0	540.0	460.0	463.3	61.3	410.0	620.0	317.0	449.0	126.7
SITE-III (B)	580.0	320.0	320.0	406.7	122.6	310.0	275.0	290.0	291.7	14.3	410.0	501.0	476.0	462.3	38.4	425.0	600.0	310.0	445.0	119.2
SITE-IV (S)	616.0	514.0	719.0	616.3	83.7	369.0	276.0	290.0	311.7	40.9	690.0	690.0	246.0	542.0	209.3	215.0	408.0	318.0	313.7	78.9
SITE-IV (B)	750.0	605.0	740.0	698.3	66.1	671.0	581.0	595.0	615.7	39.5	690.0	680.0	451.0	607.0	110.4	435.0	408.0	325.0	389.3	46.8
2001–2002	**Sep**	**Oct**	**Nov**	**Avg**	**±S.D.**	**Dec**	**Jan**	**Feb**	**Avg**	**±S.D.**	**Mar**	**April**	**May**	**Avg**	**±S.D.**	**June**	**July**	**Aug**	**Avg**	**±S.D.**
SITE-I (S)	510.0	485.0	496.0	497.0	10.2	490.0	320.0	478.0	429.3	77.5	377.0	290.0	391.0	352.7	44.7	245.0	291.0	252.0	262.7	20.2
SITE-I (B)	520.0	587.0	697.0	601.3	73.0	692.0	360.0	336.0	462.7	162.5	203.0	217.0	265.0	228.3	26.5	275.0	220.0	420.0	305.0	84.4
SITE-II (S)	460.0	449.0	500.0	469.7	21.9	487.0	320.0	470.0	425.7	75.0	370.0	390.0	390.0	383.3	9.4	345.0	421.0	405.0	390.3	32.7
SITE-II (B)	461.0	450.0	501.0	470.7	21.9	489.0	470.0	421.0	460.0	28.6	405.0	461.0	480.0	448.7	31.8	430.0	551.0	580.0	520.3	65.0
SITE-III (S)	570.0	305.0	307.0	394.0	124.5	309.0	276.0	290.0	291.7	13.5	290.0	290.0	246.0	275.3	20.7	215.0	208.0	318.0	247.0	50.3
SITE-III (B)	570.0	305.0	307.0	394.0	124.5	309.0	281.0	295.0	295.0	11.4	390.0	380.0	251.0	340.3	63.3	235.0	305.0	325.0	288.3	38.6
SITE-IV (S)	910.0	815.0	714.0	813.0	80.0	800.0	346.0	420.0	522.0	198.9	297.0	264.0	257.0	272.7	17.4	276.0	277.0	248.0	267.0	13.4
SITE-IV (B)	315.0	310.0	326.0	317.0	6.7	335.0	342.0	421.0	366.0	39.0	223.0	217.0	327.0	255.7	50.5	297.0	277.0	267.0	280.3	12.5

Avg.: Average (S): Surface
S.D.: Standard deviation (B): Bottom

Fig. 19 Monthly fluctuations in nitrate nitrogen (µg/L) at the investigated sites of the Dal lake

Site wise trend is as follows:

$$\text{Site IV} > \text{Site II} > \text{Site I} > \text{Site III}$$

The variations in the nitrate-nitrogen concentration in water at different sites of the Dal lake are due to different types of discharges received by water at different sites. Site-I is located in the lake where water does not receive any effluence. Natural weathering of rock is one source of nitrate nitrogen in water at this site. Besides, this site harbours submerged macrophytic vegetation and lacks free floating macrophytes. Nitrate-nitrogen here seems to be added into water upon death and decomposition of the macrophytes as is evident from higher nitrate values in the bottom water. Jhingran (1991b) also opined that natural waters get their nitrate through weathering of rocks while as Ellis et al. (1984) and Pandit et al. (1978) attributed it to growth of macrophytes which upon their death and decomposition, release nutrients into the water.

Site-II is located within the floating garden area used for agricultural purposes. Soil is enriched in nutrients by the use of fertilizers. The site receives nutrient loaded run-off from these gardens, thereby, increasing the concentration of nitrate-nitrogen in water. Sarwar (1991a, b) also observed that agricultural run-off increased the nitrate-nitrogen contents of water.

The site-III is located in the lake where water receives domestic sewage contributed by household effluents containing detergents from house boats which are good source of nitrate-nitrogen. The nitrate-nitrogen concentration was found to be quite high at this site but lower than site-I which is away from any sewage contamination. Fall in nitrate may be due to its uptake by macrophytes which comprise both submerged and free floating type at this site. Bath (1996), Trivedy and Goel (1986) and Das (1989) also recorded low nitrate in water due to its uptake by plants.

At site-IV where water receives domestic waste from hotels, nitrate-nitrogen was found to be maximum. This is due to addition of sewage at this site and accumulation of nitrate of water coming from site I, II and III as well. Sewage is considered to be the major nitrogen source by Thresh et al. (1944), Sylvester (1961), Jolly and Chapman (1966), Munawar (1970a), Seenayya (1971), King (1981), Trivedy and Goel (1986) and Das (1989).

According to Alikunhi (1956), presence of 60–100 μg/L of nitrate-nitrogen indicate water to be highly productive. Nitrate-nitrogen is considered to be the primary factor for enrichment of waters. In view of these statements, the Dal lake may be regarded as a highly productive and nutriently rich water body.

Earlier Sarwar and Zutshi (1989) recorded an average nitrate nitrogen value as 48 g/L while Zutshi et al. (1980) catagorised the lake to be moderatery fertile. The values recorded during present investigation are very high showing that productivity has increased due to various anthropogenic activities and the lake is highly futile.

ISI (1986) has recommended maximum permissible limit of this form of nitrogen to be 4500 μg/L. The present data reveals that nitrate- nitrogen is within the permissible limits in the Dal lake water.

Total Phosphate-Phosphorus (μg/L)

Phosphorus in the natural freshwaters is present mostly in inorganic forms such as H_2PO_4, HPO_4, PO_4. Phosphorus being an important constituent of biological systems, may also be present in the organic forms. The major sources of phosphorus are domestic sewage, detergents, agricultural effluents with fertilizers, and industrial waste waters. The higher concentration of phosphorus, therefore, is indicative of pollution (Wetzel and Likens 2000 and Esmaeili and Johal 2005).

The value of total phosphate phosphorus ranged from 120 to 850 μg/L for surface water and 180 to 851 μg/L for bottom water with slight higher values observed for bottom waters. Phosphorus in waters tends to get precipitated and lost to the sediments, resulting in the overall decrease in surface water. Wanganeo et al. (1996), however, observed a fall in the phosphate in the bottom waters in Nilnag lake of Kashmir.

The data reveals minimum total phosphate phosphorus values during summer and higher in colder months in surface water at all the studied sites of the lake. During summer, aquatic plants are found to be growing vigorously. Due to high growth rate, the total phosphate phosphorus which is one of the major constituents needed for growth is picked up from water thereby decreasing its concentration in water. Similar observations have been made by Welch (1952), Ruttner (1953), Hutchinson (1957), Michael (1969), Kaul (1977), Zutshi and Vass (1978), Harshey et al. (1982), Bath (1996), Bhatt et al. (1999) and Kumar et al. (2003).

The concentration of total phosphate phosphorus ranged from 120 to 286 μg/L for surface water and 139 to 360 μg/L for bottom water at site-I, from 195 to 850 μg/L for surface water and 315 to 851 μg/L for bottom water at site-II, from 180 to 650 μg/L for surface water and 180 to 550 μg/L for bottom water at site-III, and

Table 18 Monthly fluctuations in total phosphate phosphorus (μg/L) at the investigated sites of the Dal lake

	Months																		
2000–2001	**Sep**	**Oct**	**Nov**	**Avg**	**±S.D.**	**Dec**	**Jan**	**Feb**	**Avg**	**±S.D.**	**Mar**	**April**	**May**	**Avg**	**±S.D.**	**June**	**July**	**Aug**	**Avg**	**±S.D.**
SITE-I (S)	195	180	180	185	7.07	120	167	173	153	23.7	240	120	190	183	49.2	170	286	285	247	54.4
SITE-I (B)	210	205	210	208	2.36	315	275	185	258	54.4	255	139	210	201	47.8	195	280	270	248	37.9
SITE-II (S)	850	500	270	540	238	340	530	288	386	104	220	220	229	223	4.24	330	220	245	265	47.1
SITE-II (B)	805	495	395	565	175	335	605	391	444	116	390	325	329	348	29.7	340	340	349	343	4.24
SITE-III (S)	270	190	180	213	40.3	270	230	260	253	17	650	220	275	382	191	390	280	332	334	44.9
SITE-III (B)	275	220	210	235	28.6	295	245	290	277	22.5	550	220	295	355	141	385	220	310	305	67.5
SITE-IV (S)	489	141	232	287	147	385	390	331	369	26.7	495	434	428	452	30.3	580	328	429	446	104
SITE-IV (B)	490	342	235	356	105	480	585	330	465	105	490	630	430	517	83.8	485	328	430	414	65
2001–2002	**Sep**	**Oct**	**Nov**	**Avg**	**±S.D.**	**Dec**	**Jan**	**Feb**	**Avg**	**±S.D.**	**Mar**	**April**	**May**	**Avg**	**±S.D.**	**June**	**July**	**Aug**	**Avg**	**±S.D.**
SITE-I (S)	210	215	199	208	6.68	205	230	190	208	16.5	270	305	309	295	17.5	305	297	240	281	28.9
SITE-I (B)	210	215	199	208	6.68	207	360	360	309	72.1	253	270	255	259	7.59	180	190	211	194	12.9
SITE-II (S)	850	622	521	664	138	419	230	195	281	98.4	275	301	315	297	16.6	305	301	295	300	4.11
SITE-II (B)	851	621	520	664	139	425	315	405	382	47.8	395	340	349	361	24.1	325	390	391	369	30.9
SITE-III (S)	279	269	250	266	12	250	290	231	257	24.6	295	234	228	252	30.3	280	228	229	246	24.3
SITE-III (B)	280	269	251	267	12	250	285	230	255	22.7	290	230	230	250	28.3	285	228	230	248	26.4
SITE-IV (S)	484	389	379	417	47.3	378	480	339	399	59.4	394	435	331	387	42.8	380	220	329	310	66.7
SITE-IV (B)	289	289	381	320	43.4	375	380	320	358	27.2	390	330	231	317	65.6	285	328	363	325	31.9

Avg.: Average (S): Surface
S.D.: Standard deviation (B): Bottom

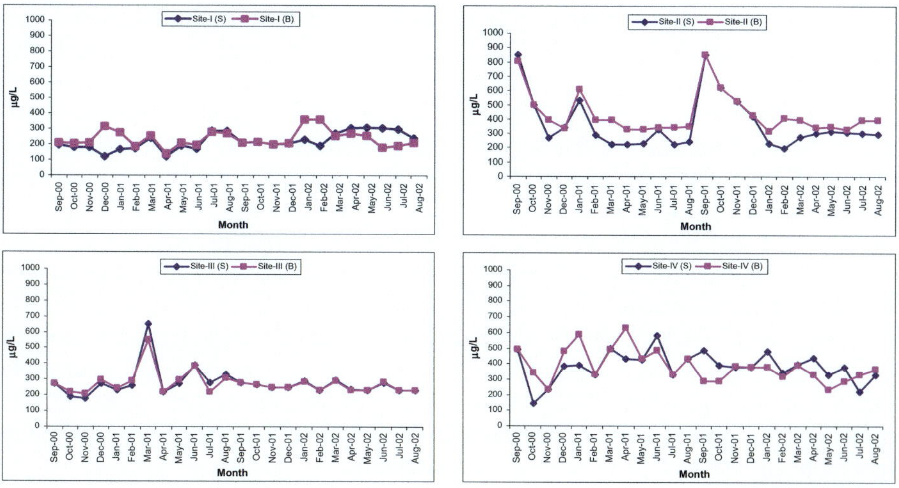

S = Surface; B = Bottom

Fig. 20 Monthly fluctuations in total phosphate phosphorus (µg/L) at the investigated sites of the Dal lake

from 141 to 580 µg/L for surface water and 231 to 630 µg/L for bottom waters at site-IV (Table 18 and Fig. 20).

Total phosphate phosphorus contents show the following trend

$$\text{Site II} > \text{Site III} > \text{Site IV} > \text{Site I}$$

Site-I, where direct source of pollution of any form is absent, the concentration of total phosphate phosphorus was recorded to be the lowest among all the sites.

Site-II which lies in floating gardens area of the Dal lake, the concentration of total phosphate phosphorus was found to be maximum. This site is characterised by luxuriant growth of macrophytes and receives substantial amount of agricultural run-off which is rich in phosphorus contents asis also suggested by Heron (1961), Singh and Roy (1990), Bandopadhyay and Gopal (1990), Sarwar (1991a, b), Thomas and Azis (1996), Bhatt et al. (1999) and Khabade et al. (2003).

At Site-III and IV where water receives domestic sewage and detergents from house boats and hotels, total phosphate phosphorus contents were found to be higher than site-I showing that domestic sewage adds total phosphate phosphorus to water. Welch (1952), Ruttner (1953), Zutshi and vass (1978), Trivedy and Goel (1986), Mishra (1988), Das (1989), Kannan (1991), Sarwar (1991a, b), Bath (1996), Bath and Kaur (1998), Khabade et al. (2003) and Prasannakumari et al. (2003) also observed increase in total phosphate phosphorus with addition of domestic sewage into water. It was further revealed that total phosphate phosphorus concentration was lower at site-III and IV as compared to site-II, showing that agricultural runoff is the richest source of total phosphate phosphorus in water of the lake.

Sawyer (1947) has set a total phosphate phosphorus concentration of 300 µg/L as the critical level beyond which algal blooms indicative of eutrophicated status can be expected. Vollenweider (1972) regarded phosphorus as a key element in the process of eutrophication. The total phosphate phosphorus contents recorded during the 2 year monitoring of the lake indicate that the lake water remained eutrophicated for most of the time at sites-II, III and IV whereas at site-I, it remained below eutrophication level for most of the period. The US department of interior division of technical support (1969) reported that in order to prevent biological nuisance, the total phosphate phosphorus concentration in the lake should not exceed 50 μg/L. In the Dal lake, the phosphorus concentration has been found to range from 120 to 850 μg/L which is well above threshold limits and thus is a cause of biological nuisance to the lake ecology.

Sarwar and Zutshi (1989) recorded an average phosphate phosphorus value of 66 μg/L in the Dal lake. The present studies show that there is a considerable increase in total phosphate phosphorus contents of the lake over the years. This is mainly due to extensive use of lake area for formation of floating gardens and more and more encroachments by hotels and houseboats.

ISI (1986) has recommended maximum permissible limit of this form of total phosphate phosphorus to be 500 μg/L. The present data reveals that values are slightly higher than the permissible limits in the Dal lake water particularly at site-II and site-IV.

Silicate (mg/L)

Silicon is the most abundant element on the earth after oxygen. Despite its over-abundance in nature, it occurs in meagre quantities in water. This is due to resistance of silica sources to chemical weathering. The solubility of silica has been found to be more at high pH and high temperature. The concentration of silica in natural waters is usually between 1 and 30 mg/L but may reach as high as 100 mg/L in hot springs. Silica concentration is also very important in regulation of growth of diatoms (Trivedy and Goel 1986).

The silicate value in the investigated lake ranged from 0.1 to 6.0 mg/L for surface water and from 0.5 to 8 mg/L for bottom water without any significant variation in the surface and bottom waters. However, Jayangoudar (1980), Thomas and Azis (1996) and Wanganeo et al. (1996) recorded higher silicate content in bottom waters in their studies.

No definite seasonal pattern in silicate concentration has been revealed. Sarwar (1991a, b) also failed to reveal any seasonal trend in silicate values. On the other hand Wanganeo et al. (1996) recorded maximum amount of silicate in bottom during summer and Kumar et al. (2003) observed high silicate during monsoon.

The concentration of silicate in the investigated lake varied from 0.4 to 5.6 mg/L for surface water and 1.0 to 3.8 mg/L for bottom waters at site-I, from 0.1 to 6.1 mg/L for surface water and 0.5 to 6.5 mg/L for bottom water at site–II, from 0.9 to 5.0 mg/L for surface waters and 2.1 to 6.0 mg/L for bottom water at site-III, and from 0.6 to 6.0 mg/L for surface water and 1.0 to 6.0 mg/L for bottom water at site-IV (Table 19 and Fig. 21).

Table 19 Monthly fluctuations in silicates (mg/L) at the investigated sites of the Dal lake

	Months																			
2000–2001	**Sep**	**Oct**	**Nov**	**Avg**	**±S.D.**	**Dec**	**Jan**	**Feb**	**Avg**	**±S.D.**	**Mar**	**April**	**May**	**Avg**	**±S.D.**	**June**	**July**	**Aug**	**Avg**	**±S.D.**
SITE-I (S)	2.5	2.9	2.1	2.5	0.3	1.9	1.5	2.0	1.8	0.2	3.1	3.4	2.9	3.1	0.2	3.8	3.7	4.2	3.9	0.2
SITE-I (B)	2.5	2.9	2.0	2.5	0.4	2.0	1.5	2.1	1.9	0.3	3.0	3.5	3.0	3.2	0.2	3.8	3.8	4.0	3.9	0.1
SITE-II (S)	3.0	2.0	1.5	2.2	0.6	0.1	0.9	1.3	0.8	0.5	2.1	2.9	3.8	2.9	0.7	4.0	6.1	5.9	5.3	0.9
SITE-II (B)	3.0	2.0	1.6	2.2	0.6	0.5	1.3	1.5	1.1	0.4	2.0	3.0	4.5	3.2	1.0	4.6	5.1	6.5	5.4	0.8
SITE-III (S)	4.0	2.0	2.5	2.8	0.8	1.0	0.9	2.5	1.5	0.7	5.0	4.0	3.0	4.0	0.8	8.0	4.0	5.0	5.7	1.7
SITE-III (B)	3.0	3.5	2.5	3.0	0.4	2.3	2.9	4.1	3.1	0.7	4.9	5.0	5.0	5.0	0.0	6.0	6.0	5.0	5.7	0.5
SITE-IV (S)	2.9	3.5	2.5	3.0	0.4	2.0	2.0	2.1	2.0	0.0	4.0	4.0	3.5	3.8	0.2	6.0	3.7	5.0	4.9	0.9
SITE-IV (B)	3.0	3.6	2.7	3.1	0.4	2.1	2.1	2.1	2.1	0.0	4.5	5.0	4.0	4.5	0.4	8.0	5.0	6.0	6.3	1.2
2001–2002	**Sep**	**Oct**	**Nov**	**Avg**	**±S.D.**	**Dec**	**Jan**	**Feb**	**Avg**	**±S.D.**	**Mar**	**April**	**May**	**Avg**	**±S.D.**	**June**	**July**	**Aug**	**Avg**	**±S.D.**
SITE-I (S)	2.0	1.9	2.1	2.0	0.1	2.5	2.0	5.6	3.4	1.6	1.6	2.0	2.6	2.1	0.4	0.4	3.4	4.0	2.6	1.6
SITE-I (B)	2.0	1.9	2.0	2.0	0.0	2.4	1.8	4.0	2.7	0.9	1.0	1.8	1.9	1.6	0.4	3.6	0.8	4.4	2.9	1.5
SITE-II (S)	3.4	2.0	1.5	2.3	0.8	1.5	2.1	5.1	2.9	1.6	1.6	2.0	2.6	2.1	0.4	0.9	2.4	3.4	2.2	1.0
SITE-II (B)	3.3	2.1	1.5	2.3	0.7	1.5	2.1	1.6	1.7	0.3	2.0	2.6	1.9	2.2	0.3	2.1	2.1	2.4	2.2	0.1
SITE-III (S)	4.0	4.5	2.5	3.7	0.8	2.5	2.0	2.1	2.2	0.2	4.0	4.0	3.5	3.8	0.2	6.0	3.7	5.0	4.9	0.9
SITE-III (B)	2.5	3.1	2.5	2.7	0.3	2.5	2.1	2.1	2.2	0.2	4.5	5.0	4.0	4.5	0.4	5.0	6.0	2.3	4.4	1.6
SITE-IV (S)	2.0	2.1	2.1	2.1	0.0	1.9	1.4	2.0	1.8	0.3	2.0	1.2	3.9	2.4	1.1	0.6	1.8	0.6	1.0	0.6
SITE-IV (B)	2.0	2.1	2.1	2.1	0.0	2.0	1.0	2.2	1.7	0.5	1.6	2.6	2.0	2.1	0.4	0.4	1.6	1.2	1.1	0.5

Avg.: Average (S): Surface
S.D.: Standard deviation (B): Bottom

Fig. 21 Monthly fluctuations in silicate (mg/L) at the investigated sites of the Dal lake

Not much difference in silicate concentration is observed at different sites. Sarwar and Majid (1997) also failed to record difference in silicate concentration at different sites of Wular lake. The present study reveals that there is no contribution of run-off from the floating gardens and sewage from houseboats and hotels towards the silicate concentration of water in the Dal lake. Only a few reports are available on the study of silicates in lake water. Main source of silicates in water is considered to be aluminosilicate present in the rocks. Contribution by sewage or agricultural run-off is not reported in the literature.

Earlier Sarwar and Zutshi (1989) reported average silicate concentration in the Dal lake to be 2.0 mg/L which is close to the level (0.1–6.0 mg/L) observed during the present investigation. Contribution of human activities towards lake water silicates is not indicated in the present studies.

Total Dissolved Solids (mg/L)

The total dissolved solid (or filterable residue) is the residue which remains after evaporation and include various kinds of minerals present in the water (Trivedy and Goel 1986 and Esmaeili and Johal 2005).

The total dissolved solids values in the Dal lake ranged from 74 to 397 mg/L for surface and from 74 to 420 mg/L for bottom water with a slightly higher values for bottom waters. The slightly higher conductivity values in bottom waters may be attributed to the solubility of nutrients at the sediment.

During the present study, no definite seasonal pattern in total dissolved solids values were observed during the whole study period.

Total dissolved solids ranged from 75 to 213 mg/L for surface water and from 74 to 213 mg/L for bottom waters at site-I, from 157 to 209 mg/L for surface waters and from 150 to 255 mg/L for bottom waters at site-II, from 153 to 297 mg/L for surface water and from 86 to 210 mg/L for bottom waters at site-III and from 80 to 397 mg/L for surface water and from 75 to 420 mg/L for bottom waters at site-IV (Table 20 and Fig. 22).

The ranges of the total dissolved solids at all the selected sites indicate the values to be higher at site-II, III and IV as compared to site-I. Sites-II receives run-off from the surrounding floating gardens and sites-III and sites-IV receive domestic sewage from the hotels and house boats. All these factors enrich the water with dissolved solids.

There are no reports of earlier studies of total dissolved solids in this lake.

ISI (1986) has suggested the maximum permissible limit to be 500 mg/L. On the basis of these recommendations, it can be concluded that the water at all the sites does not exceed the limits.

Conductivity (μS/cm)

Conductivity is the measure or numerical expression of the ability of a solution to conduct electric current. As most of the salts in the water are present in the ionic forms capable of conducting current, therefore, conductivity is a good and rapid measure of the total salt concentration (Trivedy and Goel 1986; Mishra and Saxena 1993; Jameel 1998; Bhatt et al. 1999; Esmaeili and Johal 2005).

The conductivity values in the present study of the Dal lake fluctuated between 100 and 530 μScm^{-1} for surface water and between 100 and 560 μScm^{-1} for bottom water with slightly higher values for bottom waters may be attributed to the fact that solubility of nutrients in the bottom layers increase the concentration of total dissolved solids in the form of inorganic salts which in turn reflect higher values of electrical conductivity. Wanganeo et al. (1996) also recorded higher values of conductivity in bottom waters in comparison to surface and middle.

During the present study, no definite seasonal patterns in conductivity values was observed during the whole study period.

The conductivity values ranged from 100 to 284 μScm^{-1} for surface water and from 100 to 285 μScm^{-1} for bottom water at site-I, from 210 to 265 μScm^{-1} for surface water and from 200 to 340 μScm^{-1} for bottom water at site-II, from 205 to 397 μScm^{-1} for surface water and 200 to 340 μScm^{-1} for bottom water at site-III, and from 112 to 530 μScm^{-1} for surface water and from 112 to 560 μScm^{-1} for bottom water at site-IV (Table 21 and Fig. 23).

The overall pattern of conductivity values of Dal lake follow the pattern as

$$\text{Site IV} > \text{Site III} > \text{Site II} > \text{Site I}$$

Table 20 Monthly fluctuations in total dissolved solids (mg/L) at the investigated sites of the Dal lake

	Months																		
2000–2001	**Sep**	**Oct**	**Nov**	**Avg**	**±S.D.**	**Dec**	**Jan**	**Feb**	**Avg**	**±S.D.**	**Mar**	**April**	**May**	**Avg**	**±S.D.**	**June**	**July**	**Aug**	**Avg**	**±S.D.**
SITE-I (S)	102	134	104	113	14.6	103	102	113	106	4.97	127	129	81	112	22.2	90	111	172	124	34.8
SITE-I (B)	141	132	95	123	19.9	108	107	120	112	5.91	121	149	90	120	24.1	130	114	183	142	29.5
SITE-II (S)	180	198	165	181	13.5	172	174	209	185	17	157	180	162	166	9.88	165	183	157	168	10.9
SITE-II (B)	185	189	171	182	7.72	183	185	213	194	13.7	150	185	160	165	14.7	161	160	183	168	10.6
SITE-III (S)	155	172	180	169	10.4	176	195	202	191	11	202	187	180	190	9.18	165	157	180	167	9.53
SITE-III (B)	90	104	105	99.7	6.85	106	120	206	144	44.2	131	113	93	112	15.5	86	158	195	146	45.3
SITE-IV (S)	360	337	397	365	24.7	236	232	297	255	29.7	270	194	144	203	51.8	153	165	255	191	45.5
SITE-IV (B)	367	351	339	352	11.5	249	251	301	267	24.1	285	210	155	217	53.3	150	255	420	275	111
2001–2002	**Sep**	**Oct**	**Nov**	**Avg**	**±S.D.**	**Dec**	**Jan**	**Feb**	**Avg**	**±S.D.**	**Mar**	**April**	**May**	**Avg**	**±S.D.**	**June**	**July**	**Aug**	**Avg**	**±S.D.**
SITE-I (S)	75	76	76	75.7	0.47	76	213	195	161	60.8	139	108	99	115	17.1	94	81	167	114	37.9
SITE-I (B)	75	76	74	75	0.82	75	213	198	162	61.8	140	99	101	113	18.9	75	144	174	131	41.4
SITE-II (S)	195	198	198	197	1.41	195	183	168	182	11	161	176	162	166	6.85	164	181	179	175	7.59
SITE-II (B)	255	190	191	212	30.4	192	195	183	190	5.1	171	164	182	172	7.41	161	183	191	178	12.7
SITE-III (S)	195	197	191	194	2.49	191	232	297	240	43.6	195	194	219	203	11.6	153	165	180	166	11
SITE-III (B)	120	120	118	119	0.94	118	180	157	152	25.6	210	210	161	194	23.1	158	180	195	178	15.2
SITE-IV (S)	180	176	186	181	4.11	195	188	216	200	11.9	180	117	102	133	33.8	87	155	80	107	33.8
SITE-IV (B)	180	173	180	178	3.3	189	123	325	212	84.1	175	109	78	121	40.5	75	158	159	131	39.4

Avg.: Average (S): Surface
S.D.: Standard deviation (B): Bottom

S = Surface; B = Bottom

Fig. 22 Monthly fluctuations in total dissolved solids (mg/L) at the investigated sites of the Dal lake

The range of the conductivity values at all the selected sites reveal conductivity values to be lowest at site-I for surface as well as bottom waters. This site is located in the open water area least interfered by human influence and does not receive domestic or any other waste material. The water here is crystal clear and free from any human habitation. On the other hand, water at site-III and site-IV is under intense pressure of tourism that results in the accumulation of waste disposal, flotilla of house boats and sewage from surrounding habitations whereas at site-II, agrochemicals and other nutrients added to floating gardens enter the water from the catchment area. All these factors enrich the water with dissolved solids thereby increasing the conductivity values. Trivedy and Goel (1986), Sarwar et al. (1997), Bhatt et al. (1999), Sarwar (1999), SabuThomas et al. (2001), Ara et al. (2003), Das et al. (2003), Khatri and Dhankhar (2003) and Prasannakumari et al. (2000) have also opined that higher values are related to the abundance of nutrients which are released due to decomposition of organic matter thereby increasing the electrical conductivity values.

Wilcox (1955) has made an attempt to classify water with respect to its use for irrigational purposes by taking into consideration electrical conductivity. He characterized water having conductivity of 250 μScm^{-1} as *excellent*, 250–750 μScm^{-1} as *good*, 2000–3000 μScm^{-1} as doubtful and 3000 μScm^{-1} as unsuitable and have prescribed water having conductivity more than 2000 μScm^{-1} not fit for irrigation. Trivedy and Goel (1986) supported the above classification of water. As per their suggestions, water of the Dal lake is excellent at site-I and good at rest of the sites for irrigation purposes.

Oslen (1950) classified water with conductivity values upto 200 μScm^{-1}as oligo-mesotrophic and with 200–500 μScm^{-1}as β-mesotrophic. If followed this classification, the Dal lake water falls in the category of oligotrophic at site-I and β-mesotrophic

Table 21 Monthly fluctuations in conductivity (μ Scm^{-1}) at the investigated sites of the Dal lake

	Months																		
2000–2001	**Sep**	**Oct**	**Nov**	**Avg**	**±S.D.**	**Dec**	**Jan**	**Feb**	**Avg**	**±S.D.**	**Mar**	**April**	**May**	**Avg**	**±S.D.**	**June**	**July**	**Aug**	**Avg**	**±S.D.**
SITE-I (S)	136	179	129	148	22	138	137	151	142	6.4	170	173	109	151	29	121	149	230	167	46
SITE-I (B)	189	176	127	164	27	145	143	160	149	7.6	165	199	120	161	32	174	152	245	190	40
SITE-II (S)	240	265	220	242	18	230	237	279	249	22	210	240	217	222	13	220	245	210	225	15
SITE-II (B)	250	268	229	249	16	245	249	284	259	18	200	250	215	222	21	214	244	340	266	54
SITE-III (S)	207	230	240	226	14	235	260	270	255	15	270	250	240	253	12	220	210	240	223	12
SITE-III (B)	250	268	229	249	16	245	249	284	259	18	200	250	215	222	21	214	244	340	266	54
SITE-IV (S)	480	450	530	487	33	315	310	397	341	40	360	259	192	270	69	205	220	340	255	60
SITE-IV (B)	490	470	453	471	15	332	340	410	361	35	380	280	215	292	68	210	340	560	370	144
2001–2002	**Sep**	**Oct**	**Nov**	**Avg**	**±S.D.**	**Dec**	**Jan**	**Feb**	**Avg**	**±S.D.**	**Mar**	**April**	**May**	**Avg**	**±S.D.**	**June**	**July**	**Aug**	**Avg**	**±S.D.**
SITE-I (S)	100	105	104	103	2.2	102	284	261	216	81	186	145	132	154	23	126	109	223	153	50
SITE-I (B)	100	105	104	103	2.2	102	285	264	217	81	187	132	135	151	25	105	193	232	177	53
SITE-II (S)	260	265	264	263	2.2	260	244	224	243	15	215	235	216	222	9.2	219	242	239	233	10
SITE-II (B)	260	265	264	263	2.2	260	244	229	244	13	219	243	215	226	12	214	244	255	238	17
SITE-III (S)	260	260	255	258	2.4	255	310	397	321	58	260	259	292	270	15	205	220	240	222	14
SITE-III (B)	260	265	264	263	2.2	260	244	229	244	13	219	243	215	226	12	214	244	255	238	17
SITE-IV (S)	240	235	249	241	5.8	260	251	289	267	16	240	157	136	178	45	117	207	112	145	44
SITE-IV (B)	240	235	249	241	5.8	260	165	475	300	130	234	146	123	168	48	112	211	208	177	46

Avg.: Average (S): Surface
S.D.: Standard deviation (B): Bottom

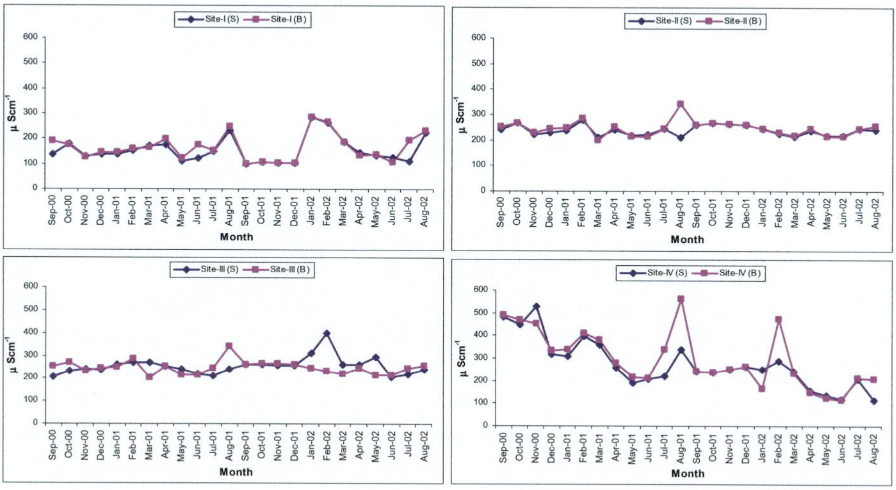

S = Surface; B = Bottom

Fig. 23 Monthly fluctuations in conductivity (µ Scm^{-1}) at the investigated sites of the Dal lake

at sites-II, III and IV. It been seen to cross β-mesotrophic level at site-IV only rarely. However, Dunn (1962) categorized water to be oligotrophic if conductivity values are less than 200 µScm^{-1} and eutrophic if values are more than 200 µScm^{-1}. According to this classification, water of the Dal lake is oligotrophic at site-I and eutrophic at all other sites.

A previous study on the Dal lake by Sarwar and Zutshi (1989) shows average conductivity value of the Dal lake to be 133 µS/cm, much lower than the values recorded during the present study, thereby showing that ionic concentration has considerably increased in water during these years.

ISI (1986) prescribed the maximum permissible limits of conductivity for drinking water to be not more than 750 µS/cm. On the basis of this value, the water at site-II and site-IV shows increased values of conductivity.

Biological Parameters

Phytoplankton Phytoplankton comprise of bacillariophyceae, chlorophyceae, cyanophyceae, dinophyceae, euglenophyceae and chrysophyceae.

Bacillariophyceae (Diatoms)

This class outnumbered the phytoplankton species both in terms of species composition and population density in the Dal lake. Mir and Kachroo (1982), Sarwar and Zutshi (1988) and Sarwar et al. (1996) also reported the dominance of diatoms in the valley lakes. Shetty et al. (1961), Richardson (1968), Crayton and Sommerfeld (1978), Venkateshwarlu and Reddy (1985), Varma (1998) and Katti et al. (2002) reported similarly for other freshwater bodies of India.

In the 2 year study of the Dal lake, 75 diatom taxa distributed at four sites of the lake with 73 at site-I, 72 at site-II, 68 at site-III and 71 at site-IV were recorded (Table 22).

In the present study, the maximum population density of bacillariophyceae was observed during autumn at the sites I, III, and IV, but in summer at site-II. The species diversity was recorded to be highest during autumn at site- I and II, during autumn and winter at site-III and during winter at site-IV. In terms of population diversity and density, the diatoms reflect their affinity for cold temperature. Lund (1965) reported abundance of diatom in colder months and opined that they grow under the conditions of weak light and low temperature. Zafar (1967) also observed colder months to be more favourable for multiplication of diatoms in freshwater bodies. Similar observations have also been made by Roa (1955), Venkateshwarlu (1969), Vass et al. (1978), Manikya (1984), Sarwar (1985), Singh and Srivastava (1991), Mir (1995), Sarwar et al. (1996), Krishnamurthy and Bharati (1996), Gujarathi and Kanhere (1998), Pandey et al. (2002) and Sedamkar and Angadi (2003). However, Bhowmick and Singh (1985) recorded diatom peak in summer while Mallick and Padhi (1999) and Khanna and Bhutiani (2003) recorded the peak during monsoon season (Table 23, Figs. 24 and 25).

The taxa (73) recorded at site-I include *Achnanthes sp., Amphora bitumida, Amphora constrictum, Amphora falcatus, Amphora normanii, Amphora ovalis, Amphora proteus, Amphora sp., Anomoeoneis sp., Aphanocapsa sp., Asterionella formosa, Caloneis sp., Chaetoceros sp., Cocconeis placentula, Cocconeis sp., Coscinodiscus sp., Cyclotella sp., Cymbella cistula, Cymbella lanceolata, Cymbella microporum, Cymbella prostrata, Cymbella sp., Cymbella tumida, Cymbella turgida, Cymbella ventricosa, Diatoma elongatum, Diatomella sp., Diploneis sp., Epithemia sp., Epithemia turgida, Eunotia diodon, Eunotia gracilis, Eunotia pectinalis, Eunotia sorex, Eunotia sp., Eunotia triodon, Eunotia zebra, Fragilaria acus, Fragilaria capucina, Fragilaria construens, Fragilaria crotonensis, Fragilaria sp., Fragilaria vaucheriae, G.subventricosum, Gomhonema augur, Gomphonema accuminatum, Gomphonema geminatum, Gomphonema olivaceum, Gomphonema sp., Gomphonema constrictum, Melosira granulate, Melosira sp., Meridion sp., N. rhyncocephala, Navicula constricta, Navicula radiosa, Navicula sp., Nitzschia accicularis, Nitzschia epiculata, Nitzschia longissima, Nitzschia plana, Nitzschia sp., Pinnularia sp., Pleurosigma sp., Rhopalodia gibba, Rhopalodia sp., Stauronies sp., Suriella sp., Synedra lingus, Synedra sp., Synedra ulna, Tabellaria fenestrata,* and *Tabellaria sp.* The average population density of bacillariophyceae was recorded lowest at this site ranging from 1300 ind/mL (spring) to 2080 ind/mL (autumn). The species diversity was high during autumn with 54 taxa. This site is located in the open water area least interfered by human influence and does not receive domestic or any other waste material. The water here is crystal clear and free from any human habitation (Table 24).

Table 22 List of Bacillariophyceae taxa recorded at the investigated sites of the Dal lake

S. No.	Bacillariophyceae taxa	Site-I	Site-II	Site-III	Site-IV
1	*Achnanthes sp.*	+	+	+	+
2	*Amphora bitumida* Prowse	+	+	+	+
3	*Amphora constrictum* Donk	+	+	+	+
4	*Amphora falcatus* Kutz	+	+	+	+
5	*Amphora normanii* Donk	+	+	+	+
6	*Amphora ocellata* Kutz	–	+	+	+
7	*Amphora ovalis* Kutz	+	+	+	+
8	*Amphora proteus* Greg	+	+	–	+
9	*Amphora sp.*	+	+	+	+
10	*Anomoeoneis sp.*	+	+	+	+
11	*Aphanocapsa sp.*	+	–	–	–
12	*Asterionella formosa* Hassal	+	+	+	+
13	*Caloneis sp.*	+	+	+	+
14	*Ceratoneis arcus* (Ehr) Kutz	-	+	+	+
15	*Chaetoceros sp.*	+	+	+	+
16	*Cocconeis placentula* Her	+	+	+	+
17	*Cocconeis sp.*	+	+	+	+
18	*Coscinodiscus sp.*	+	+	+	+
19	*Cyclotella sp.*	+	+	+	+
20	*Cymbella cistula* Hempr	+	+	+	+
21	*Cymbella lanceolata* Her	+	+	+	+
22	*Cymbella microporum* Kutz	+	+	+	+
23	*Cymbella prostrata* Ralfs	+	+	–	+
24	*Cymbella sp.*	+	+	+	+
25	*Cymbella tumida* Breb	+	+	+	+
26	*Cymbella turgida* Greg	+	+	+	+
27	*Cymbella ventricosa* Kutz	+	+	+	+
28	*Diatoma elongatum* Agardh	+	+	+	+
29	*Diatomella sp.*	+	+	–	–
30	*Diploneis sp.*	+	+	–	–
31	*Epithemia sp.*	+	+	+	+
32	*Epithemia turgida* (Ehr) Kutz	+	+	+	+
33	*Eunotia diodon* Her	+	+	+	+
34	*Eunotia gracilis* (Ehr) Rab	+	+	+	+
35	*Eunotia pectinalis* (Kutz) Raben	+	+	+	+
36	*Eunotia sorex* Kutz	+	+	+	+
37	*Eunotia sp.*	+	+	+	+
38	*Eunotia triodon* Her	+	+	+	+
39	*Eunotia zebra* (Ehr) Kutz	+	+	+	+
40	*Fragilaria acus* Desmaz	+	+	+	+
41	*Fragilaria capucina* Desmaz	+	+	+	+
42	*Fragilaria construens* (Ehr) Grun	+	+	+	+
43	*Fragilaria crotonensis* (Ed) Kitt	+	+	+	+

(continued)

Table 22 (continued)

S. No.	Bacillariophyceae taxa	Site-I	Site-II	Site-III	Site-IV
44	*Fragilaria sp.*	+	+	+	+
45	*Fragilaria vaucheriae* (Kutz)Peterson	+	+	+	+
46	*Gomphonema constrictum* Her	+	+	+	+
47	*G. subventricosum* Her	+	+	+	+
48	*G. augur* Her	+	+	+	+
49	*G. accuminatum* Her	+	+	+	+
50	*G. geminatum* (Lyngb) Ag	+	+	+	+
51	*G. olivaceum* Kutz	+	-	+	+
52	*Gomphonema sp.*	+	+	+	+
53	*Melosira granulata* Kutz	+	+	+	+
54	*Melosira sp.*	+	+	+	+
55	*Meridion sp.*	+	+	+	+
56	*Navicula rhyncocephala* Kutz	+	+	+	+
57	*N. constricta* Bory	+	+	+	+
58	*N. radiosa* Kutz	+	+	+	+
59	*Navicula sp.*	+	+	+	+
60	*Nitzschia accicularis* W.Sm.	+	+	+	+
61	*N. epiculata* (Greg) Grun	+	+	+	+
62	*N. longissima* (Breb) Ralfs	+	+	+	+
63	*N. plana* W.Sm.	+	+	+	+
64	*Nitzschia sp.*	+	+	+	+
65	*Pinnularia sp.*	+	+	+	+
66	*Pleurosigma sp.*	+	+	+	+
67	*Rhopalodia gibba* (Ehr) O.Mull	+	+	+	+
68	*Rhopalodia sp.*	+	+	–	+
69	*Stauronies sp.*	+	+	+	+
70	*Suriella sp.*	+	+	+	+
71	*Synedra lingus* Her	+	+	+	+
72	*Synedra sp.*	+	+	+	+
73	*Synedra ulna* (Nitz) Her	+	+	+	+
74	*Tabellaria fenestrata* (Lyng)	+	+	+	+
75	*Tabellaria sp.*	+	–	–	–

The taxa (72) recorded at site-II include *Achnanthes sp., Amphora butimida, A. constrictum, A. falcatus, A. normanii, A. ocellata, A. ovalis, A. proteus, Amphora sp., Anomoeoneis sp., Asterionella formosa, Caloneis sp., Ceratoneis arcus, Chaetoceros sp., Cocconeis placentula, Cocconeis sp., Coscinodiscus sp., Cyclotella sp., Cymbella cistula, C. lanceolata, C. microporum, C. prostrata, C. tumida, C.turgida, C. ventricosa, Cymbella sp., Diatoma elongatum, Diatomella sp., Diploneis sp., Epithemia turgida, Epithemia sp., Eunotia sorex, E. diodon, E.gracilis, E. pectinalis, E. triodon, E. zebra, Eunotia sp., Fragilaria acus, F. capucina, F. construens, F. crotonensis, F.vaucheriae, Fragilaria sp., Gomphonema accuminatum,*

Table 23 Seasonal variations in the population density (Ind/mL) of the Bacillariophyceae at the investigated sites of the Dal lake

Sites	Year	Autumn	Winter	Spring	Summer
SITE-I	**2000–2001**	2096	1572	1456	2200
	2001–2002	2064	1584	1144	1296
	Average	2080	1578	1300	1748
SITE-II	**2000–2001**	2328	1936	2142	2704
	2001–2002	1292	1836	1880	1880
	Average	1812	1886	2011	2292
SITE-III	**2000–2001**	1956	2132	2248	2120
	2001–2002	2368	2024	2064	1838
	Average	2162	2078	2156	1978
SITE-IV	**2000–2001**	2365	1816	2528	2304
	2001–2002	2576	2056	2248	2048
	Average	2470	1936	2388	2176

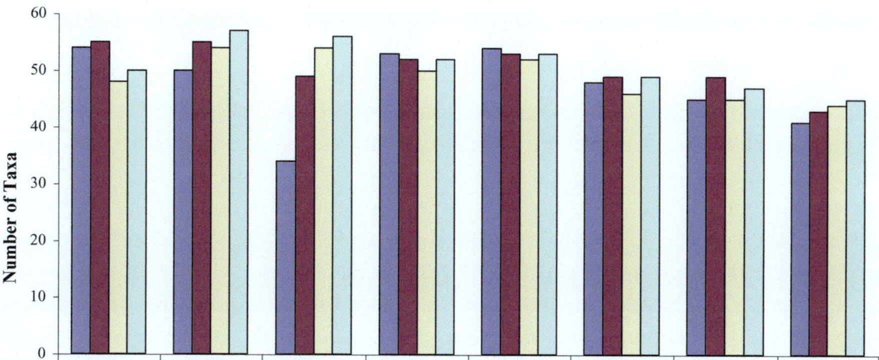

	Autumn	Winter	Spring	Summer	Autumn	Winter	Spring	Summer
Site-I	54	50	34	53	54	48	45	41
Site-II	55	55	49	52	53	49	49	43
Site-III	48	54	54	50	52	46	45	44
Site-IV	50	57	56	52	53	49	47	45

Season

Fig. 24 Seasonal variations in the species diversity of the Bacillariophyceae at the investigated sites of the Dal lake

*G. constrictum. G. geminatum, G.subventricosum, G. augur, Gomphonema sp., Melosira granulata, Melosira sp., Meridion sp., Navicula constricta, N. rhyncocephala, N.radiosa, Navicula sp., Nitzchia longissima, N.accicularis, N. epiculata, N. plana, Nitzschia sp., Pinnularia sp., Pleurosigma sp., Rhopalodia gibba, Rhopalodia sp., Stauronies sp., Suriella sp., Synedra lingus, S. ulna, Synedra sp.,*and *Tabellaria fenestrata.* The population density of bacillariophyceae was recorded higher than site-I but lower than site-IV. The average population density ranged from 1812 ind/mL (autumn) to 2292 ind/mL (summer) at this site. The maximum average species diversity at this site was recorded in autumn with 54 taxa. This site

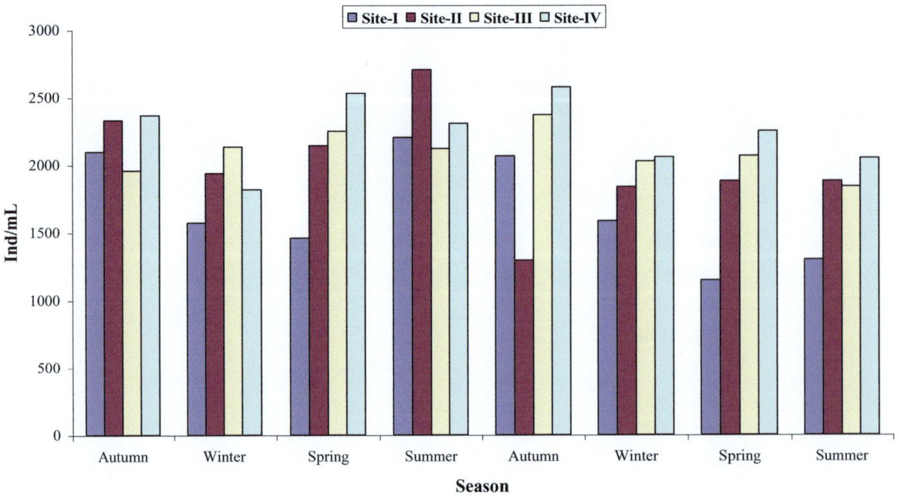

Fig. 25 Seasonal variations in the population density (Ind/mL) of the Bacillariophyceae at the investigated sites of the Dal lake

is located in the lake where floating gardens cover a large area of the lake. This site is characterized by rich growth of macrophytes and receives agricultural run-off from the floating gardens and dead organic matter by death and decay of macrophytes. Sarwar (1991a, b) attributed maximum population density and diversity to diverse macrophytic assemblage and to the nutrient enrichment caused by run-off and leakage from adjoining vegetable fields (Table 25).

The taxa (68) recorded at site-III include *Achnanthes sp., Amphora bitumida, A. constrictum, A. falcatus, A. normani, A. ocellata, A. ovalis, Amphora sp., Anomoeoneis sp., Asterionella formosa, Caloneis sp., Ceratoneis arcus, Chaetoceros sp., Cocconeis placentula, Cocconeis sp., Coscinodiscus sp., Cyclotella sp., Cymbella cistula, C. lanceolata, C. microporum, C. tumida, C.turgida, C. ventricosa, Cymbella sp., Diatoma elongatum, Epithemia turgida, Epithemia sp., Eunotia diodon, E. gracilis, E. pectinalis, E. sorex, E. triodon, E. zebra, Eunotia sp., Fragilaria acus, F. capucina, F. construens, F. crotonensis, F. vaucheriae, Fragilaria sp., Gomphonema accuminatum, G. augur, G. constrictum. G. geminatum, G.subventricosum, G. olivaceum, Gomphonema sp., Melosira granulata, Melosira sp., Meridion sp., Navicula constricta, N. rhyncocephala, N. radiosa, Navicula sp., Nitzschia accicularis, N. epiculata, N. longissima, N. plana, Nitzschia sp., Pinnularia sp., Pleurosigma sp., Rhopalodia gibba, Stauronies sp., Suriella sp., Synedra lingus, S. ulna, Synedra sp., and Tabellaria fenestrata.* The average population density of bacillariophyceae at this site ranged from 1978 ind/mL (summer) to 2162 ind/mL (autumn). The species diversity at this site was maximum in autumn and winter with average value of 50 diatom taxa. This site is located where the lake water receives effluence from house boats (Table 26).

Table 24 Monthly variations in the population density (Ind/mL) of the Bacillariophyceae at the study site-I of the Dal lake

Taxa	Sep	Oct	Nov	Dec	Jan	Feb	Mar	Apr	May	Jun	Jul	Aug	Sep	Oct	Nov	Dec	Jan	Feb	Mar	Apr	May	Jun	Jul	Aug
Amphora sp.	−	−	+	+	−	−	−	+	+	−	+	−	+	−	+	+	−	+	−	+	−	+	−	−
Achnanthes sp.	−	+	+	−	+	−	+	+	+	−	+	+	+	−	+	+	+	+	+	+	+	+	+	+
Amphora bitumida	+	+		−	+	−	+	+	−	+	+	+	−	−	+	+	+	−	+	−	−	−	−	+
Amphora constrictum	−	−	−	−	−	+	−	−	−	−	+	+	−	−	−	−	−	−	−	−	−	−	+	+
Amphora falcatus	−	−	−	−	+	−	−	−	−	−	−	+	−	+	−	−	−	−	−	−	+	+	+	−
Amphora normanii	+	+		+	−	−	−	+	+	+	−	+	−	+	+	+	−	−	−	+	−	−	−	−
Amphora ovalis	+	+	+	+	+	−	+	+	+	+	+	+	−	+	+	+	+	+	+	+	+	+	+	−
Amphora proteus	+	+	+	+	+	+	−	−	−	+	+	+	+	−	+	+	+	+	+	+	+	−	−	−
Anomoneis sp.	+	+	+	−	+	−	+	+	+	+	+	−	+	+	+	−	+	+	+	−	−	−	+	−
Aphanocapsa sp.	−	−	−	−	+	−	−	−	−	−	+	−	+	−	−	−	−	−	−	−	−	−	−	−
Asterionella formosa	−	−	−	−	−	−	−	−	−	−	−	−	+	−	−	−	−	−	−	−	−	−	−	−
Caloneis sp.	−	−	−	−	+	−	−	+	+	−	−	+	−	−	−	+	+	−	−	+	−	−	−	+
Chaetoceros sp.	+	+	−	+	−	−	−	−	−	+	−	−	+	−	−	+	+	−	−	+	+	+	+	−
Cocconeis placentula	−	−	+	+	+	−	+	+	+	−	+	−	+	+	+	+	−	−	−	+	−	+	−	−
Cocconeis sp.	−	−	−	−	−	−	−	−	−	−	−	+	+	−	+	−	+	−	+	−	−	−	−	−
Coscinodiscus sp.	+	+	−	+	+	+	+	−	+	+	+	−	+	−	−	−	+	+	+	−	+	+	−	−
Cyclotella sp.	−	−	+	+	+	+	−	−	+	−	+	+	+	−	+	+	+	+	−	−	+	+	−	+
Cymbella cistula	+	+	−	+	+	−	+	+	+	+	−	+	+	+	−	+	+	+	−	−	−	+	−	+
Cymbella tumida	+	+	+	−	+	−	+	−	−	−	−	−	+	+	+	+	+	−	+	−	−	−	−	−
Cymbella lanceolata	−	+	+	−	−	−	−	−	−	−	+	+	+	+	−	−	−	−	−	−	−	+	+	−
Cymbella microporum	+	+	+	−	−	−	−	−	−	−	+	+	+	+	−	−	−	−	−	−	−	+	+	−
Cymbella prostrata	+	+	−	−	−	+	+	+	+	+	+	+	+	−	−	−	+	+	+	−	−	−	+	−
Cymbella sp.	−	−	−	−	−	−	−	−	−	+	+	−	−	−	−	−	−	−	−	−	−	−	−	−
Cymbella turgida	−	−	−	−	−	−	−	−	−	−	+	−	−	−	−	−	−	−	−	−	−	−	−	−
Cymbella ventricosa	+	+	+	+	+	+	+	+	+	+	+	+	+	+	+	+	+	+	+	+	+	+	+	+

(continued)

Table 24 (continued)

Taxa	Sep	Oct	Nov	Dec	Jan	Feb	Mar	Apr	May	Jun	Jul	Aug	Sep	Oct	Nov	Dec	Jan	Feb	Mar	Apr	May	Jun	Jul	Aug
Diatoma elongatum	+	+	+	–	+	–	–	+	+	+	–	–	–	+	–	–	+	–	–	+	+	–	+	–
Diatomella sp.	–	+	–	–	–	–	–	–	–	+	+	+	–	–	–	+	–	–	+	–	–	+	–	–
Diploneis sp.	+	+	+	–	+	–	+	–	+	+	–	–	+	–	+	–	–	–	–	–	–	+	+	–
Epithemia sp.	–	–	–	–	–	–	–	–	–	–	–	–	–	+	–	–	–	–	–	–	–	–	–	–
Epithemia turgida	+	+	–	–	+	–	–	–	+	–	+	–	+	–	+	–			+	–	+	–	–	–
Eunotia sorex	+	–	–	–	+	–	+	–	–	+	–	+	+	+	–	–	+	–	–	+	–	–	–	–
Eunotia gracilis	+	+	–	+	–	–	–	–	–	+	+	+	+	–	+	–	–	+	+	+	–	+	–	–
Eunotia pectinalis	+	+	–	–	+	–	–	+	–	+	–	+	+	+	–	+	+	–	–	+	+	+	–	+
Eunotia sp.	–	+	+	+	+	–	+	+	+	–	–	–	+	+	+	+	–	+	+	–	+	+	–	+
Eunotia triodon	–	+	–	–	+	+	–	–	–	–	–	–	+	+	+	–	–	+	–	–	+	+	–	+
Eunotia zebra	+	+	–	–	+	–	+	–	–	+	–	+	+	+	–	+	–	–	–	–	+	+	–	+
Eunotia diodon	–	+	–	+	+	–	+	–	+	–	+	+	+	+	+	+	+	+	–	+	+	+	–	+
Fragilaria acus	–	+	+	+	–	–	–	–	–	–	–	–	–	–	–	–	+	–	–	–	–	–	–	–
Fragilaria capucina	+	+	–	+	+	–	+	–	+	+		+	–	–	–				+		+		+	
Fragilaria construens	+	+	–	–	–	+	–	+	–	+	+	–	+	+	+	–	–	+	–	+	–	+	+	+
Fragilaria crotonensis	–	–	–	+	+	–	–	–	+	–	–	–	+	–	+	–	–	–	–	–	–	+	–	–
Fragilaria sp.	–	+	–	–	–	–	–	–	–	–	–	–	–	–	–	–	–	–	–	–	–	–	–	–
Fragilaria vaucheria	+	+	+	+	+	+	+	+	+	+	–	+	+	+	+	+	+	–	+	+	+	+	+	+
G. constrictum	+	+	–	+	–	–	–	–	–	+	–	–	–	+	–	–	–	–	–	–	–	–	+	–
G.subventricosum	–	+	–	–	–	–	–	–	–	–	–	–	–	–	–	–	–	+	–	+	–	–	–	–
Gomhonema augur	–	+	–	+	–	–	–	–	–	+	–	+	–	+	–	+	–	–	–	–	+	–	+	–
Gomphonema geminatum	–	–	+	–	+	–	–	–	–	–	–	–	–	–	–	–	–	–	+	–	–	–	–	–
Gomphonema accuminatum	–	–	–	–	–	–	–	–	–	+	+	–	–	–	–	–	–	–	–	–	–	–	–	–
Gomphonema olivaceum	+	+	+	+	+	+	+	+	++	+	+	+	+	+	+	+	+	+	+	–	–	+	+	–

Gomphonema sp.	+	+	+	−	−	−	−	−	−	−	−	−	+	+	−	−	+	−	−	−	−	−	−	−
Melosira granulata	+	+	−	+	+	+	+	+	−	+	−	+	+	+	+	+	+	+	−	+	+	+	+	+
Melosira sp.	−	+	−	+	−	+	−	−	−	−	+	−	+	++	−	−	−	−	−	−	−	−	−	−
Meridion sp.	−	−	−	−	−	−	−	−	−	−	−	−	−	−	−	−	−	−	−	−	+	−	−	−
Navicula constricta	−	−	−	+	−	+	−	−	−	−	−	−	+	+	−	+	−	+	−	−	−	−	+	+
N. rhyncocephala	+	−	−	+	+	+	−	+	+	+	−	−	+	−	+	−	+	−	+	+	+	−	−	−
N. radiosa	+	+	−	+	+	+	−	+	+	+	−	+	+	−	−	+	+	+	+	−	+	−	−	+
Navicula sp.	−	−	−	−	−	−	−	−	−	−	+	−	+	−	−	−	−	−	−	−	−	−	−	−
Nitzschia longisima	−	+	+	+	+	−	+	−	−	−	−	+	+	−	−	+	−	+	−	+	−	+	−	−
Nitzschia accicularis	+	+	+	+	+	+	−	−	+	+	+	+	+	+	+	+	+	+	+	−	+	+	−	−
Nitzschia epiculata	−	−	−	−	+	−	+	−	−	−	−	−	−	−	−	−	−	−	−	−	+	−	−	−
Nitzschia plana	+	−	−	−	−	+	−	−	−	+	+	−	−	+	−	−	+	+	−	−	−	−	−	−
Nitzschia sp.	−	+	−	−	+	−	−	−	−	−	−	−	−	−	−	+	−	−	−	−	+	−	−	−
Pinnularia sp.	+	−	+	−	−	−	−	−	−	−	−	+	−	−	−	+	−	−	−	−	+	−	−	−
Pleurosigma sp.	−	+	−	−	−	−	−	−	−	−	−	−	−	−	+	−	−	+	−	−	−	−	−	−
Rhopalodia gibba	+	+	−	+	−	−	−	+	+	+	−	+	+	+	−	−	+	+	+	−	−	−	−	−
Rhopalodia sp.	−	−	−	−	−	−	−	−	−	−	+	−	−	−	−	−	−	−	−	−	−	−	−	−
Stauronies sp.	+	−	−	−	+	−	−	−	−	+	+	−	−	−	−	−	+	−	−	−	−	−	−	−
Suriella sp.	−	−	−	−	−	−	−	−	−	−	−	−	−	−	−	−	+	−	−	−	−	+	−	−
Synedra lingus	−	+	+	−	−	−	−	−	−	−	+	+	+	−	+	−	−	−	−	−	−	+	−	+
Synedra sp.	+	+	+	−	−	−	−	−	−	−	+	−	+	−	+	−	+	−	−	+	−	−	−	−
Synedra ulna	−	−	−	−	−	−	−	−	−	−	−	−	+	−	−	−	−	−	−	−	−	−	−	−
Tabellaria fenestrata	+	+	+	−	+	−	−	−	−	+	−	−	+	−	−	+	−	+	−	−	+	+	+	+
Tabellaria sp.	+	−	+	−	−	−	−	−	−	−	−	+	−	−	−	+	−	−	−	−	+	−	−	−

0–5 ind/mL = +
6–10 ind/mL = ++
11–15 ind/mL = +++
and above = ++++

Table 25 Monthly variations in the population density (Ind/mL) of the Bacillariophyceae at study site-II of the Dal lake

Taxa	Sep	Oct	Nov	Dec	Jan	Feb	Mar	Apr	May	Jun	Jul	Aug	Sep	Oct	Nov	Dec	Jan	Feb	Mar	Apr	May	Jun	Jul	Aug
Amphora sp.	+	−	+	−	−	+	−	+	−	+	−	++	−	+	−	+	−	+	−	+	−	++	−	−
Amphora normanii	−	−	++	−	+	−	+	+	−	+	−	++	++	−	++	−	−	−	−	+	−	−	−	−
Amphora ovalis	+	+	−	+	++	+	+	++	+	+	++	+	++	+	++	+	++	+	++	+	+	++	++	−
Amphora proteus	−	−	−	−	+	+	−	+	−	−	−	+	++	−	+	+	+	+	+	+	−	−	−	−
Amphora butimida	−	−	+	−	+	−	+	−	−	+	−	+	+	+	+	+	++	−	−	−	−	−	−	+
Achnanthes sp.	+	+	++	−	+	+	−	++	+	+	+	+	−	+	+	++	++	+	++	++	+	+	++	++
Anomoneis sp.	+	+	−	−	−	+	+	+	+	+	+	+	++	−	−	−	++	++	+	−	+	−	+	−
Amphora constrictum	−	−	−	−	−	−	+	−	−	−	++	−	−	−	−	−	−	−	−	−	−	−	−	−
Asterionella formosa	−	−	−	+	−	−	−	−	−	−	+	−	+	−	−	−	−	−	+	−	−	−	−	+
Amphora ocellata	−	−	−	−	−	−	−	+	+	−	−	−	−	−	++	−	−	−	−	−	+	−	−	−
Amphora falcatus	−	−	+	−	−	−	−	−	+	+	+	−	−	+	+	−	−	−	−	−	+	−	−	−
Chaetoceros sp.	+	−	−	−	+	−	−	+	+	+	−	−	+	−	−	+	++	−	−	+	+	+	+	+
Caloneis sp.	−	−	+	−	+	−	−	+	−	−	−	−	−	−	+	+	+	−	−	+	−	−	−	+
Cymbella ventricosa	++	++	++	++	+	+	+	++	++	++	++	+	+	+	++	+	++	+	+	++	++	++	+	+
Coscinodiscus sp.	+	+	++	+	+	+	−	−	+	+	+	−	+	+	−	−	++	+	+	−	+	+	−	+
Cyclotella sp.	−	−	−	+	−	−	−	−	−	−	−	−	−	−	−	−	+	+	−	−	+	−	−	−
Cymbella cistula	−	+	+	+	+	+	−	−	−	+	+	−	+	−	+	+	+	+	−	−	−	++	−	+
Cymbella turgida	−	+	−	−	+	−	−	+	−	−	+	−	−	+	−	−	−	−	−	−	−	−	−	−
Cocconeis placentula	+	+	++	+	−	+	++	−	+	++	++	+	+	+	++	−	++	+	+	++	+	++	+	−
Cymbella tumida	+	++	+	+	−	+	+	+	+	+	+	+	++	+	++	+	−	+	++	+	+	−	+	−
Cocconeis sp.	−	+	−	−	−	+	−	−	−	−	+	+	−	−	−	−	+	+	+	−	−	−	−	−
Cymbella prostrata	+	+	−	−	++	−	+	−	+	+	−	−	+		+	+					+	−	−	−
Cymbella lanceolata	+	+	+	−	−	−	−	+	+	−	−	−	−	+	++	−	++	−	−	+	+	−	−	−
Cymbella microporum	+	−	+	−	−	−	−	−	−	+	++	−	+	−	−	−	−	−	−	−	+	+	+	−
Cymbella sp.	−	−	−	−	−	−	−	−	−	−	−	−	+	+	−	−	−	−	−	−	−	−	−	−
Ceratoneis arcus	−	−	−	−	−	−	+	−	−	−	−	−	−	−	−	−	−	−	−	−	−	+	−	−

Diatoma elongatum	+	+	+	+	–	–	–	+	+	–	+	–	+	+	–	–	+	–	–	+	+	–	+	+
Diploneis sp.	–	+	++	–	–	+	–	+	–	–	–	++	++	–	–	–	–	–	–	+	–	+	+	–
Diatomella sp.	–	+	–	–	–	–	–	–	–	–	–	–	–	–	–	–	+	–	–	–	–	–	–	–
Eunotia zebera	+	+	–	+	++	–	–	–	+	+	+	–	+	–	++	+	–	–	–	–	+	+	–	+
Eunotia gracilis	–	+	+	–	+	+	–	+	–	+	++	+	++	+	+	–	–	+	++	+	–	+	–	–
Eunotia pectinalis	+	+	–	–	+	–	–	+	+	–	–	–	+	–	++	–	++	–	–	+	+	+	–	+
Eunota sorex	–	–	–	–	+	–	–	+	–	–	–	–	+	–	+	–	+	–	–	+	–	–	–	–
Eunotia sp.	++	++	–	–	–	+	+	–	++	+	+	+	–	+	–	+	–	+	+	–	++	+	–	+
Eunotia triodon	++	+	+	–	–	+	–	+	++	+	–	+	–	–	–	–	–	+	–	–	++	+	–	+
Eunotia diodon	+	–	–	+	–	+	+	+	+	–	–	+	–	+	++	+	+	+	–	+	+	+	–	+
Epithemia sp.	–	–	–	+	–	–	–	–	–	–	–	–	–	–	–	–	–	–	–	–	–	–	–	–
Epithemia turgida	–	–	+	+	–	–	–	–	–	–	–	+	++	–	+	–	–	–	+	–	–	–	–	+
Fragilaria vaucheria	++	+	++	++	++	+	++	+	++	+	–	++	++	–	–	+	+	–	+	+	+	++	++	+
Fragilaria sp.	–	+	+	++	+	+	+	–	–	++	–	+	+	–	+	++	+	++	–	++	–	–	++	++
Fragilaria acus	–	–	–	–	+	–	–	–	–	–	–	–	–	–	–	–	–	–	–	–	–	–	–	–
Fragilaria crotonensis	–	++	+	–	+	–	–	+	–	+	–	+	–	–	+	+	++	–	–	+	–	+	–	+
Fragilaria capucina	–	–	–	–	–	–	+	–	–	–	–	–	–	–	–	–	–	–	–	–	–	–	–	–
Fragilaria construens	–	+	–	+	–	–	–	++	–	++	–	+	+	–	+	–	–	–	–	–	–	+	+	+
Gomphonema augur	+	–	–	–	–	–	–	–	+	+	–	–	–	+	–	+	–	–	–	–	+	–	+	–
G. accuminatum	+	+	–	+	–	–	–	–	–	–	+	–	–	–	–	–	–	–	–	–	–	–	–	+
G.subventricosum	–	–	+	–	–	+	–	+	–	+	–	–	–	–	–	–	+	–	+	+	–	–	–	–
G. geminatum	–	–	–	+	–	–	+	–	–	–	–	–	–	–	+	–	–	–	–	–	–	–	–	–
G. constrictum	–	+	+	–	–	–	–	–	–	+	+	–	+	–	–	–	–	–	–	–	–	–	+	–
Gomphonema sp.	+	–	+	–	++	–	–	–	–	+	++	+	–	–	++	–	–	+	–	–	–	–	–	–
Melosira granulata	+	–	+	+	–	+	+	+	+	+	+	+	++	–	+	+	+	+	–	+	+	+	+	+
Melosira sp.	–	–	–	–	–	–	+	–	–	–	–	–	–	–	–	+	+	–	–	–	–	–	–	–
Meridion sp.	–	–	–	–	–	–	–	+	–	–	–	–	–	–	–	–	–	–	–	+	–	–	–	–

(continued)

Table 25 (continued)

Taxa	Sep	Oct	Nov	Dec	Jan	Feb	Mar	Apr	May	Jun	Jul	Aug	Sep	Oct	Nov	Dec	Jan	Feb	Mar	Apr	May	Jun	Jul	Aug
Navicula radiosa	+	–	+	+	++	+	–	–	+	–	–	–	+	–	++	+	+	+	+	–	+	–	–	+
N. rhyncocephala	+	++	++	+	++	–	+	+	+	–	+	+	+	–	–	+	+	–	++	+	+	–	–	+
N. constricta	–	+	+	+	+	+	+	–	–	–	+	+	–	–	–	++	–	+	–	–	–	–	++	+
Nitzschia accicularis	+	–	+	+	++	++	+	–	+	+	+	+	++	+	++	+	++	++	+	–	+	+	–	–
Nitzschia sp.	+	–	+	++	+	–	–	–	+	+	+	+	–	–	+	++	++	–	–	–	+	+	–	+
Navicula sp.	+	–	–	–	+	–	–	–	–	–	–	–	–	+	–	–	–	–	+	–	–	–	–	–
Nitzchia longisima	+	+	+	+	++	+	+	+	–	+	–	–	–	–	+	+	–	+	–	+	–	+	–	–
Nitzschia plana	+	–	–	++	–	+	+	+	+	+	–	–	+	–	–	–	+	+	–	+	+	–	–	+
Nitzschia epiculata	+	–	–	–	–	–	–	–	–	–	–	–	–	–	–	–	–	–	–	–	+	–	–	–
Pinnularia sp.	–	+	–	+	+	–	–	–	–	–	–	–	–	–	–	–	++	–	–	–	–	–	–	–
Pleurosigma sp.	–	–	–	–	–	+	–	–	–	+	–	+	–	–	–	–	–	+	–	–	–	–	–	–
Rhopalodia gibba	–	+	–	+	+	+	–	–	–	+	+	–	++	–	+	–	+	+	+	–	–	–	–	+
Rhopalodia sp.	–	–	–	–	–	–	–	–	–	–	–	–	–	+	–	–	–	–	–	–	–	–	–	–
Stauronies sp.	–	+	+	–	+	–	–	–	–	–	+	–	+	–	–	–	+	–	–	–	–	–	–	–
Synedra ulna	+	+	+	–	+	++	–	++	+	+	++	++	+	+	+	+	+	++	+	++	+	+	+	+
Suriella sp.	–	–	–	–	–	–	–	–	–	–	+	–	–	–	–	–	+	–	–	–	–	–	–	–
Synedra sp.	+	–	–	+	–	–	–	–	–	–	–	+	–	+	–	+	+	–	–	+	–	–	–	–
Synedra lingus	–	+	+	–	–	–	–	–	–	+	–	+	–	–	–	–	–	–	–	–	–	–	–	+
Tabellaria fenestrata	+	–	–	+	–	+	–	–	+	+	+	–	+	–	–	++	–	+	–	–	+	+	+	+

0–5 ind/mL = +
6–10 ind/mL = ++
11–15 ind/mL = +++
and above = ++++

Table 26 Monthly variations in the population density (Ind/mL) of the Bacillariophyceae at study site-III of the Dal lake

Taxa	Sep	Oct	Nov	Dec	Jan	Feb	Mar	Apr	May	Jun	Jul	Aug	Sep	Oct	Nov	Dec	Jan	Feb	Mar	Apr	May	Jun	Jul	Aug
Amphora sp.	−	+	−	−	+	−	+	+	+	+	−	+	−	+	+	+	+	−	−	+	+	+	+	+
Amphora normani	−	−	+	−	+	−	+	−	+	−	+	+	−	+	−	+	−	+	−	−	+	−	+	+
Amphora ovalis	+	+	+	+	+	+	+	+	+	+	+	+	+	+	+	+	−	+	−	+	+	+	+	+
Amphora bitumida	−	−	+	−	+	−	+	−	+	−	+	−	−	+	+	+	−	−	++	+	++	−	−	−
Achnanthes sp.	+	+	+	+	+	+	+	−	+	+	−	+	+	−	+	+	+	−	−	++	++	+	+	+
Anomoneis sp.	+	+	+	+	+	+	+	−	−	+	+	+	−	+	+	−	+	+	+	−	+	+	+	−
Amphora constrictum	−	−	−	−	−	++	−	−	−	−	+	−	−	−	−	+	−	−	+	−	−	−	−	−
Asterionella formosa	−	−	+	−	−	+	−	+	−	−	−	−	−	−	−	−	−	−	+	−	−	−	−	−
Amphora ocellata	+	−	−	−	−	−	−	−	−	−	−	+	+	+	−	−	−	−	+	−	−	−	−	+
Amphora falcatus	−	−	−	−	+	+	−	+	−	−	−	−	+	−	−	+	−	+	−	−	−	−	−	−
Chaetoceros sp.	+	+	−	−	+	−	−	−	+	−	−	+	+	+	+	−	+	−	+	+	++	−	−	+
Caloneis sp.	−	−	+	−	−	−	−	−	+	−	−	+	−	−	−	+	−	−	+	+	+	−	−	+
Cymbella ventricosa	+	+	+	+	++	++	+	+	+	+	+	+	+	+	+	+	+	+	+	+	+	+	++	+
Coscinodiscus sp.	+	+	+	−	+	+	−	+	+	+	−	−	+	+	+	−	+	−	−	−	+	+	+	+
Cyclotella sp.	−	−	+	+	−	−	−	−	−	−	−	−	+	−	−	+	+	−	+	+	+	+	−	−
Cymbella cistula	−	−	+	−	+	+	−	+	−	+	−	−	−	+	+	+	+	+	+	+	+	+	−	−
Cymbella turgida	−	−	−	−	−	+	−	−	+	−	−	+	−	−	−	−	−	−	−	−	−	−	−	−
Cocconeis placentula	++	++	+	+	++	++	+	+	−	+	+	−	−	−	−	−	+	++	−	+	−	−	−	+
Cymbella tumida	+	+	−	−	++	+	−	+	−	+	++	+	−	−	−	−	−	−	−	−	−	−	−	−
Cocconeis sp.	−	−	−	−	+	+	−	+	+	+	−	−	−	−	−	−	−	−	−	−	+	+	−	−
Cymbella lanceolata	+	+	+	+	−	−	−	−	+	−	−	+	+	+	+	+	−	−	+	+	+	−	−	+
Cymbella microporum	−	+	−	−	++	+	+	−	−	−	−	−	+	+	+	−	+	+	−	−	+	−	−	−
Cymbella sp.	−	−	−	−	−	+	−	−	−	−	−	+	+	+	+	−	−	−	−	−	+	−	−	−
Ceratoneis arcus	−	+	−	−	−	−	−	−	−	−	−	+	−	−	−	−	−	−	−	−	−	−	−	+
Diatoma elongatum	+	++	+	−	−	+	−	+	+	−	−	+	+	+	+	++	−	+	+	−	−	−	−	+

(continued)

Table 26 (continued)

Taxa	Sep	Oct	Nov	Dec	Jan	Feb	Mar	Apr	May	Jun	Jul	Aug	Sep	Oct	Nov	Dec	Jan	Feb	Mar	Apr	May	Jun	Jul	Aug
Eunotia zebra	+	++	+	+	+	+	−	+	−	−	−	−	+	+	+	−	+	−	+	+	+	−	−	−
Eunotia gracilis	−	++	+	+	+	+	+	−	−	+	++	+	+	+	++	+	++	++	−	−	−	−	−	+
Eunotia pectinalis	+	++	−	++	−	−	−	−	+	−	−	+	+	+	+	−	+	−	+	+	+	−	−	+
Eunota sorex	−	++	−	+	−	−	−	−	+	−	−	+	−	−	−	−	−	−	−	−	+	−	−	+
Eunotia sp.	+	−	+	−	+	+	+	−	−	+	+	−	+	+	+	+	+	−	+	+	−	+	−	−
Eunotia triodon	+	−	−	−	+	−	+	−	−	+	−	−	+	+	+	+	+	−	+	−	−	+	−	+
Eunotia diodon	+	−	+	+	−	−	+	+	+	+	−	+	+	+	−	−	+	−	+	+	−	+	+	+
Epithemia sp.	−	−	−	−	−	−	−	+	−	−	−	−	−	−	−	−	−	−	−	−	−	−	−	−
Epithemia turgida	+	−	−	−	−	−	−	−	−	−	−	−	−	−	+	+	−	−	−	−	−	−	−	−
Fragilaria vaucheria	++	++	−	−	++	+	++	++	++	−	++	+	+	++	++	+	+	+	+	+	+	+	+	++
Fragilaria sp.	−	+	−	+	+	+	+	++	+	++	−	++	++	+	+	−	−	++	++	+	++	++	+	++
Fragilaria acus	−	−	−	−	−	−	−	−	−	−	+	−	−	−	−	−	−	−	−	−	−	−	+	−
Fragilaria crotonensis	−	−	−	+	+	−	+	−	++	−	−	+	−	−	++	+	+	−	++	+	+	−	−	+
Fragilaria capucina	+	−	−	−	−	−	−	−	−	−	+	+	+	−	−	−	−	−	−	−	−	−	−	+
Fragilaria construens	−	++	−	−	++	−	+	+	−	−	+	++	−	−	+	−	+	+	+	−	−	−	−	+
Gomphonema olivaceum	+	++	++	+	+	+	+	−	−	+	+	++	+	−	−	−	+	−	−	−	−	+	+	+
G. augur	+	−	+	−	+	−	−	−	−	−	+	−	+	+	−	−	−	+	−	−	−	−	−	−
G. accuminatum	−	−	−	−	−	+	−	+	−	−	−	−	−	+	+	−	−	−	+	−	−	−	−	−
G.subventricosum	−	−	−	−	+	−	−	−	+	−	+	+	−	−	−	++	−	−	−	−	−	+	−	+
G. geminatum	−	−	−	−	−	−	−	−	+	−	−	−	−	−	−	−	+	−	−	−	−	−	+	−
G. constrictum	−	++	−	−	+	+	−	−	−	−	−	−	−	−	+	++	−	+	−	−	−	−	−	−
Gomphonema sp.	−	−	−	++	++	+	+	−	−	+	−	−	−	+	−	++	−	−	−	−	++	−	−	−

Melosira granulata	−	−	−	−	+	+	++	+	−	+	−	−	−	−	+	−	−	−	−	−	−	+	−	−
Melosira sp.	+	+	−	−	−	−	−	+	−	+	−	+	+	−	−	+	−	−	−	−	−	+	−	+
Meridion sp.	−	−	−	−	−	−	−	−	−	−	−	−	−	−	−	−	−	−	−	−	−	−	−	+
Navicula radiosa	+	+	−	+	−	−	−	+	+	+	+	−	+	+	−	+	−	−	+	+	−	+	−	−
N rhyncocephala	+	+	−	−	−	+	+	+	+	−	+	+	+	+	++	+	−	−	+	+	++	−	+	+
N constricta	−	−	−	−	−	+	+	+	−	+	+	−	−	−	+	+	−	+	+	+	+	−	+	−
Nitzschia accicularis	+	+	+	++	+	+	+	+	+	+	−	−	−	+	−	+	+	−	−	+	++	+	+	−
Nitzschia sp.	+	−	−	+	+	+	+	+	+	−	−	−	−	+	−	+	+	−	−	+	+	−	−	−
Navicula sp.	−	−	−	−	−	−	−	−	−	−	+	−	−	+	−	−	−	−	−	−	+	−	−	−
Nitzschia longisima	−	−	+	+	+	−	−	+	−	+	−	+	−	+	+	+	+	−	−	+	+	+	+	+
Nitzschia plana	+	+	−	−	+	−	−	+	+	+	−	−	−	+	−	−	−	−	−	−	−	+	+	+
Nitzschia epiculata	+	−	−	−	−	−	−	−	−	−	−	−	−	+	−	−	−	−	−	−	−	−	−	−
Pinnularia sp.	−	−	−	−	−	+	−	−	−	−	−	−	−	−	−	−	−	−	−	−	−	−	−	+
Pleurosigima sp.	−	−	−	−	−	−	+	−	−	−	−	−	−	+	−	−	−	−	−	−	−	−	−	−
Rhopalodia gibba	−	+	−	+	+	+	−	+	+	+	+	+	++	+	++	−	−	−	−	−	−	−	−	+
Stauronies sp.	−	++	−	−	−	+	−	−	+	−	−	−	−	−	+	+	−	−	−	−	+	−	−	−
Synedra ulna	+	+	+	+	+	+	++	−	+	+	+	+	+	+	+	+	+	+	+	+	−	+	−	+
Suriella sp.	−	−	−	−	−	+	−	−	+	−	−	−	−	−	−	−	−	−	−	−	−	−	−	−
Synedra sp.	−	−	+	−	−	−	+	−	+	−	−	+	−	+	−	−	−	−	−	+	−	−	−	−
Synedra lingus	−	−	−	−	+	−	+	−	−	−	−	−	−	−	+	+	−	−	+	−	−	−	−	−
Tabellaria fenestrata	+	+	−	−	+	+	−	+	−	+	−	−	+	+	−	−	+	+	+	+	−	−	−	−

0–5 ind/mL = +
6–10 ind/mL = ++
11–15 ind/mL = +++
and above = ++++

The taxa (71) recorded at site-IV include *Achnanthes sp., Amphora butimida, A. falcatus, A. normanii, A. ocellata, A. ovalis, A. proteus, A. constrictum, Amphora sp., Anomoeoneis sp., Asterionella formosa, Caloneis sp., Ceratoneis arcus, Chaetoceros sp., C. placentula, Cocconeis sp., Coscinodiscus sp., Cyclotella sp., Cymbella cistula, C. lanceolata, C. microporum, C. prostrate, C. tumida, C. turgida, C. ventricosa, Cymbella sp., Diatoma elongatum, Eunotia diodon, E. gracilis, E. pectinalis, E. triodon, E. zebera, Eunotia sorex, Eunotia sp., Epithemia turgida, Epithemia sp., Fragilaria acus, F. capucina, F. construens, F. crotonensis, F. vaucheriae, Fragilaria sp., Gomhonema constrictum,G. acuminatum, G. augur, G. geminatum, G. olivacium, G.subventricosum, Gomphonema sp., Melosira granulate, Melosira sp., Meridion sp., Navicula radiosa, N. constricta, N. rhyncocephala, Navicula sp., Nitzschia accicularis, N. epiculata, N. longissima, N. plana, Nitzschia sp., Pinnularia sp., Pleurosigma sp., Rhopalodia gibba, Rhopalodia sp., Stauronies sp., Suriella sp., Synedra lingus, Synedra ulna, Synedra sp., and Tabellaria fenestrata* (Table 27). The population density of bacillariophyceae was recorded to be highest ranging from 1936 ind/mL (winter) to 2470 ind/mL (autumn) at this site. The species diversity at this site was maximum in winter with average 53 diatom taxa. This site is located in the lake where water receives waste from the hotels. According Hentschel (1928), Gunther (1936), Pearsall (1932), Zafar (1964), Munawar (1970a, b), Singh (1995) and Sedamkar and Angadi (2003) in polluted streams the total number of species may be reduced however, those that survive grow aggressively as there is little competition. Kar et al. (1987), Khan et al. (1998) and Biswas and Konar (2000) and Katti et al. (2002) found greater abundance of diatoms and related it to the higher concentration of plant nutrients. Kumar (1998) found maximum growth of diatoms at the polluted sampling stations in his study.

The overall distribution pattern of bacillariophyceae indicates that the species composition is not much variable at different sites. Out of total of 75, 65 have been found to be present at all the sites.

The taxa *Aphanocapsa* sp. and *Tabellaria* sp. were found at site-I only. The presence of these two indicates that they cannot tolerate any level of pollution. Kolkwitz and Marsson (1908), Jolly and Chapman (1966) and Das (1989) also found *Tabellaria* sp. in clean waters only. However, Bulusu et al. (1967), Verma and Shukla (1969) and Syal (1996) found it from clean waters to highly polluted waters. No taxon was exclusively recovered from site-II, III and site-IV.

In all, 5 species belonging to genera *Amphora* were found to be present at all the four sites. These include *A. bitumida, A. constrictum, A. falcatus, A. normanii* and *A. ovalis*. The *A. proteus* was present at site-I, II and IV and *A. ocellata* was recovered from site-II, III and site-IV only. Sarwar (1991a, b) found *A.ovalis* from clean to polluted waters. Bhatt et al. (1999) recorded *Amphora* sp. in waters used for bathing, washing clothes, and containing chemical fertilizers in Taudaha lake.

In all, 6 species belonging to genus *Cymbella* were found to be present at all the four sites. These include *C. cistula, C. tumida, C. lanceolata, C. microporum, C. turgida* and *C. ventricosa*. In the present study, the density of *Cymbella ventricosa* was seen to increase at sites II, III and IV. The highest density was recorded at site-II where water receives good amount of agricultural run-off. According to Dickman

Table 27 Monthly variations in the population density (Ind/mL) of the Bacillariophyceae at study site-IV of the Dal lake

Taxa	Sep	Oct	Nov	Dec	Jan	Feb	Mar	Apr	May	Jun	Jul	Aug	Sep	Oct	Nov	Dec	Jan	Feb	Mar	Apr	May	Jun	Jul	Aug
Amphora sp.	–	+	–	–	+	–	++	+	+	+	–	+	–	+	+	+	+	–	–	+	+	+	+	+
A. butimida	–	–	+	–	+	–	+	–	+	–	+	–	–	+	+	+	–	–	+	+	++	–	–	–
A. normanii	–	–	+	–	+	–	++	–	+	–	+	+	–	+	–	+	–	+	–	–	+	–	+	+
A. ovalis	+	+	++	+	+	++	+	+	+	+	+	++	+	++	+	+	–	+	–	+	+	+	+	+
A. proteus			++	+	+	+	+		+	+		+	+	+	+	+	+		+	+	+	+	+	+
Achnanthes sp.	+	+	+	+	+	++	+	–	+	+	–	+	+	–	+	+	+	–	–	++	++	+	++	+
Amphora constrictum	–	–	–	–	–	++	–	–	–	–	+	–	–	–	–	+	–	–	+	–	–	–	–	–
A. falcatus	–	–	–	–	+	+	–	+	–	–	–	–	+	–	–	+	–	+	–	–	–	–	–	–
A. ocellata	+	–	–	–	–	–	–	–	–	–	–	+	+	+	–	–	–	–	+	–	–	–	–	+
Anomoneis sp.	+	+	+	+	+	+	+	–	–	+	+	+	–	+	+	–	+	+	+	–	++	+	+	–
Asterionella formosa	–	–	+	–	–	+	–	+	–	–	–	–	–	–	–	–	–	–	+	–	–	–	–	–
Caloneis sp.	–	–	+	–	–	–	–	–	+	–	–	+	–	–	–	+	–	–	+	+	+	–	–	+
Ceratoneis arcus	–	+	–	–	–	–	–	–	–	–	–	+	–	–	–	–	–	–	–	–	–	–	–	+
Chaetoceros sp.	+	+	–	–	+	–	–	–	+	–	–	+	+	+	+	–	+	–	+	+	++	–	–	+
Cocconeis placentula	++	++	+	+	++	++	+	+	–	+	++	–	–	–	–	–	+	++	–	+	–	–	–	+
Cymbella prostrata	+	–	+	+	+	–	–	–	+	–	+	–	–	–	–	–	+	+	–	+	–	–	–	–
Cocconeis sp.	–	–		–	+	+	–	+	+	+	–	–	–	–	–	–	–	–	–	–	+	+	–	–
Coscinodiscus sp.	+	+	++	–	+	+	–	+	+	+	–	–	+	+	+	–	+	–	–	–	+	+	+	+
Cyclotella sp.	–	–	+	+	–	–	–	–	–	–	–	–	+	–	–	+	+	–	+	++	+	+	–	–
Cymbella cistula	–	–	+	–	++	+	–	+	–	+	–	–	–	+	+	+	+	+	+	+	+	+	–	–
Cymbella tumida	+	+		–	++	+	–	+	–	+	++	+	–	–	–	–	–	–	–	–	–	–	–	–
Cymbella lanceolata	+	+	+	+	–	–	–	–	+	–	–	+	+	+	+	+	–	–	+	+	+	–	–	+
Cymbella microporum	–	+	–	–	++	+	+	–	–	–	–	–	+	+	+	–	+	+	–	–	+	–	–	–
Cymbella sp.	–	–	–	–	–	+	–	–	–	–	–	+	+	+	+	–	–	–	–	–	+	–	–	–
Cymbella turgida	–	–	–	–	–	+	–	–	+	–	–	+	–	–	–	–	–	–	–	–	–	–	–	–

(continued)

Table 27 (continued)

Taxa	Sep	Oct	Nov	Dec	Jan	Feb	Mar	Apr	May	Jun	Jul	Aug	Sep	Oct	Nov	Dec	Jan	Feb	Mar	Apr	May	Jun	Jul	Aug
Cymbella ventricosa	+	+	+	+	++	++	+	+	+	+	+	++	+	+	+	+	+	+	+	++	++	+	++	+
Diatoma elongatum	+	++	+	–	–	+	–	+	++	–	–	+	+	+	+	++	–	+	+	–	–	–	–	+
Epithemia sp.	–	–	–	–	–	–	–	+	–	–	–	–	–	–	–	–	–	–	–	–	–	–	–	–
Epithemia turgida	+	–	–	–	–	–	–	–	–	–	–	–	–	–	+	+	–	–	–	–	–	–	–	–
Eunotia sorex	–	++	–	+	–	–	–	–	+	–	–	+	–	–	–	–	–	–	–	–	+	–	–	+
E. gracilis	–	++	+	+	++	++	+	–	–	+	++	+	+	++	++		++	++	–	–	–	–	–	+
E. pectinalis	+	+	–	++	–	–	–	–	++	–	–	+	+	+	+	–	+	–	+	+	+	–	–	+
Eunotia sp.	+	–	+	–	+	+	+	–	–	+	+	–	+	+	+	+	+	–	+	+	–	+	–	–
E. triodon	+	–	–	–	+	–	+	–	–	+	–	–	+	+	+	+	+	–	+	–	–	+	–	+
E. zebera	+	++	+	+	+	+	–	+	–	–	–	–	+	+	+	–	+	–	+	+	+	–	–	–
E. diodon	+	–	+	+	–	–	+	+	+	+	–	+	+	+	–	–	+	–	+	+	–	+	+	+
Fragilaria acus	–	–	–	–	–	–	–	–	–	–	+	–	–	–	–	–	–	–	–	–	–	–	+	–
F. capucina	+	–	–	–	–	–	–	–	–	–	+	+	+	–	–	–	–	–	–	–	–	–	–	+
F. construens	–	++	–	–	++	–	++	+	–	–	+	++	–	–	+	–	+	+	+	–	–	–	–	++
F. crotonensis	–	–	–	+	+	–	+	–	++	–	–	+	–	–	++	+	+	–	++	+	+	–	–	+
Fragilaria sp.	–	+	–	+	+	+	+	++	+	++	–	++	++	+	+	–	–	++	++	+	++	++	+	++
F. vaucheria	++	++	–	–	++	+	++	++	++	–	++	+	+	++	++	+	+	+	+	+	+	+	+	++
Gomhonema constrictum	–	++	–	–	+	+	–	–	–	–	–	–	–	–	+	++	–	+	–	–	–	–	–	–
G. geminatum	–	–	–	–	–	–	–	–	+	–	–	–	–	–	–	–	+	–	–	–	–	–	+	–
G.subventricosum	–	–	–	–	+	–	–	–	+	–	+	+	–	–	–	++	–	–	–	–	–	+	–	+
G. augur	+	–	+	–	+	–	–	–	–	–	+	–	+	+	–	–	–	+	–	–	–	–	–	–
G. acuminatum	–	–	–	–	–	+	–	+	–	–	–	–	–	+	+	–	–	–	+	–	–	–	–	–
G. olivacium	+	++	++	+	+	+	+	–	–	+	+	++	+	–	–	–	+	–	–	–	–	+	+	+
Gomphonema sp.	–	–	–	++	++	++	+	–	–	+	–	–	–	+	–	++	–	–	–	–	++	–	–	–
Melosira granulata	–	–	–	–	+	+	++	+	–	+	–	–	–	–	+	–	–	–	–	–	–	+	–	–

Melosira sp.	+	+	−	−	−	−	−	+	−	+	−	+	+	−	−	+	−	−	−	−	−	+	−	+
Meridion sp.	−	−	−	−	−	−	−	−	−	−	−	−	−	−	−	−	−	−	−	−	−	−	−	+
Navicula constricta	−	−	−	−	−	+	+	+	−	+	+	−	−	−	+	+	−	+	+	+	+	−	+	−
N. rhyncocephala	+	+	−	−	−	+	+	+	+	−	+	+	+	+	++	+	−	−	+	+	++	−	+	+
N. radiosa	+	+	−	+	−	−	−	+	+	+	+	−	+	+	−	+	−	−	+	+	−	+	−	−
Navicula sp.	−	−	−	−	−	−	−	−	−	−	+	−	−	+	−	−	−	−	−	−	+	−	−	−
Nitzschia sp.	+	−	−	+	+	+	+	++	+	−	−	−	−	+	−	+	+	−	−	+	+	−	−	−
N. longisima	−	−	+	+	+	−	−	+	−	+	−	+	−	+	+	+	+	−	−	+	+	+	+	+
N. accicularis	+	++	+	++	+	+	+	+	+	+	−	−	−	+	−	+	+	−	−	+	++	+	+	−
N. epiculata	+	−	−	−	−	−	−	−	−	−	−	−	−	+	−	−	−	−	−	−	−	−	−	−
N. plana	+	+	−	−	+	−	−	+	+	+	−	−	−	+	−	−	−	−	−	−	−	+	+	+
Pinnularia sp.	−	−	−	−	−	+	−	−	−	−	−	−	−	−	−	−	−	−	−	−	−	−	−	+
Pleurosigma sp.	−	−	−	−	−	−	+	−	−	−	−	−	−	+	−	−	−	−	−	−	−	−	−	−
Rhopalodia gibba	−	+	−	+	+	+	−	+	+	+	+	+	++	++	++	−	−	−	−	−	−	−	−	+
Rhopalodia sp.	−	−	−	−	+	−	−	−	−	−	−	−	−	−	−	−	−	+	−	−	−	−	−	−
Stauronies sp.	−	++	−	−	−	+	−	−	+	−	−	−	−	−	+	+	−	−	−	−	+	−	−	−
Suriella sp.	−	−	−	−	−	+	−	−	+	−	−	−	−	−	−	−	−	−	−	−	−	−	−	−
Synedra lingus	−	−	−	−	+	−	+	−	−	−	−	−	−	−	+	+	−	−	+	−	−	−	−	−
Synedra sp.	−	−	+	−	−	−	+	−	+	−	−	+	−	+	−	−	−	−	−	+	−	−	−	−
Synedra ulna	+	+	+	+	+	+	++	−	+	+	+	+	+	+	+	+	+	+	+	+	−	+	−	+
Tabellaria fenestrata	+	+	−	−	+	+	−	+	−	+	−	−	+	+	−	−	+	+	+	+	−	−	−	−

0–5 ind/mL = +
6–10 ind/mL = ++
11–15 ind/mL = +++
and above = ++++

(1975), *Cymbella ventricosa* is commonly found in organically polluted waters. Sarwar (1991a, b) recorded this species from clean as well as polluted waters. Syal (1996) found *C.cistula* inhabiting both clean and contaminated water with sewage. Kolkwitz and Marsson (1908) however, reported *C. cistula* only in clean waters. The *C.prostrata* was recorded at site-I, II and IV and absent from site-III. In the present study the density of *C. tumida* was seen to increase at site-II as compared to other sites, thereby indicating that this species can tolerate wide range of agricultural run-off. *Cymbella* sp. has been reported in a variety of waters. Zutshi et al. (1980) found it in clean and highly polluted waters. Bulusu et al. (1967), Singh et al. (1969) and Munawar (1970a) found it in waters polluted with sewage and industrial wastes while Verma and Shukla (1969, 1970) and Govindan and Sundaresan (1979) found it only in clean waters. Varma (1998) recorded *Cymbella* sp. from industrially polluted waters. Kawoosa (1985) related the nutrient enrichment of the Dal lake to the presence of *Cymbella* sp.

In all 6 species belonging to genus *Eunotia* were found to be present at all the four sites. These include *E. sorex, E. gracilis, E. pectinalis, E. triodon, E. zebra,* and *E. diodon.* In this study the density of *E. sorex* was seen to be lower at site-II as compared to site-I, II and III, thus indicating that this species cannot thrive well in waters which are polluted with agricultural run-off while as *E. gracilis* increased at site-III and IV thereby indicating that this species can thrive well in waters which are contaminated with sewage. Siver (1977) found *Eunotia* sp. among the most dominant in Wheelwright pond in New Hampshire containing organic waste.

In all, 5 species belonging to genus *Fragilaria* were found to be present at all the four sites. These include *F. acus, F. capucina, F. construens, F. crotonensis,* and *F. vaucheriae.* In the present study, the density of *F. capucina* was seen to be lowest at site-II thereby indicating that this species cannot flourish well in waters contaminated with agricultural run-off. David et al. (1981) opined that this species is favoured by moderate to highly enriched waters and eliminated from extremely polluted habitats. However, Dickman (1975) found this species in organically polluted waters. This species has been labelled as indicator of eutrophy by Jarnefelt (1952), Stockner and Benson (1971) and Vollenweider (1972). Khan and Bhat (2000) recorded this species in Manasbal lake harbouring rich growth of aquatic weeds and the water receiving multiple organic wastes from anthropogenic perturbations. Shetty et al. (1961), Singh (1960), Stockner and Benson (1971), Williams (1969), Conner (1977), Cassie (1979) and Valecha and Bhatnagar (1988) have also treated this species as pollution indicator. The density of *F.vaucheria* in the present study has been seen to increase at site-II, III and IV which are receiving agricultural run-off from the floating gardens and sewage from the houseboats and hotels Sarwar (1991a, b) recorded *F. vaucheria* in clean as well as polluted waters. The density of *F.crotonensis* and *Fragilaria* sp. in the present study has been seen to increase at site-II, III and IV, thereby indicating that these can thrive well in waters contaminated with agricultural run-off and sewage. Williams (1969) considered the presence of *Fragilaria* sp. as indication of sewage pollution. Chakraborty and Konar (2000) recorded *Fragilaria* sp. in water with huge organic materials. Syal (1996) reported *Fragilaria* sp. in clean as well as waters receiving sewage effluence. David and Ray (1966), Zutshi et al. (1980), Paramasivam and Sreenivasan (1981) and Das (1989) found it in both clean and highly polluted waters, whereas, George et al. (1966), Bulusu et al. (1967) and Verma et al. (1978) found it only in highly polluted waters contaminated with sewage and industrial wastes. Kawoosa (1985) related the nutrient enrichment of water of the Dal lake to the presence of *Fragilaria* sp. On the other hand, Kolkwitz and Marsson (1908) and Verma and Shukla (1970) found it only in clean waters.

In all, 5 species belonging to the genus *Gomphonema* were found to be present at all the four sites. These include *G. constrictum, G. subventricosum, G. augur, G. geminatum,* and *G. accuminatum. G. olivaceum* was recorded from site-I, III and IV only. Dickman (1975) found *G. olivaceum* in organically polluted waters. The density of *Gomphonema* sp. was observed to increase at the sites receiving agricultural run-off from the floating gardens and sewage from the houseboats and hotels. David et al. (1981) opined that *Gomphonema* sp. is adapted to withstand physical disturbance and benefit from organic enrichment. Bhatt et al. (1999) recorded this taxon to be present in sewage polluted water. Syal (1996) opined that *Gomphonema* sp. can inhabit clean as well as water receiving sewage effluence. Kolkwitz and Marsson (1908), David and Ray (1966), Jolly and Chapman (1966) and Zutshi et al. (1980) found this in clean waters only.

In all, 3 species belonging to genus *Navicula* were found to be present at all the four sites. These include *N. constricta, N. rhyncocephala,* and *N. radiosa.* Among these species, the density of *N. constricta* was found to increase at the sites receiving agricultural run-off from floating gardens and sewage from houseboats and hotels. *N.radiosa* was found to be the dominating species at all the sites. Khan and Bhat (2000) recorded *N.radiosa* where the lake basin is infested with aquatic weeds and is under multiple anthropogenic pressures. The density of *Navicula* sp. in the present study was seen to be low at all the sites. Bhatt et al. (1999) recorded *Navicula* sp. in places where the water is used for bathing and washing and carries chemical fertilizers and insecticides. Chakraborty and Konar (2000) found this taxon where there is huge storage of organic materials. On the other hand, Sedamkar and Angadi (2003) considered *Navicula* sp. as the indicator of clean water. David and Ray (1966), Jolly and Chapman (1966), Govindan and Sundaresan (1979) and Zutshi et al. (1980) found *Navicula* sp. both in clean and highly polluted waters, whereas, George et al. (1966), Bulusu et al. (1967), Munawar (1970b), Das (1989) and Verma et al. (1997) found it in waters contaminated with sewage and industrial effluence. Kolkwitz and Marsson (1908), Verma and Shukla (1969, 1970) and Paramavsivam and Sreenivasan (1981) found it both in clean and moderately polluted waters.

In all, 4 species belonging to genus *Nitzschia* were found to be present at all the four sites. These include *N.longissima, N. accicularis, N. epiculata,* and *N. plana.* In the present study the density of *N.accicularis* was seen to increase at site-II, III and IV, thereby meaning that this species can thrive well in such kind of waters. Khan and Bhat (2000) recorded this species where the lake basin is infested with aquatic weeds and is under multiple anthropogenic pressures. Shetty et al. (1961), Singh (1961) and Valecha and Bhatnagar (1988) in their investigations have also impressed upon the utility of *N. accicularis* as the potential indicator of pollution and eutrophy. Stockner and Benson (1971), Williams (1969), Conner (1977) and Cassie (1979) have also treated this species as pollution indicator. At site-I, the density of *Nitzschia* sp. was recorded to be very low as compared to site-II, III and site-IV. Richardson (1968) considered the presence of this taxon to be characteristics of organically rich waters and labeled it a typical eutrophic species. Weilgolaski (1975), Ahmad (1996) and Varma (1998) labeled *Nitzschia* sp. as pollution indicator.

The presence of *Pinnularia* sp. at all the sites indicates that it can thrive well in clean as well as organically rich waters. Bhatt et al. (1999) found this taxon present throughout in sewage water also.

In all, 2 species belonging to genus *Synedra* were found to be present at all the sites in the present investigation. These include *S. lingus* and *S. ulna*. The density of *S.ulna* at the site-II, III and IV has been seen to increase to a larger proportions as compared to site-I. The out break of this species at the sites receiving agricultural run-off from floating gardens and sewage from houseboats and hotels indicated that the species can thrive well in such waters and to be tolerant to such kind of pollution. David et al. (1981) recorded *S. ulna* to be abundant in shallow and enriched zones of waters. Khan and Bhat (2000) recorded this in the places of water body rich in macrophytes and where the water receives good amount of organic waste. Dickman (1975) found this species commonly in organically polluted waters. Weilgolaski (1975), Ahmad (1996) and Varma (1998) attributed the presence of *S. ulna* to be the indication of pollution where the density was high. Lower (1974) labeled it as eutrophic species. Syal (1996) found *S.ulna* at the sites receiving sewage as well as free of pollution. Kolkwitz and Marsson (1908), Jolly and Chapman (1966), Zutshi et al. (1980), Paramasivam and Sreenivasan (1981) and Das (1989) recorded *S.ulna* in clean waters only. Chakraborty and Konar (2000) found *Synedra* sp. in water containing high organic matter.

The genus *Tabellaria* is represented by *T.fenestrata* and *Tabellaria* sp., *T. fenestrata* was present at all the sites whereas *Tabellaria* sp. was recorded from site-I only. Kolkwitz and Marsson (1908), Jolly and Chapman (1966) and Das (1989) also found *Tabellaria* sp. in clean waters only. However, Bulusu et al. (1967), Verma and Shukla (1969) and Syal (1996) found it from clean waters to highly polluted waters.

Achnanthes sp. was present at all the sites in the present study. However, at site-II, III and IV, the density of this taxon was high as compared to site-I. The highest density was recorded at site-II where water receives agricultural run-off from the adjoining floating gardens. Sarwar (1991a, b) found this taxon to be dominant at all the places receiving sewage. Siver (1977) found this taxon to be most abundant in moderately to polluted waters. Kawoosa (1985) observed presence of *Achnanthes* sp. as indicator of eutrophy in the Dal lake. David et al. (1981) opined that this taxon is adapted to moderately to highly enriched environments, able to survive anywhere within the low nutrient conditions also. Blinn et al. (1980) and Stevenson (1984) have found low abundance of this taxon due to nutrient limitation.

Anomoeoneis sp. was found to be present at all the sites in the present study. David et al. (1981) recorded this taxon in Smith mountain lake throughout the study period receiving major municipal and industrial discharges and also under low nutrient conditions. Siver (1977) found this taxon occasionally abundant in Wheelwright pond in New Hampshire receiving sewage.

Cocconeis placentula was found to be present at all the study sites. The high density of *C. placentula* at the site-II indicates that this can thrive well in waters receiving agricultural run-off. At site-III and IV also, the density was seen to be higher as compared to site-I. Siver (1977) found *C. placentula* abundant in Wheelwright pond in New Hampshire receiving sewage. According to Dickman (1975), *C. placentula* is commonly found in organically polluted waters. The presence of *Cocconeis* sp. also indicates that it can thrive well in waters contaminated with agricultural run-off and sewage. Syal (1996) found *Cocconeis* sp. to be present at polluted and pollution free sites. Jolly and Chapman (1966) and Zutshi et al. (1980) found *Cocconeis sp.* only in clean waters.

Cyclotella sp. was present at all the sites. However, the density of this taxon was found to be low at the site-II, thereby indicating that this cannot withstand water contaminated with agricultural run-off. David et al. (1981) found this taxon abundant in eutrophicated waters. *Diatoma elongatum* was present at all the sites with higher density at site-IV which is receiving sewage from hotels. According to Dickman (1975), this species is commonly found in organically polluted waters. Bhatt et al. (1999) found *Diatoma* sp. in water polluted with bathing, washing and chemical fertilizers.

Melosira was represented by *M. granulata* and *Melosira* sp. Both the taxa were recorded from all the sites. However, the density of *Melosira* sp. was found to be low at site-II which is receiving agricultural run-off from floating gardens. Kolkwitz and Marsson (1908) found it both in clean and moderately polluted waters whereas, Gaur and Khan (1993) found *Melosira* sp. in least numbers after receiving effluents from a thermal power plant. David and Ray (1966), Jolly and Chapman (1966), Zutshi et al. (1980), Das (1989) and Syal (1996) found *Melosira* sp. both in clean and highly polluted waters, However, Pearsall (1932), Verma and Shukla (1968, 1970) and Gunale and Balakrishnan (1981) found it only in clean waters.

Epithemia is represented by *E. turgida* and *Epithemia* sp., both present at all the sites. However, the total contribution of *Epithemia* sp. was found to be low as compared to *E.turgida*. Syal (1996) found *Epithemia* sp. to be present only in highly polluted waters at the sites receiving paper mill wastes. David and Ray (1966) and Verma et al. (1978) found it to be present in highly polluted waters contaminated with sewage.

Asterionella formosa in the present study was found to be present at all the sites. Sarwar (1991a, b) also found this species in less polluted to highly polluted waters in the marshlands around Srinagar. *Coscinodiscus* sp. was found to be present at all the investigated sites in the present study. Sarwar (1991a, b) found this taxon at the places where the water is greatly overgrown with macrophytes. *Pleurosigma* sp. was found to be present at all the study sites in the present investigation. Sarwar (1991a, b) also found this taxon to be present at all the sites in marshlands around Srinagar. *Rhopalodia gibba* was also present at all the study sites in the present investigation. Sarwar (1991a, b) also found this taxon to be present at all the study sites. He observed increased abundance of this taxon at sites receiving sewage and agricultural wastes which is also true for the present study. *Stauronies* sp. and *Suriella* sp. were found to be present at all the sites in the present study. Sarwar (1991a, b) however, recorded *Stauronies anceps* to be the frequent species and *Suriella ovalis* to be present only at the sites receiving sewage and agricultural wastes. *Caloneis* sp. was found to be present at all the sites.

Chlorophyceae (Green algae)

The Chlorophyceae constituted the second largest group in terms of population density and species diversity after diatoms in the present study. Biswas and Konar (2000) and Yousuf et al. (2002) also reported Chlorophyceae to be the second largest group of phytoplankton in the river Ganga and the Anchar lake respectively.

In the 2 year study of the Dal lake, 34 chlorophyceae taxa distributed at four sites of the Dal lake with 33 at site I, 28 at site-II and 25 each at site-III and site-IV (Table 28) are *Ankistrodesmus spiralis, Ankistrodesmus* sp., *Characium* sp., *Chlorella* sp., *Chlorococcum* sp., *Cladophora* sp., *Clostridium* sp., *Coelastrum* sp., *Cosmarium,* sp., *Desmidium* sp., *Euastrum* sp., *Gonatozygon* sp., *Mougeotia* sp., *Oedogonium,* sp., *Oocystis* sp., *Pediastrum boryanum, P. duplex, P. ovatum, P. spinosum, P. tetras, P. simplex, Pediastrum* sp., *Scenedesmus bijugates, S. dimorphus, Scendesmus* sp., *Selenastrum gracile, Selenastrum* sp., *Sphaerocystis, Spirogyra* sp., *Staurastrum* sp., *Tetraedon, Ulothrix* sp., *Volvox* sp., *Zygnema* sp.,

In the present study, the maximum population density was observed during summer at site-I, II, and IV only. The maximum species diversity was observed during summer at all the sites. Kant and Kachroo (1974; 1977); Zutshi and Vass (1982); Gujarathi and Kanhere (1988), Bhatt et al. (1999), Yousuf et al. (2002) and Sedamkar and Angadi (2003) also recorded population density to be highest during summer (Table 29, Figs. 26 and 27).

The taxa recorded at site-I include *Ankistrodesmus spiralis, Ankistrodesmus* sp., *Characium* sp., *Chlorella* sp., *Chlorococcum* sp., *Cladophora* sp., *Clostridium* sp., *Coelastrum* sp., *Cosmarium,* sp., *Desmidium* sp., *Euastrum* sp., *Gonatozygon* sp., *Mougeotia* sp., *Oedogonium,* sp., *Oocystis* sp., *Pediastrum boryanum, P. duplex, P. ovatum, P. spinosum, P. tetras, P. simplex, Pediastrum* sp., *Scenedesmus bijugates, S. dimorphus, Scendesmus* sp., *Selenastrum* sp., *Sphaerocystis, Spirogyra* sp., *Staurastrum* sp., *Tetraedon, Ulothrix* sp., *Volvox* sp. and *Zygnema* sp. (Table 30).

The taxa recorded at site-II include *Ankistrodesmus spiralis, Characium* sp., *Chlorella* sp., *Chlorococcum* sp., *Cladophora* sp., *Clostridium* sp., *Cosmarium,* sp., *Desmidium* sp., *Euastrum* sp., *Gonatozygon* sp., *Mougeotia* sp., *Oedogonium,* sp., *Oocystis* sp., *Pediastrum boryanum, P. duplex, P. tetras, P. simplex, Pediastrum* sp., *Scenedesmus bijugates, S. dimorphus, Scendesmus* sp., *Selenastrum* sp., *Sphaerocystis, Spirogyra* sp., *Staurastrum* sp., *Ulothrix* sp., *Volvox* sp.and *Zygnema* sp. (Table 31).

Table 28 List of Chlorophyceae taxa recorded at the investigated sites of the Dal lake

S. No.	Chlorophyceae taxa	Site-I	Site-II	Site-III	Site-IV
1	*Ankistrodesmus spiralis (*Turn.) Lemm	+	+	–	–
2	*Ankistrodesmus sp.*	+	–	+	+
3	*Characium sp.*	+	+	+	+
4	*Chlorella sp.*	+	+	+	+
5	*Chlorococcum sp.*	+	+	+	+
6	*Cladophora sp.*	+	+	+	+
7	*Clostridium sp.*	+	+	+	+
8	*Coelastrum sp.*	+	–	–	–
9	*Cosmarium sp.*	+	+	+	+
10	*Desmidium sp.*	+	+	+	+
11	*Euastrum sp.*	+	+	–	–
12	*Gonatozygon sp.*	+	+	+	+
13	*Mougeotia sp.*	+	+	+	+
14	*Oedogonium sp.*	+	+	+	+
15	*Oocystis sp.*	+	+	+	+
16	*Pediastrum boryanum* (Turp) Menegh	+	+	+	+
17	*P. duplex*	+	+	+	+
18	*P. ovatum*	+	–	–	–
19	*P. spinosum*	+	–	–	–
20	*P. tetras* (Ehr) Ralfs	+	+	–	–
21	*P. simplex* Mey. Rabenh	+	+	+	+
22	*Pediastrum sp.*	+	+	–	–
23	*Scenedesmus bijugates* (Turp) Kutz.	+	+	+	+
24	*Scenedesmus dimorphus* (Turp) Kutz.	+	+	+	+
25	*Scendesmus sp.*	+	+	+	+
26	*Selenastrum gracile* Reinsch	–	–	+	+
27	*Selenastrum sp.*	+	+	–	–
28	*Sphaerocystis*	+	+	+	+
29	*Spirogyra sp.*	+	+	+	+
30	*Staurastrum sp.*	+	+	+	+
31	*Tetraedon*	+	–	–	–
32	*Ulothrix sp.*	+	+	+	+
33	*Volvox sp.*	+	+	+	+
34	*Zygnema sp.*	+	+	+	+

The taxa recorded at site-III include *Ankistrodesmus* sp., *Characium* sp., *Chlorella* sp., *Chlorococcum* sp., *Cladophora* sp., *Clostridium* sp., *Cosmarium,* sp., *Desmidium* sp., *Gonatozygon* sp., *Mougeotia* sp., *Oedogonium,* sp., *Oocystis* sp., *Pediastrum boryanum, P. duplex, P. simplex, Scenedesmus bijugates, S. dimorphus, Scendesmus* sp., *Selenastrum gracile, Sphaerocystis, Spirogyra* sp., *Staurastrum* sp., *Ulothrix* sp., *Volvox* sp., and *Zygnema* sp. (Table 32).

Table 29 Seasonal variations in the population density (Ind/mL) of the Chlorophyceae at the investigated sites of the Dal lake

Sites	Year	Autumn	Winter	Spring	Summer
Site-I	**2000–2001**	71	81	84	121
	2001–2002	42	47	43	121
	Average	56.5	64	63.5	121
Site-II	**2000–2001**	83	47	48	115
	2001–2002	26	63	55	163
	Average	54.5	55	51.5	139
Site-III	**2000–2001**	73	46	66	41
	2001–2002	32	51	44	57
	Average	52.5	48.5	55	49
Site-IV	**2000–2001**	83	22	50	50
	2001–2002	35	81	61	33
	Average	59	51.5	55.5	41.5

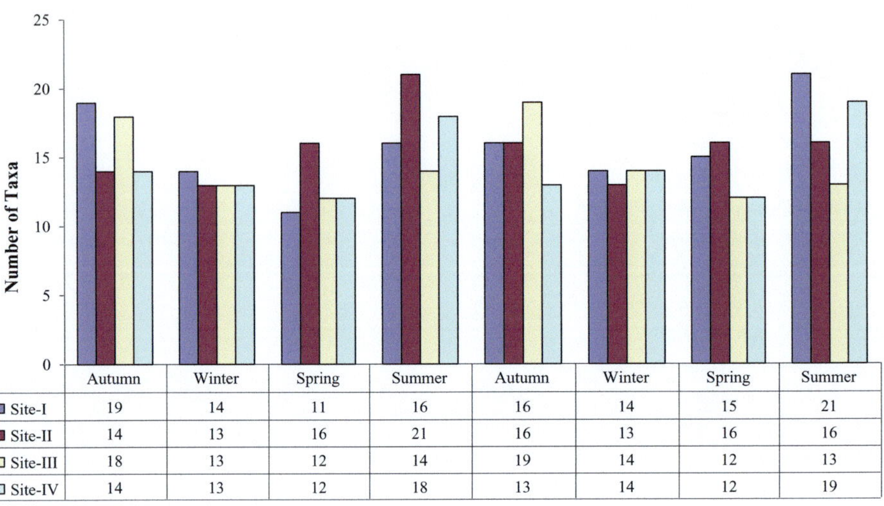

	Autumn	Winter	Spring	Summer	Autumn	Winter	Spring	Summer
Site-I	19	14	11	16	16	14	15	21
Site-II	14	13	16	21	16	13	16	16
Site-III	18	13	12	14	19	14	12	13
Site-IV	14	13	12	18	13	14	12	19

Fig. 26 Seasonal variations in the species diversity of the Chlorophyceae at the investigated sites of the Dal lake

The taxa recorded at site-IV include *Ankistrodesmus* sp., *Characium* sp., *Chlorella* sp., *Chlorococcum* sp., *Cladophora* sp., *Clostridium* sp., *Cosmarium,* sp., *Desmidium* sp., *Gonatozygon* sp., *Mougeotia* sp., *Oedogonium,* sp., *Oocystis* sp., *Pediastrum boryanum, P. duplex, P. simplex, Scenedesmus bijugates, S. dimorphus, Scendesmus* sp., *Selenastrum gracile, Sphaerocystis, Spirogyra* sp., *Staurastrum* sp., *Ulothrix* sp., *Volvox* sp., and *Zygnema* sp. (Table 33).

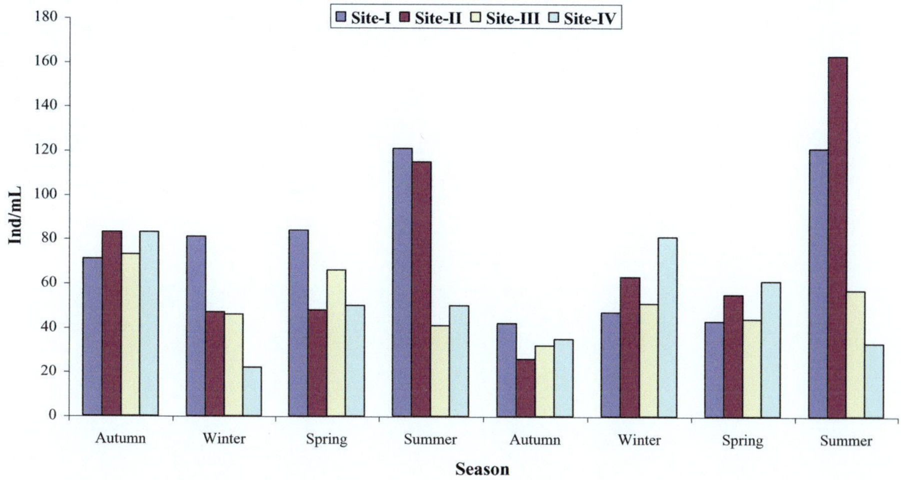

Fig. 27 Seasonal variations in the population density (Ind/mL) of the Chlorophyceae at the investigated sites of the Dal lake

The overall distribution pattern of Chlorophyceae in the present study indicates that the species composition is not much variable at different sites. Out of 34, 23 taxa were recovered from all the sites.

The taxa exclusively recorded at site-I include *Coelastrum* sp., *Pediastrum spinosum, P. ovatum* and *Tetraedon* sp. The presence of *P. ovatum* in open water only where the water is comparatively less interfered by human habitation is also reported by Sarwar et al. (1996). *Tetraedon* sp. was found to be distributed in clean as well as waters receiving sewage and paper mill effluence by Zutshi et al. (1980) and Syal (1996), whereas, Munawar (1970b) and Govindan and Sundaresan (1979) found it to be distributed only in highly polluted waters enriched with sewage and industrial wastes. Sarwar (1991a, b) recorded *Tetraedon* sp. to be present from the marshy lands which receive sewage. Absence of all these including *Coelastrum* sp. and *Pediastrum spinosum,* at site-II, III and IV indicate their aversion to organic pollution in the present study of the Dal lake. No taxa was exclusively recovered from site-II, III and site-IV.

Ankistrodesmus spiralis and *Ankistrodesmus* sp. were recovered in the present study from the Dal lake. *Ankistrodesmus spiralis* was found to be present at site-I and site-II and absent from site-III and IV. The presence of this species indicate that they can withstand water contaminated with agricultural run-off. Sarwar (1991a, b) also recorded this species from the polluted sites in the marsh lands around Srinagar. The another taxon *Ankistrodesmus* sp. was found to be present at the site- I, III and IV and absent from site-II. Bhatt et al. (1999) found *Ankistrodesmus* sp. (without mentioning its species) in waters used for bathing, washing clothes and carrying fertilizers. Venkateshwarlu et al. (1990), Reddy et al. (1991) and Suvarna and Somashekar (2000) opined that this taxon can survive in waters deficient in dis-

Table 30 Monthly variations in the population density (Ind/mL) of the Chlorophyceae at the study site-I of the Dal lake

Taxa	Sep	Oct	Nov	Dec	Jan	Feb	Mar	Apr	May	Jun	Jul	Aug	Sep	Oct	Nov	Dec	Jan	Feb	Mar	Apr	May	Jun	Jul	Aug
Ankistrodesmus spiralis	−	+	−	−	−	−	−	−	−	−	−	−	+	++	+	+	−	−	−	−	−	−	−	−
Characium sp.	−	−	+	−	−	−	+	+	+	−	+	−	−	−	+	−	−	−	+	+	+	−	−	+
Chlorella sp.	+	−	+	+	++	+	+	−	−	++	++	+	++	−	−	+	−	−	+	−	−	+	−	−
Chlorococcuim sp.	+	+	−	−		++		−	+	−	−	−	+	+	−	+	+	+	+	−	+	−	−	+
Cladophora sp.	+	−	+	−	−	−	−	−	−		+	+	+	−	−	−	−	−	−	−	−	+	+	−
Clostridium sp.	−	+	−	+	−	−	+	−		+	+	+	+	−	−	+	−	−	−	−	+	−	++	−
Coelastrum sp.	+	−	−	−	−	−	−	−	−	−	−	−	−	−	−	−	−	−	−	−	−	−	−	−
Cosmaruim sp.	−	−	−	−	+	+	−	−	−	+	−	−	+	+	+	+	+	+	−	−	−	−	−	−
Desmidium sp.	−	+	−	−	+	−	+	−	−	+	+	+	−	+	+	+	+	−	+	−	−	+	+	+
Euastrum sp.	−	−	−	−	−	−	−	−			+	−	−	−		−	−	−	−	−	−	−	−	−
Gonatozygon	−	−	+	−	−	−	−	−	−	−	−	−	−	−	−	−	−	−	−	−	−	−	−	−
Mougeotia sp.	+	−	+	+	+	+	+	+	+	++	+	−	−	−	+	+	+	−	−	+	+	+	−	++
Oedogonium sp.	−	−	−	+	+	+	+	++	+	++	+	+	−	−	−	−	−	−	+	+	+	+	+	+
*Oocystis**	+	+	+	−	+	−	+	−	−	+	+	+	−	+	−	−	−	−	+	−	−	+	+	+
Pediastrum boryanum	−	−	−	−	−	−		−	−	−	+	−	−	−	−	−	−	−	−	−	−	−	−	+
Pediastrum ovatum		−	−	−	−	−		−			+	−	−	−	−	−	−	−	+	−	−	−	−	−

Pediastrum simplex	–	+	+	+	+	–	–	–	–	–	–	+	+	–	–	–	+	–	–	–	–	–	–	–
Pediastrum sp.	–	–	–		–	–	–	–		–	–	–	–	+	+	–	–	–	–	–	+	–	–	–
Pediastrum spinosum	+	–	–	–	–	–	–	–	–	–	–	–	–	–	–	–	–	–	–	–	–	++	–	–
Pediastrum tetras	–	–	–	–	–	–		–	–		–	+	–	–	–	–	–	–	–	–	–	–	–	–
Pediastum duplex	–	+	+	–		–	–			+	–	–	–	–	+	+	+	–	–	+	+	++	+	+
Scendesmus bijugates	+	+	+	+	+	–	–	–	–	+	–	+	+	–	–	+	–	–	–	–	–	++	+	+
Scendesmus sp.	+	–	–	–	–	–	–	–			–	–	–	–	–	–	–	–	–	–	+	+	+	+
Scenedesmus dimorphus	–	–	+	–	–	–		–	–	–	–	–	–	–	–		–	–	+	–	–	–	–	+
Selenastrum sp.	–	–	–	–	–	–	–	–	–	++	–	–	–	–	–	–	–	–	–	–	–	–	+	+
Sphaerocystis	+	–	+	–	+	+	–		+	++	+	+	+	–	–	–	+		–	+	–	+	–	+
Spirogyra sp.	–	–	–	+	–	–	++	++	++	–	–	–	–	–	–	–	–	–	–	–	–	–	–	+
Staurastrum sp.	–	–	–	–	–	–	–		+	–	–	–	–	–	–	–	–	–	+	–	–	–	–	–
Tetraedon	–	–	–	–	–		–	–	–	–	–	+	–	–	–	–	–	+	–	–	–	–	+	–
Ulothrix sp.	+	+	+	+	–	–	–	–	–	–	–	–	–	+	+	–	–	–	–	–	–	–	–	–
Volvox sp.	–	–	–	–	–	–	–	–	–	+	–	–	–	–	–	–	–	–	–	–	–	–	+	–
Zygnema sp.	–	–	–	+	+			–	+	++	–	–	–	–	–	–	–	+	+	–	+	–	+	–

0–5 ind/mL=+
6–10 ind/mL=++
11–15 ind/mL=+++
and above=++++

Table 31 Monthly variation in the population density (Ind/mL) of the Chlorophyceae at the study site-II of the Dal lake

Taxa	Sep	Oct	Nov	Dec	Jan	Feb	Mar	Apr	May	Jun	Jul	Aug	Sep	Oct	Nov	Dec	Jan	Feb	Mar	Apr	May	Jun	Jul	Aug
Ankistrodesmus spiralis	–	+	–	–	–	–	–	–	–	–	–	–	–	–	–	–	–	–	–	–	–	–	–	–
Characium sp.	+	–	+	–	–	–	–	–	–	–	–	+	–	–	–	–	–	–	–	–	–	–	–	–
Chlorella sp.	++	+	++	++	+	–	+	–	–	–	–	+	+	–	–	+	–	–	++	–	–	++	–	–
Chlorococcum sp.	–	+	–	–	++	+	+	–	+	+	–	++	+	+	–	+	+	+	+	–	+	–	–	++
Cladophora sp.	–	–	–	–	–	+	–	–	–	–	–	–	–	–	–	–	–	–	–	–	–	–	–	–
Clostridium sp.	–	–	–	+	–	–	–	–	+	+	+	++	+	–	–	–	–	–	–	–	+	–	++	–
Cosmarium sp.	–	–	–	–	–	–	–	–	–	–	–	–	+	–	–	+	–	–	–	–	–	–	–	–
Desmidium sp.	+	+	–	+	+	–	–	–	–	–	–	++	–	+	++	+	+	–	+	–	–	+	+	++
Euastrum sp.	–	–	–	–	–	–	–	+	–	+	–	+	–	–	–	–	–	–	–	+	–	–	–	–
Gonatozygon sp.	+	+	–	–	–	–	–	–	–	–	+	–	–	–	+	–	–	–	–	–	+	–	–	–
Mougeotia sp.	+	+	+	+	+	–	–	+	+	+	+	–	–	–	+	+	+	–	–	+	+	++	–	+
Oedogonium sp.	+	–	–	–	+	–	+	+	+	+	–	–	–	+	+	+	–	–	+	+	+	++	+	+
Oocystis	–	–	–	–	–	–	–	–	–	–	–	–	–	–	–	–	–	–	–	–	–	–	–	–
Pediastrum boryanum	–	–	–	–	–	–	–	+	–	–	+	–	–	–	–	–	+	–	–	+	–	–	+	–
P. duplex	+	–	+	–	–	–	–	+	+	+	–	+	–	–	+	+	+	–	–	+	+	++	+	++
P. tetras	–	–	–	–	–	–	–	+	–	–	–	++	–	–	+	–	–	–	–	+	–	++	–	–
Pediastnum sp.	–	–	–	–	–	–	–	–	–	+	–	–	–	+	–	–	–	–	–	–	–	–	–	–
P. simplex	–	–	+	–	+	–	–	–	–	–	–	++	+	–	–	–	+	–	–	–	–	–	–	–

Scenedesmus bijugates	−	+	+	+	+	−	−	+	−	−	+	−	+	−	+	+	−	−	−	−	−	++	+	++
Scenedesmus dimorphus	−	−	+	−	−	−	+	−	−	−	−	++	−	−	−	−	−	+	−	−	−	+	++	−
Scenedesmus sp.	−	−	−	−	−	−	−	+	−	−	+	+	−	−	−	−	−	−	−	+	+	++	+	−
Selenastrum sp.	−	−	−	−	−	−	−	−	−	−	−	−	−	−	−	−	−	−	−	−	−	++	+	−
Sphaerocystis	+	++	+	−	+	−	+	+	+	+	+	++	+	−	−	−	+	−	−	+	+	+	−	+
Spirogyra sp.	−	−	−	−	−	−	−	−	−	−	+	++	−	−	−	−	−	−	−	−	−	−	−	+
Staurastrum sp.	−	−	−	−	+	−	−	+	−	−	−	−	−	−	−	−	−	−	−	+	−	−	−	−
Ulothrix sp.	−	+	+	−	−	−	−	−	−	−	−	−	+	−	−	−	−	−	−	−	−	−	−	−
Vovlox sp.	−	−	−	−	+	−	−	−	+	+	−	−	−	−	−	−	−	−	−	−	+	−	−	−
Zygnema sp.	−	−	−	+	+	+	−	+	+	+	−	−	−	−	+	+	++	+	−	+	+	++	−	−

0–5 ind/mL = +
6–10 ind/mL = ++
11–15 ind/mL = +++
and above = ++++

Table 32 Monthly variation in the population density (Ind/mL) of the Chlorophyceae at the study site-III of the Dal lake

Taxa	Sep	Oct	Nov	Dec	Jan	Feb	Mar	Apr	May	Jun	Jul	Aug	Sep	Oct	Nov	Dec	Jan	Feb	Mar	Apr	May	Jun	Jul	Aug
Ankistrodesmus sp.	–	–	–	–	+	–	–	–	–	–	–	–	–	+	–	+	–	–	–	–	–	–	–	–
Characium sp.	–	–	+	–	+	–	+	–	–	–	–	–	+	–	–	–	–	–	+	–	–	–	+	+
Chlorella sp.	–	+	–	+	+	+	+	+	+	–	+	–	–	+	+	–	+	–	+	+	–	–	+	+
Chloroccum sp.	+	+	–	–	+	+	–	–	+	+	+	–	+	–	–	+	+	+	–	+	+	+	+	+
Cladophora sp.	–	–		–	+	–	–	–	–	+	–	–	–	–	–	–	–	–	–	+	–	–	+	–
Clostridium sp.	+	+	+	+	–	–	+	+	–	–	–	–	+	–	–	–	–	+	–	+	–	–	–	–
Cosmarium sp.	–	+	–	–	–	–	–	–	–	+	–	–	+	–	–	–	–	–	–	–	–	–	–	–
Desmidium sp.	–	+	+	+	+	+	–	+	+	–	+	+	+	–	–	+	–	+	+	+	+	–	–	–
Gonatozygon sp.	+	+	+	–	–	–	–	–	–	–	–	–	–	+	+	–	–	–	–	–	–	–	–	–
Mougeotia sp.	–	–	+	–	–	–	+	+	+	–	–	+	–	+	–	+	–	–	–	–	–	–	–	+
Oedogonium sp.	+	–	+	+	+	–	–	–	–	–	+	+	+	–	–	–	+	–	+	+	+	–	–	+
Oocystis sp.	–		–	–	–	–	–	–	–	–	–	–	+	+	+	+	–	–	–	–	–	–	–	–
Pediastrum boryanum	–	–	–	–	–	–	–	–	–	+	–	–	–	–	–	–	–	–	–	–	–	–	–	–
P. duplex	+	–	–	+	–	+	+	–	+	–	–	+	+	+	–	–	–	–	+	–	–	–	–	+
P. simplex	–	–	–	–	–	+	+	+	+	–	–	–	–	–	–	–	+	–	–	–	+	–	+	+

Scendesmus bijugates	–	++	+	+	–	+	++	–	–	–	–	–	–	–	+	–	+	–	+	+	–	–	–	+
Scendesmus dimorphus	–	+	++	–	–	–	+	–	–	+	–	–	–	–	–	++	–	–	–	–	–	–	+	–
Scendesmus sp.	+	++	+	–	–	–	–	–	–	–	–	+	–	–	+	+	–	–	–	–	–	–	–	+
Selenastrum gracile	–	+	+	–	–	–	–	–	–	–	–	–	–	–	–	–	–	–	–	–	–	–	–	–
Sphaerocystis sp.	+		–	+	++	++	+	–	+	–	–	+	+	+	+	+	+	–	–	–	+	–	+	+
Spirogyra sp.	–	–	–	+	–	–	–	–	–	–	–	–	–	–	+	+	–	–	–	–	–	–	–	–
Staurastrum sp.	–	–	–	–	–	–	–	–	–	–	–	+	–	–	–	–	–	–	–	–	+	–	–	+
Ulothrix sp.	+	+	–	–	–	–	–	–	–	–	–	–	–	+	–	–	–	–	–	–	–	–	–	–
Volvox sp.	–	–	+	–	–	–	–	–	+	–	–	–	–	+	–	–	–	–	–	–	–	–	–	–
Zygnema sp.	+	–	–	–	–	–	–	–	–	–	–	+	+	+	–	–	–	–	–	–	–	–	–	–

0–5 ind/mL = +
6–10 ind/mL = ++
11–15 ind/mL = +++
and above = ++++

Table 33 Monthly variation in the population density (Ind/mL) of the Chlorophyceae at the study site-IV of the Dal lake

Taxa	Sep	Oct	Nov	Dec	Jan	Feb	Mar	Apr	May	Jun	Jul	Aug	Sep	Oct	Nov	Dec	Jan	Feb	Mar	Apr	May	Jun	Jul	Aug
Ankistrodesmus sp.	−	−	−	−	+	−	−	−	−	−	−	−	−	−	−	+	−	−	−	−	−	−	+	−
Chloroccum sp.	+	+	−	−	+	+	−	−	+	+	+	−	+	+	+	+	+	+	−	+	+	+	−	−
Chlorella sp.	−	+	−	+	++	+	+	+	++	−	+	−	−	++	+	−	+	−	+	+	−	−	+	+
Characium sp.	−	−	−	−	+	−	+	−	−	−	−	+	−	+	+	−	−	−	+	−	−	+	−	−
Clostridium sp.	−	−	−	+	−	−	+	+	−	+	+	+	−	−	−	−	−	+	−	++	−	+	−	−
Cladophora sp.	+	−	−	−	+	−	−	−	−	−	−	−	−	+	−	−	−	−	−	+	−	−	−	−
Cosmarium sp.	+	−	−	−	−	−	−	−	−	−	+	−	−	−	−	−	−	−	−	−	−	+	−	−
Desmidium sp.	−	+	+	+	+	+	−	+	+	−	+	+	−	−	−	+	−	+	+	+	+	+	−	−
Gonatozygon sp.	−	−	−	−	−	−	−	−	−	+	+	+	−	−	−	−	−	−	−	−	−	−	+	+
Mougeotia sp.	−	−	+	−	−	−	+	+	+	−	−	+	−	−	+	+	−	−	−	−	−	−	+	−
Oedogonium sp.	−	+	+	+	++	−	−	−	−	+		+	−	−	+	−	+	−	+	+	+	+	−	−
Oocystis sp.	−	−	−	−	−	−	−	−	−	−		−	−	−	−	+	−	−	−	−	−	+	+	+
Pediastrum simplex	−	−	−	−	−	+	+	+	+	−	−	−	−	+	+	−	+	−	−	−	+	−	−	−
P. duplex	−	−	+	+	−	+	+	−	+	+	−	−	−	−	+	−	−	−	+	−	−	+	+	−
P. boryanum	+	−	−	−	−	−	−	−	−	−	−	−	−	−	−	−	−	−	−	−	−	−	−	−

Sphaerocystis sp.	−	−	+	+	++	++	+	−	+	+		−	−	+	+	+	+	−	−	−	+	+	+	+
Staurastrum sp.	−	−	+	−	−	−	−	−	−	−	−	−	−	−	+	−	−	−	−	−	+	−	−	−
Scendesmu bijugates	−	−	−	+	−	+	++	−	−	−	++	+	−	−	+	−	+	−	+	+	−	−	−	+
Scendesmus sp.	−	−	+	−	−	−	−	−	−	+	++	+	−	−	+	+	−	−	−	−	−	−	−	+
Spirogyra sp.	−	−	−	+	−	−	−	−	−	−	−	−	−	−	−	+	−	−	−	−	−	−	−	+
Senedesmus dimorphus	+	−	−	−	−	−	+	−	−	−	+	++	−	+	−	++	−	−	−	−	−	−	−	−
Selenastrum gracile	−	−	−	−	−	−	−	−	−	−	+	+	−	−	−	−	−	−	−	−	−	−	−	−
Ulothrix sp.	−	−	−	−	−	−	−	−	−	+	+	−	−	−	−	−	−	−	−	−	−	−	+	−
Volvox sp.	−	−	−	−	−	−	−	−	+	−	−	+	−	−	−	−	−	−	−	−	−	−	+	−
Zygnema sp.	−	−	+	−	−	−	−	−	−	+	−	−	−	−	−	−	−	−	−	−	−	+	+	−

0–5 ind/mL = +
6–10 ind/mL = ++
11–15 ind/mL = +++
and above = ++++

solved oxygen. Sedamkar and Angadi (2003), on the other hand, considered this taxon as the indicator of clean water.

Characium sp. was found to be present at all the sites. Syal (1996) found this taxon to inhabit clean waters to moderately polluted waters with sewage.

Chlorella sp. was recorded from all the sites in the present study. Bhatt et al. (1999) without mentioning its species, found it in waters used for bathing, washing clothes and carrying fertilizers. On the other hand Varma (1998) labelled *Chlorella vulgaris* as pollution indicator.

Cladophora sp. was recovered from all the sites in the present study. Kolkwitz and Marsson (1908) and Syal (1996) also found that this taxon can inhabit all types of waters. Verma and Shukla (1970) and Das (1989) found it only in highly polluted waters. Sladecek (1961) and Govindan and Sundaresan (1979) found it only in moderately polluted waters, while Jolly and Chapman (1966) found it only in clean waters. Sarwar (1991a, b) found this taxon in the marshy lands around Srinagar

Clostridium sp. was found to be present at all the sites. Jolly and Chapman (1966), Kolkwitz and Marrson (1908) and Syal (1996) found it in waters ranging from clean to moderately polluted. Venkateshwarlu and Jayanti (1968), Singh et al. (1969) and Munawar (1970b) found it in waters contaminated with sewage and industrial wastes while David and Ray (1966), Verma and Shukla (1968, 1970) and Paramavsivam and Sreenivasan (1981) found it only in clean waters. Gaur and Khan (1993) reported this taxon to be the most dominant and present at all the sites. Sarwar (1991a, b) found it in marshy lands receiving sewage. Bhatt et al. (1999) found it in waters used for bathing, washing clothes and carrying fertilizers.

Cosmarium sp. was found to be present at all the sites. This taxon can, therefore, withstand agriculture run-off and sewage effluence. Jolly and Chapman (1966), Zutshi et al. (1980) and Sarwar (1991a, b) reported its presence in waters contaminated with sewage. However, David and Ray (1966), Verma and Shukla (1969, 1970) and Paramavsivam and Sreenivasan (1981) reported its presence only in clean waters.

Chlarococcum sp., *Gonatozygon* sp. and *Desmidium* sp. were found to be present at all the sites. Out of these, site-II receives the agricultural run-off from the floating gardens, site-III and site-IV receive sewage effluence from house boats and hotels. Presence of these taxa in clean or moderately or polluted waters has not been reported yet.

Euastrum sp. was found to be present at site-I and site-II only. Site-I is located in the lake where the water is least interfered by the human habitation. Site-II receives agriculture effluence from the floating gardens. This taxon therefore, can withstand less polluted to polluted waters. Sarwar (1991a, b) recorded this from the marshy lands receiving sewage.

Mougeotia sp. was found to be present at all the sites. This taxon was reported from clean waters by Kolkwitz and Marsson (1908), David and Ray (1966), Bulusu et al. (1967), Gunale and Balakrishnan (1981) and Syal (1996). Sarwar (1991a, b) found this from all the sites of the marshy lands receiving sewage.

Oedogonium sp. was found to be present at all the sites. Sarwar (1991a, b) found this from all the sites of the marshy lands receiving sewage and agricultural run-off. Gaur and Khan (1993) also recorded this taxon from all the sites but the abundance was least. Syal (1996) recorded this from pollution free site to the sites receiving fertilizer wastes. Verma and Shukla (1968, 1970) reported this from clean waters.

In all, 7 species belonging to genus *Pediastrum* were found to be present in the Dal lake. Out of these, *P. boryanum*, *P.duplex* and *P. simplex* were found to be present at all the sites. Kolkwitz and Marsson (1908) and Syal (1996) recorded *P. simplex* both in clean and moderately polluted waters while George et al. (1966), Munawar (1970b), Verma et al. (1978) and Gunale and Balakrishnan (1981) and Sarwar (1991a, b) found it in waters receiving sewage and industrial effluence. However, Venkateshwarlu and Jayanti (1968), Zutshi et al. (1980) and Paramasivam and Sreenivasan (1981) recorded its presence both in clean and highly polluted waters. Jolly and Chapman (1966), Verma and Shukla (1969) and Govindan and Sundaresan (1979) recorded its presence only in clean waters. *P.duplex* was found to be the dominant species by Khan and Bhat (2000) in Manasbal lake receiving sewage and the sites infested with abundance of macrophytes. *P. boryanum* was reported to be present at all the sites in marshy lands receiving sewage by Sarwar (1991a, b). *P.ovatum* and *P.spinosum* were found to be present at site-I only in the present study. Sarwar et al. (1996) also reported *P.ovatum* to be present in open waters least interfered by human interference and free from direct source of pollution. Absence of *P.ovatum* and *P.spinosum* from other sites indicate their aversion to organic pollution in the present study. *P.tetras* and *Pediastrum* sp. were recorded from site-I and site-II only thereby indicating they can withstand waters contaminated by agricultural run-off. Sarwar (1991a, b) recorded *P.tetras* at all the sites of marshy lands receiving sewage. Bhatt et al. (1999) found *Pediastrum* sp. in waters used for bathing, washing clothes and carrying fertilizers. Gujarathi and Kanhere (1998) found this taxon to be highest in the Barwani pond. Chakraborty and Konar (2000) found this from sites polluted with organic matter in Damodar river.

The genus *Scenedesmus* was represented by *S. bijugates, S. dimorphus* and *Scenedesmus sp.* All the three were found to be present at all the sites thereby showing wide range of tolerance. Sarwar (1991a, b) found *S. bijugates* and *S. dimorphus* in the marshy lands receiving sewage. Khan and Bhat (2000) found *Scenedesmus sp.* to be the dominant taxon in the lake Manasbal. Bhatt et al. (1999) found *Scenedesmus sp.* in waters used for bathing, washing clothes and water carrying fertilizers. Weilgolaski (1975) and Ahmad (1996) labeled this taxon as pollution indicator. Jolly and Chapman (1966), Venkateshwarlu and Jayanti (1968), Zutshi et al. (1980) and Syal (1996) found it in both clean as well as highly polluted waters with sewage. David and Ray (1966), Munawar (1970b), Verma et al. (1978), Govindan and Sundaresan (1979) and Gunale and Balakrishnan (1981) reported its presence in waters highly contaminated with sewage and industrial wastes. However, Kolkwitz and Marsson (1908) reported its presence only in moderately polluted waters.

The genus *Selenastrum* was represented by *S.gracile* and *Selenastrum* sp. which were recovered in the Dal lake during the present study. Among these *S.gracile* was found to be present at site-III and site-IV only, thereby meaning that this species can withstand sewage pollution. Sarwar (1991a, b) recorded this species to be present from the marshy lands which receive sewage. *Selenastrum* sp. was found to be present at site-I and site-II only.

Sphaerocystis sp. was found to be present at all the four sites. This can, therefore, thrive well in clean waters as well as waters receiving effluence from floating gardens, house boats and hotels. Khan and Bhat (2000) found *Sphaerocystis schroeteri* to be present in the phytoplankton of the lake Manasbal which is under the impact of multiple anthropogenic perturbations.

Spirogyra sp. was found to be present at all the four sites. This can, therefore, thrive well in clean waters as well as waters receiving effluence from floating gardens, house boats and hotels. David and Ray (1966), Jolly and Chapman (1966), and Paramasivam and Sreenivasan (1981) also found it both in clean and highly polluted waters. Kolkwitz and Marsson (1908) found it both in clean and moderately polluted waters. Bulusu et al. (1967), Govindan and Sundarsan (1979) and Das (1989) found it in waters enriched with sewage and industrial wastes, while Verma and Shukla (1968, 1969, 1970) and Gunale and Balkrishnan (1981) found this species only in clean waters. Sedamkar and Angadi (2003) recorded high density of this taxon at less polluted sites. Sarwar (1991a, b) recorded this species to be present from the marshy lands which receive sewage.

Staurastrum sp. was found to be present at all the sites. This taxon can thus thrive well in clean as well as waters receiving agricultural run-off and sewage. Zutshi et al. (1980) also reported this taxon from the Dal lake. Sarwar (1991a, b) recorded this species to be present from the marshy lands which receive sewage. Gaur and Khan (1993) found this taxon in water receiving effluents from a thermal power plant near Aligarh. Chakraborty and Konar (2000) found this taxon in polluted streams in river Damodar.

Ulothrix sp., *Volvox* sp. and *Zygnema* sp. were found to be present at all the sites thereby indicating that they can thrive well in clean waters as well as waters receiving effluence from floating gardens, house boats and hotels. Among these *Ulothrix* sp. was found in clean waters by Kolkwitz and Marsson (1908), Verma and Shukla (1968, 1969, 1970), Govindan and Sundaresan (1979), Paramasivam and Sreenivasan (1981) and Syal (1996).

Cyanophyceae (blue green algae)

In the 2 year study of the Dal lake, 13 cyanophyceae taxa have been recovered with 10 at site-I, 11 each at site-II and IV and 12 at site-IV (Table 34).

In the present study, the maximum population density of blue green algae was observed during summer at the sites I, II and III and during winter at site-IV. Higher population density in summer is also observed by Gonzalves and Joshi (1946), Zafar (1967), Munawar (1970a, b), Nandan and Aher (2002) and Sedamkar and Angadi (2003). Sabater and Isabel (1990) reported abundance of cyanophyceae from late spring to summer while Biswas and Konar (2000) recorded increased

Table 34 List of Cyanophyceae Taxa recorded at the investigated sites of the Dal lake

S. No.	Cyanophyceae Taxa	Site-I	Site-II	Site-III	Site-IV
1	*Anabaena sp.*	+	+	+	+
2	*Anacystis*	+	+	+	+
3	*Gleotrichia sp.*	+	+	+	+
4	*Gomphosphaera sp.*	+	+	+	+
5	*Lyngbia sp.*	+	+	+	+
6	*Microcystis sp.*	+	+	+	+
7	*M.aeroginosa*	+	+	+	+
8	*Merismopedia formosa* A.Br.	–	+	+	+
9	*M.glauca Meyen*	+	–	–	–
10	*Nostoc sp.*	+	+	+	+
11	*Oscillatoria sp.*	+	+	+	+
12	*O.princeps*	–	+	+	+
13	*Spirulina sp.*	–	–	–	+

Table 35 Seasonal variations in the population density (Ind/mL) of the Cyanophyceae at the investigated sites of the Dal lake

Sites	Year	Autumn	Winter	Spring	Summer
Site-I	**2000–2001**	51	23	15	58
	2001–2002	23	33	27	34
	Average	37	28	21	46
Site-II	**2000–2001**	35	20	25	33
	2001–2002	50	17	27	76
	Average	42.5	18.5	26	54.5
Site-III	**2000–2001**	32	39	14	52
	2001–2002	32	49	56	45
	Average	32	44	35	48.5
Site-IV	**2000–2001**	36	63	62	32
	2001–2002	50	50	16	41
	Average	43	56.5	39	36.5

population in rainy season. The species diversity was recorded highest in summer at sites I, II and III and during autumn at site-IV (Table 35, Figs. 28 and 29).

Out of total of 13, 9 cyanophycean taxa were found to be present at all the sites. These include *Anabaena* sp., *Anacystis* sp., *Gleotrichia* sp., *Gomphosphaera* sp., *Lyngbia* sp., *Microcystis* sp., *M. aeroginosa, Nostoc* sp., *and Oscillatoria* sp. Two species *Merismopedia formosa* and *Oscillatona princeps* were found absent from site – I and present at other three sites (Tables 36, 37, 38 and 39). *Merismopedia*

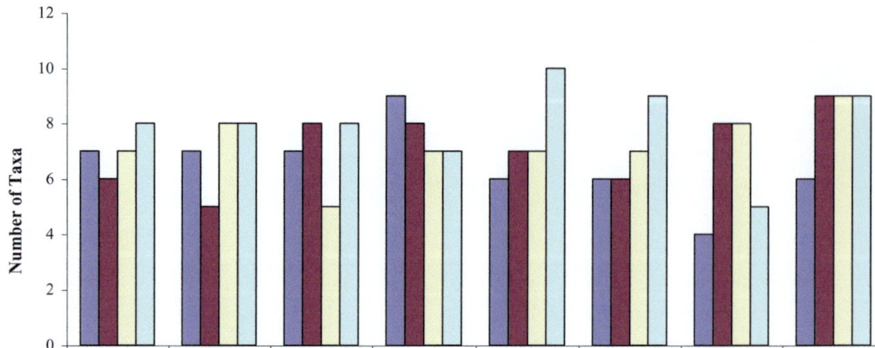

	Autumn	Winter	Spring	Summer	Autumn	Winter	Spring	Summer
Site-I	7	7	7	9	6	6	4	6
Site-II	6	5	8	8	7	6	8	9
Site-III	7	8	5	7	7	7	8	9
Site-IV	8	8	8	7	10	9	5	9

Season

Fig. 28 Seasonal variations in species diversity of the Cyanophyceae at the investigated sites of the Dal lake

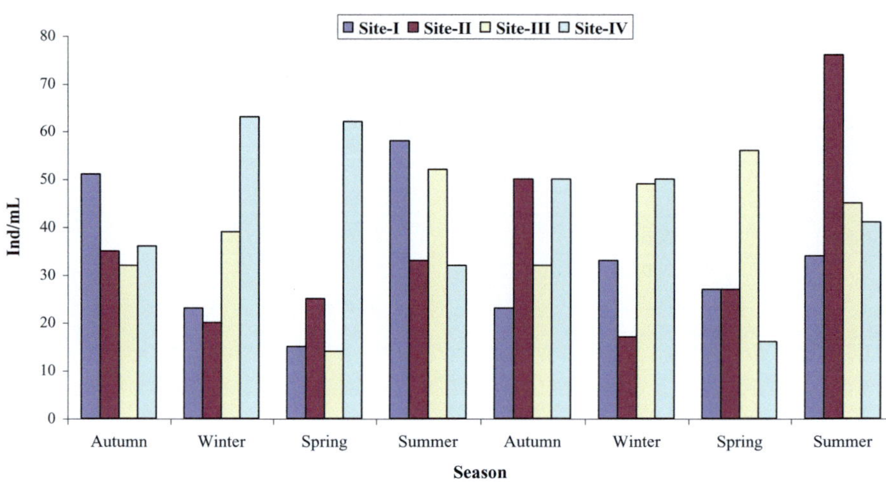

Fig. 29 Seasonal variations in population density (Ind/mL) of the Cyanophyceae at the investigated sites of the Dal lake

glauca was recorded from site-I only. Sarwar et al. (1996) also recorded *M.glauca* in less polluted and open water area of this lake. Kolkwitz and Marsson (1908), Jolly and Chapman (1966), Govindan and Sundaresan (1979), Zutshi et al. (1980) and Parmasivam and Sreenivasan (1981) found *Merismopedia* sp. in clean waters only. *Spirulina* sp. was found to be present at site-IV only. Bhatt et al. (1999) recorded its presence at the sites where water is used for bathing, washing and car-

Table 36 Monthly variation in the population density (ind/mL) of the Cyanophyceae at the study site-I of the Dal lake

Taxa	Sep	Oct	Nov	Dec	Jan	Feb	Mar	Apr	May	Jun	Jul	Aug	Sep	Oct	Nov	Dec	Jan	Feb	Mar	Apr	May	Jun	Jul	Aug
Anabaena sp.	+	−	−	−	−	−	+	−	−	−	+	−	−	−	−	−	−	−	−	−	−	−	−	−
Anacystis	−	−	+	−	+	+	+	−	−	−	+	+	−	+	−	−	+	+	+	−	−	+	+	+
Gleotrichia sp.	+	+	−	−	−	+	−	+	−	−	+	+	−	+	−	+	−	−	−	−	−	−	−	+
Gomphosphaera sp.	−	−	−	−	+	−	−	−	−	−	+	−	−	+	−	−	−	−	−	−	+	−	−	−
Lyngbia sp.	−	+	−	−	−	−	−	−	−	−	+	−	−	−	−	−	−	−	−	−	−	−	−	−
Microcystis aeroginosa	−	−	−	−	+	+	−	+	+	−	+	−	−	+	−	−	−	+	−	−	−	−	−	+
Merismopedia glauca	−	−	−	−	+	−	−	−	−	−	+	−	−	−	−	−	+	−	−	−	−	+	−	−
Microcystis sp.	−	−	+	−	−	−	+	−	+	−	−	−	−	−	−	−	−	+	−	−	−	−	+	−
Nostoc sp.	+	−	−	+	+	−	−	+	−	+	+	+	+	−	+	−	−	−	−	+	+	+	+	+
Oscillatoria sp.	−	+	−	−	+	−	+	−	−	+	+	−	+	−	−	−	−	+	+	+	+	+	−	−

0–5 ind/mL = +
6–10 ind/mL = ++
11–15 ind/mL = +++
and above = ++++

Table 37 Monthly variations in the population density (Ind/mL) of the Cyanophyceae at the study site-II of the Dal lake

Taxa	Sep	Oct	Nov	Dec	Jan	Feb	Mar	Apr	May	Jun	Jul	Aug	Sep	Oct	Nov	Dec	Jan	Feb	Mar	Apr	May	Jun	Jul	Aug
Anabaena sp.	+	−	−	−	−	−	+	−	−	−	−	−	−	−	−	−	−	−	−	−	−	−	+	−
Anacystis	−	++	+	−	−	−	+	+	−	−	−	+	+	+	+	−	−	−	+	+	−	++	+	++
Gleotrichia sp.	+	++	−	−	++	−	−	+	+	+	−	−	−	++	−	+	−	−	−	−	+	−	−	++
Gomphosphaera sp.	−	−	−	−	−	−	−	−	−	+	−	+	−	−	−	−	−	−	−	−	+	−	−	−
Lyngbya sp.	−	−	−	−	−	−	−	+	−	−	+	−	+	−	−	+	−	−	−	+	−	+	−	++
Microcystis sp.	−	−	−	−	−	+	−	−	−	−	−	−	−	−	−	−	−	+	−	−	−	−	+	−
M aeroginosa	−	−	−	−	−	+	−	+	+	+	−	−	−	++	−	−	−	+	−	+	+	−	−	++
Merismopedia formosa	−	−	−	−	−	−	−	−	−	−	+	−	−	−	−	−	−	−	−	−	−	−	−	−
Nostoc sp.	+	−	−	+	+	+	+	+	+	+	+	−	++	−	+	−	−	−	−	+	+	+	+	++
Oscillatoria sp.	+	−	++	−	−	−	+	−	−	−	−	−	−	++	++	+	−	−	+	−	−	−	++	−
O. princeps	−	−	+	−	−	+	+	+	−	−	−	++	+	+	+	+	−	+	++	+	−	+	−	++

0–5 ind/mL = +
6–10 ind/mL = ++
11–15 ind/mL = +++
and above = ++++

Table 38 Monthly variation in the population density (ind/mL) of the Cyanophyceae at the study site-III of the Dal lake

Taxa	Sep	Oct	Nov	Dec	Jan	Feb	Mar	Apr	May	Jun	Jul	Aug	Sep	Oct	Nov	Dec	Jan	Feb	Mar	Apr	May	Jun	Jul	Aug
Anabaena sp.	–	–	–	–	–	–	–	–	+	+	+	–	–	–	–	–	–	–	–	–	–	–	+	+
Anacystis	–	–	+	–	+	+	+	–	–	–	+	++	–	+	+	+	+	+	+	–	–	–	+	–
Gleotrichia sp.	+	–	–	–	–	+	–	–	–	++	+	+	–	–	+	–	–	+	–	+	+	+	+	–
Gomphosphaera sp.	+	–	–	–	–	–	–	–	–	–	–	–	–	–	–	–	–	–	–	–	–	+	+	–
Lyngbia sp.	–	–	–	–	–	–	–	–	–	–	+	–	–	–	–	–	–	–	–	–	–	–	–	–
Microcystis sp.	+	+	+	–	–	++	–	–	–	–	–	–	–	–	+	+	+	+	+	+	+	+	++	–
Microcystis aeroginosa	+	+	–	+	+	+	–	–	–	+	–	+	+	–	+	+	+	+	+	+	+	+	+	–
Merismopedia formosa	–	–	–	–	+	–	–	–	–	–	–	–	–	–	–	–	–	–	–	–	+	–	–	–
Nostoc sp.	+	+	+	–	–	+	–	+	–	+	+	–	+	+	+	–	+	–	+	–	+	+	+	+
Osciltoria sp.	–		–	–	+	–	+	–	–	+	+	–	–	–	+	+	–	–	+	+	+	+	–	–
O. princeps	–		+	+	+	+	+	–	–	–	–	–	–	+	+	+	+	+	+	+	+	+	–	–

0–5 ind/mL = +
6–10 ind/mL = ++
11–15 ind/mL = +++
and above = ++++

Table 39 Monthly variation in the population density (ind/mL) of the Cyanophyceae at the study site-IV of the Dal lake

Taxa	Sep	Oct	Nov	Dec	Jan	Feb	Mar	Apr	May	Jun	Jul	Aug	Sep	Oct	Nov	Dec	Jan	Feb	Mar	Apr	May	Jun	Jul	Aug
Anabaena sp.	–	–	–	–	–	–	–	–	+	+	+	–	–	+	–	–	–	–	–	–	–	–	+	–
Anacystis	++	+	+	+	–	–	–	–	–	+	–	+	+	+	+	–	–	–	+	+	–	+	+	++
Gleotrichia sp.	–	–	–	–	–	+	–	+	–	++	+	++	+	+	–	–	–	+	–	–	–	–	–	++
Gomphosphaera sp.	+	–	–	–	–	–	–	–	–	–	–	–	+	+	–	–	+	–	–	–	–	–	–	–
Lyngbia sp.	+	++	–	–	–	–	–	+	+	+	+	+	–	–	–	+	–	–	–	–	–	+	–	+
Microcystis aeroginosa	+	+	–	+	+	+	–	–	–	+	–	+	+	++	–	+	-	+	+	+	++	++	+	+
Merismopaedia formosa	–	–	–	–	+	–	–	+	–	–	–	–	–	–	–	–	+	–	–	–	–	–	–	–
Microcystis sp.	+	+	+	–	–	++	–	–	–	–	–	–	+	++	–	–	–	+	+	+	+	+	+	+
Nostoc sp.	+	+	+	–	–	+	–	+	–	+	+	–	+	+	+	+	+	+	+	–	+	–	+	–
Osciltoria princeps	–	–	+	+	+	+	+	–	–	–	–	–	+	–	–	–	+	+	+	+	+	+	+	+
Osciltoria sp.	+	–	+	++	–	+	–	–	+	–	–	+	–	+	+	+	–	–	–	–	–	–	+	–
Spirulina sp.	–	–	+	–	–	–	–	+	–	–	–	–	–	+	–	–	–	–	–	–	–	–	–	–

0–5 ind/mL = +
6–10 ind/mL = ++
11–15 ind/mL = +++
and above = ++++

ries chemical fertilizers. Sarwar (1991a, b) marked its presence in the marshy lands receiving sewage. Khan and Bhat (2000) recorded it in the waters which are contaminated with multiple anthropogenic activities. Bulusu et al. (1967) and Syal (1996) found it both in clean and moderately polluted waters, whereas, David and Ray (1966) found it only in moderately polluted waters. Paramasivam and Sreenivasan (1981) found it in waters contaminated with sewage and effluence from industries and Verma and Shukla (1968) found it only in clean waters. *Anabaena* sp. was found to be present at all the sites. Sedamkar and Angadi (2003) considered this as the indicator of polluted waters. Sarwar (1991a, b) marked its presence as the indication of fertility status. Bhatt et al. (1999) recorded its presence at the sites where water is used for bathing, washing and carries chemical fertilizers. Syal (1996) found this in both clean and polluted waters. David and Ray (1966), Bulusu et al. (1967) and Govindan and Sundaresan (1979) reported it in moderately polluted waters while Kolkwitz and Marsson (1908) in clean waters.

Oscillatoria genus is represented by *O. princeps* and *Oscillatoria* sp. Among these *Oscillatoria* sp. was found to be present at all the sites and *O. princes* was found to be present at site-II, III and IV and absent from site-I. David and Ray (1966), Jolly and Chapman (1966), Zutshi et al. (1980), Paramasivam and Sreenivasan (1981) and Syal (1996) found *Oscillatoria* sp. in clean and highly polluted waters. Kolkwitz and Marsson (1908) and Sladecek (1961) found it both in clean and moderately polluted waters. George et al. (1966), Venkateshwarlu and Jayanti (1968), Singh et al. (1969), Verma et al. (1978), Gunale and Balakrishnan (1981), Mahadevan and Krishnaswamy (1983) and Das (1989) found it in waters highly polluted with sewage and industrial wastes. Sarwar (1991a, b), Bhatt et al. (1999) and Khan and Bhat (2000) recorded *Oscillatoria* sp. in the waters which are contaminated with multiple anthropogenic activities, the chief being sewage. Anjana and Kanhere (1998) recorded its highest density in a pond. Verma and Shukla (1968, 1969, 1970) found it only in clean waters.

Lyngbia sp. was found to be present at all the sites. Bhatt et al. (1999) found this to be present at the sites where water was used for bathing, washing and carries chemical fertilizers Verma and Shukla (1970), Paramasivam and Sreenivasan (1981) and Syal (1996) found this taxon in clean water.

Anacystis was found to be present at all the sites. Chakraborty and Konar (2000) found this at the sites where water carries huge storage of organic materials.

Microcystis is represented by *M. aeroginosa* and *Microcystis* sp. Both of these were present at all the sites. *M. aeroginosa* was observed to be common in the floating garden site and open water site by Sarwar et al. (1996). Bhatt et al. (1999) found *Microcystis* sp. to be present at the sites where water was used for bathing, washing and carries chemical fertilizers. Khan and Bhat (2000) found *M. aeroginosa* to be the dominant species where the lake basin is infested with aquatic weeds and under the impact of multiple anthrogenic perturbations.

Dinophyceae (Dinoflagellates)

Only two Dinophyceae taxa were recorded at the four investigated sites of the Dal lake which include *Gymnodium* sp. and *Peridinium* sp. Dinoflagellates do not contribute much to the total species composition of the Phytoplankton. Vareethiah and Haniffa (1998) also recorded low contribution of Dinophyceae group towards phytoplankton in their study. *Gymnodium* sp. was recorded to be dominant throughout the study period. Tilzer (1973) also found *Gymnodium* sp. to be dominant throughout the year in the Vorderer Finstertaler high mountain lake while as Yousuf et al. (2002) recorded *Peridinium* sp. being the dominant in their study.

In the present study, no marked or definite seasonal trend was observed in the population density and species diversity (Figs. 30 and 31).

Peridinium sp. and *Gymnodium* sp. were recorded to be present at all the sites. Sarwar and Zutshi (1989) found *Peridinium* sp. frequently in the Dal lake and Sarwar (1991a, b) recovered this taxon in the marshy lands receiving sewage (Table 40).

Euglenophyceae

The two Euglenophycean taxa recorded at the four investigated sites of the Dal lake include *Euglena* sp. and *Phacus* sp. The population density of euglenophyceae was recorded to be highest in summer at site-I, II and IV while as at site-III no marked seasonal trend was observed. The species diversity also do not show any definite pattern. Chakraborty and Konar (2000) also recorded high density during in summer in river Damodar and Yousuf et al. (2002) recorded high density of this group during cooler months in the Anchar lake (Tables 41 and 42, Figs. 32 and 33).

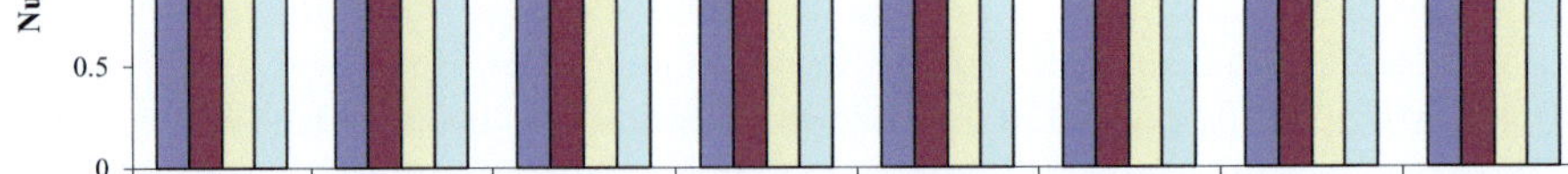

	Autum	Winter	Spring	Summer	Autum	Winter	Spring	Summer
Site-I	2	2	2	2	2	1	2	2
Site-II	2	2	2	2	2	2	2	2
Site-III	1	1	2	2	1	2	1	1
Site-IV	1	2	1	1	1	2	2	2

Season

Fig. 30 Seasonal variations in the species diversity of the Dinophyceae at the investigated sites of the Anchar lake

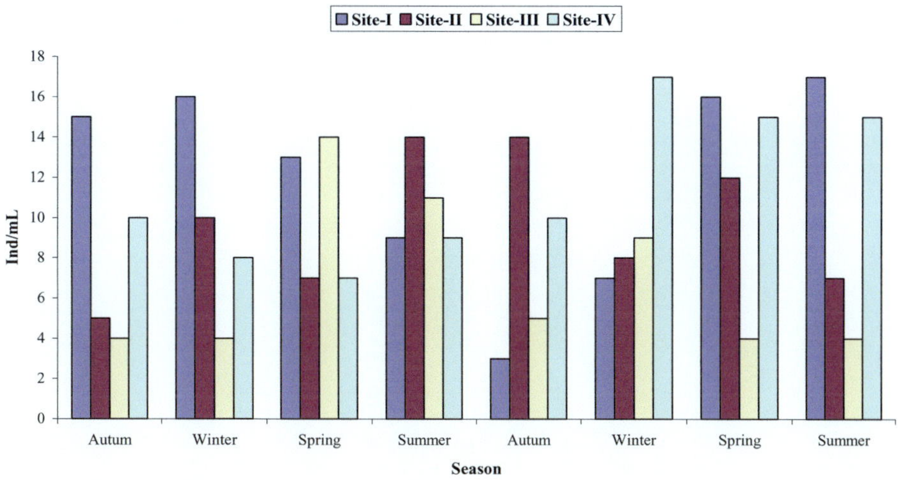

Fig. 31 Seasonal variations in the population density (Ind/mL) of the Dinophyceae at the investigated sites of the Anchar lake

Euglena sp. and *Phacus* sp. were found to be present at all the sites in the present investigation. Carter (1956b) reported the occurrence of *Euglena* sp. in fresh water pools in the vicinity of Bombay. Khan and Bhat (2000) attributed the red water phenomenon in lake Manasbal to this taxon. Sedamkar and Angadi (2003) considered this as an indicator of pollution. Sarwar and Zutshi (1989) categorized this as a rare taxon in the Dal lake. Munawar (1974b), Rai and Kumar (1977), Rai (1978) and Syal (1996) recorded this both in clean and highly polluted waters. Zafar (1959a, b), Munawar (1970b, 1972, 1974a) and Jhingran (1991a) found it only in waters contaminated with sewage. Palmer (1969), Rishi and Kachroo (1981) have described this as pollution tolerant. *Phacus* sp. has also been considered as indicator of polluted waters by Sedamkar and Angadi (2003). Khan and Bhat (2000) found this taxon in waters receiving sewage. Hedge and Bharati (1984) observed that high pH favoured the growth of euglenophyceae. From the study it seems that alkaline pH favours the growth of euglenoids.

Chrysophyceae

The only Chrysophyceae taxon recorded in the Dal lake is *Dinobryon* sp. The population density of this in the lake is extremely low and its contribution towards algal population is almost negligible (Table 43).

In the present study, no seasonal trend in the population density of *Dinobryon* sp. was recorded. Skuja (1964) however, recorded that its high growth from April to May.

At Site-I no chrysophycean taxon was recorded. The only taxon *Dinobryon* sp. was found to be present at site-II, III and IV. Sarwar and Zutshi (1989), however, found this taxon rarely in the Dal lake. Khan and Bhat (2000) recorded its presence

Table 40 Monthly variations in the population density (Ind/mL) of the Dinophyceae at the study sites of the Dal lake

Site-I	**Sep**	**Oct**	**Nov**	**Dec**	**Jan**	**Feb**	**Mar**	**Apr**	**May**	**Jun**	**Jul**	**Aug**	**Sep**	**Oct**	**Nov**	**Dec**	**Jan**	**Feb**	**Mar**	**Apr**	**May**	**Jun**	**Jul**	**Aug**
Gymnodium sp.	+	+	–	++	+	+	++	+	–	+	+	+	+	–	–	+	+	–	+	+	+	++	–	–
Peridinium sp.	+	+	–	+	–	–	+	+	+	++	–	+	+	+	–	–	–	–	+	–	–	+	–	–
Site-II	**Sep**	**Oct**	**Nov**	**Dec**	**Jan**	**Feb**	**Mar**	**Apr**	**May**	**Jun**	**Jul**	**Aug**	**Sep**	**Oct**	**Nov**	**Dec**	**Jan**	**Feb**	**Mar**	**Apr**	**May**	**Jun**	**Jul**	**Aug**
Gymnodium sp.	–	+	–	+	+	–	–	+	+	+	–	+	+	+	–	+	+	–	+	+	+	+	–	–
Peridinium sp.	+	–	–	–	+	–	+	–	–	–	+	+	+	+	–		+	–	+	–	–	++	–	–
Site-III	**Sep**	**Oct**	**Nov**	**Dec**	**Jan**	**Feb**	**Mar**	**Apr**	**May**	**Jun**	**Jul**	**Aug**	**Sep**	**Oct**	**Nov**	**Dec**	**Jan**	**Feb**	**Mar**	**Apr**	**May**	**Jun**	**Jul**	**Aug**
Gymnodium sp.	+	+	–	–	–		–	+	+	–	+	+	+	+	–	+	+	+	–	+	+	–	–	+
Peridinium sp.	–	–	–	–	+	+	–	+	+	–	+	–	–	–	–	+	+	+	–	–	–	–	–	–
Site-IV	**Sep**	**Oct**	**Nov**	**Dec**	**Jan**	**Feb**	**Mar**	**Apr**	**May**	**Jun**	**Jul**	**Aug**	**Sep**	**Oct**	**Nov**	**Dec**	**Jan**	**Feb**	**Mar**	**Apr**	**May**	**Jun**	**Jul**	**Aug**
Gymnodium sp.	+	+	–	–	–	++	–	++	+	–	+	+	+	+	–	+	+	+	–	+	+	–	–	+
Peridinium sp.	–	–	–	–	+	+	–	–	–	–	–	–	–	–	–	+	+	+	–	+	–	–	+	–

0–5 ind/mL = +
6–10 ind/mL = ++
11–15 ind/mL = +++
and above = ++++

Table 41 Seasonal variations in the population density (Ind/mL) of the Euglenophyceae at the investigated sites of the Dal lake

Sites	Year	Autumn	Winter	Spring	Summer
Site-I	**2000–2001**	6	5	5	6
	2001–2002	3	5	1	7
	Average	4.5	5	3	6.5
Site-II	**2000–2001**	10	11	12	11
	2001–2002	6	10	14	19
	Average	8	10.5	13	15
Site-III	**2000–2001**	7	6	7	7
	2001–2002	4	5	1	7
	Average	5.5	5.5	4	7
Site-IV	**2000–2001**	8	14	8	22
	2001–2002	3	2	7	14
	Average	5.5	8	7.5	18

in the Manasbal lake where the lake water was under the impact of multiple anthropogenic perturbations. Zutshi and Vass (1982) attributed its presence to high levels of nitrogen and phosphorus in the Dal lake. Rodhe (1948) and Mc Murry and Olive (1975) have found high levels of nitrogen and phosphorous to have inhibitory influence on the chrysophycean growth.

Macrophytes

The presence and abundance of macrophytes in aquatic habitats is governed by a set of environmental features. Macrophytes being ecological dominants of freshwater bodies especially lakes, have received great deal of attention both from limnologists and ecologists. Even though elaborate ecological studies on freshwater bodies of Kashmir Himalaya commenced after 1960s, the earliest reference of macrophytes of Kashmir is that of Hooker (1875) who listed some aquatic plants of Kashmir in his 'Flora of British India'.

A total of (45) species of macrophytes were recorded from various sites in the Dal lake. The maximum number of (34) species were recorded in August as against the minimum of (11) species in April (Tables 44 and 45, Fig. 34).

All the life-form classes of macrophytes viz, emergents, rooted floating-leaf types, submergeds and free floating types were present with varying number of species. The emergents recorded a maximum number of (29) species which included *Alisma plantago aquatica*; *Berula sp.*, *Bidens cernua; Bromus japonica, Cyperus glomeratus; Carex sp.; Echinocloa crusgatti, Epilobium hersutam, Eliocharis palustris, Gallum sp., Lathyrus salicaria, Lycopus europus, Marsilea quadrifolia, Menyanthese trifolata, Mentha longifolia, Myriophyllum verticillatum, Nasturtium officinale, Paspalum sp., Polygonum minus, Paspalum sp., Phragmites australis, Polygonum amphibium, Ranunculus scleratus, Rorripa islandica,Ranunculus*

Table 42 Monthly variations in the population density (Ind/mL) of the Euglenophyceae at the study sites of the Dal lake

Site-I	**Sep**	**Oct**	**Nov**	**Dec**	**Jan**	**Feb**	**Mar**	**Apr**	**May**	**Jun**	**Jul**	**Aug**	**Sep**	**Oct**	**Nov**	**Dec**	**Jan**	**Feb**	**Mar**	**Apr**	**May**	**Jun**	**Jul**	**Aug**
Euglena sp.	+	–	+	+	–	+	–	–	–	–	–	–	–	–	–	–	–	–	+	+	+	+	+	–
Phacus sp.	+	+	–	+	–	–	–	+	+	++	–	+	+	+	–	–	+	–	–	–	–	+	+	+
Site-II	**Sep**	**Oct**	**Nov**	**Dec**	**Jan**	**Feb**	**Mar**	**Apr**	**May**	**Jun**	**Jul**	**Aug**	**Sep**	**Oct**	**Nov**	**Dec**	**Jan**	**Feb**	**Mar**	**Apr**	**May**	**Jun**	**Jul**	**Aug**
Euglena sp.	+	–	+	–	+	+	+	+	+	+	+	+	–	–	–	–	+	+	+	+	+	+	+	+
Phacus sp.	–	+	–	–	+	+	–	+	+	+	–	–	+	+	–	–	+	+	+	+	+	+	+	++
Site-III	**Sep**	**Oct**	**Nov**	**Dec**	**Jan**	**Feb**	**Mar**	**Apr**	**May**	**Jun**	**Jul**	**Aug**	**Sep**	**Oct**	**Nov**	**Dec**	**Jan**	**Feb**	**Mar**	**Apr**	**May**	**Jun**	**Jul**	**Aug**
Euglena sp.	–	+	+	–	–	–	–	–	–	+	+	–		+	+	–	–	–	–	–	–	+	+	–
Phacus sp.	–	+	+	+	+	+	–	+	+	–	–	–	–	–	–	+	+	+	–	–	+	–	+	–
Site-IV	**Sep**	**Oct**	**Nov**	**Dec**	**Jan**	**Feb**	**Mar**	**Apr**	**May**	**Jun**	**Jul**	**Aug**	**Sep**	**Oct**	**Nov**	**Dec**	**Jan**	**Feb**	**Mar**	**Apr**	**May**	**Jun**	**Jul**	**Aug**
Euglena sp.	–	+	+	–	+	–	–	–	–	+	+	–		+	+	–	–	–	–	+	–	+	+	–
Phacus sp.	–	+	+	+	+	+	–	+	+	–	+	–	–	–	–	+	+	+	–	–	+	–	+	–

0–5 ind/mL = +
6–10 ind/mL = ++
11–15 ind/mL = +++
and above = ++++

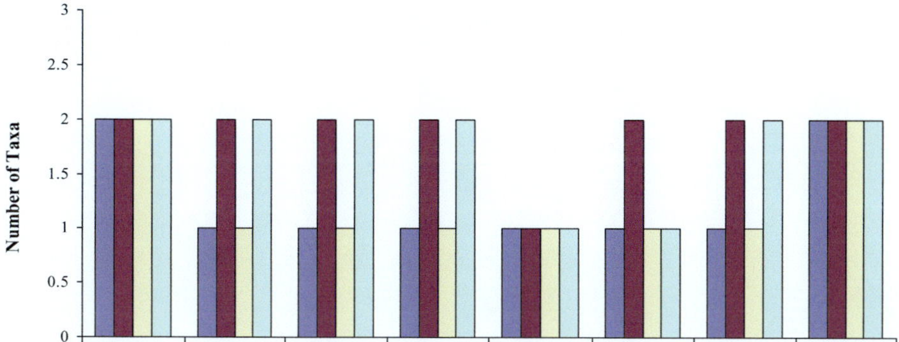

	Autumn	Winter	Spring	Summer	Autumn	Winter	Spring	Summer
Site-I	2	1	1	1	1	1	1	2
Site-II	2	2	2	2	1	2	2	2
Site-III	2	1	1	1	1	1	1	2
Site-IV	2	2	2	2	1	1	2	2

Season

Fig. 32 Seasonal variations in the species diversity of the Euglenophyceae at the investigated sites of the Dal lake

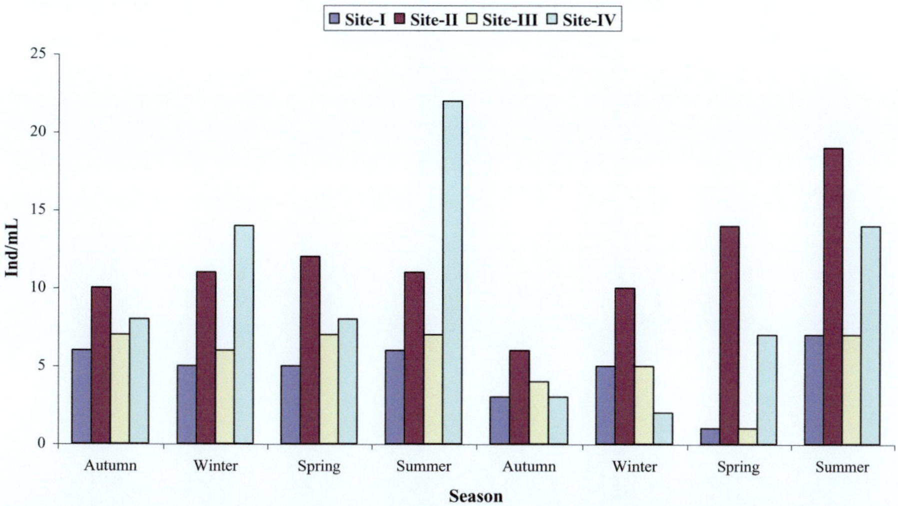

Fig. 33 Seasonal variations in the population density (Ind/mL) of the Euglenophyceae at the investigated sites of the Dal lake

Table 43 Monthly variations in the population density (Ind/mL) of the Chrysophyceae at the study sites of the Dal lake

Site-I	**Sep**	**Oct**	**Nov**	**Dec**	**Jan**	**Feb**	**Mar**	**Apr**	**May**	**Jun**	**Jul**	**Aug**	**Sep**	**Oct**	**Nov**	**Dec**	**Jan**	**Feb**	**Mar**	**Apr**	**May**	**Jun**	**Jul**	**Aug**
Dinobryon sp.	**No Chrysophyceae was recorded from this Site**																							
Site-II	**Sep**	**Oct**	**Nov**	**Dec**	**Jan**	**Feb**	**Mar**	**Apr**	**May**	**Jun**	**Jul**	**Aug**	**Sep**	**Oct**	**Nov**	**Dec**	**Jan**	**Feb**	**Mar**	**Apr**	**May**	**Jun**	**Jul**	**Aug**
Dinobryon sp.	–	–	–	+	–	–	–	–	–	–	–	–	–	–	+	–	–	–	–	–	–	+	–	–
Site-III	**Sep**	**Oct**	**Nov**	**Dec**	**Jan**	**Feb**	**Mar**	**Apr**	**May**	**Jun**	**Jul**	**Aug**	**Sep**	**Oct**	**Nov**	**Dec**	**Jan**	**Feb**	**Mar**	**Apr**	**May**	**Jun**	**Jul**	**Aug**
Dinobryon sp.	–	+	–	–	–	–	–	+	–	–	–	–	–	–	–	–	–	–	–	–	–	–	+	–
Site-IV	**Sep**	**Oct**	**Nov**	**Dec**	**Jan**	**Feb**	**Mar**	**Apr**	**May**	**Jun**	**Jul**	**Aug**	**Sep**	**Oct**	**Nov**	**Dec**	**Jan**	**Feb**	**Mar**	**Apr**	**May**	**Jun**	**Jul**	**Aug**
Dinobryon sp.	–	+	–	–	–	–	–	+	–	–	–	–	–	–	–	–	–	–	–	–	–	–		–

0–5 ind/mL = +
6–10 ind/mL = ++
11–15 ind/mL = +++
and above = ++++

Table 44 Importance Value Index (IVI) of macrophytes in the Dal lake

Taxa	April	May	June	July	August	September	October
A. Emergents							
Alisma plantago aquatica	–	–	–	–	1.31	2.09	–
Bidens cernua	–	–	–	2.23	1.07	0.93	–
Cyperus glomeratus	–	–	–	3.35	4.02	5.01	6.31
Eliocharis palustris	–	–	–	1.27	1.81	2	–
Echinocloa crusgalli	–	–	–	1.06	0.83	1.31	–
Gallium sp.	–	–	–	0.89	0.75	1.21	–
Myriophyllum verticillatum	55.45	55.31	42.31	45.1	41.21	35.13	40.51
Lycopus europaeus	–	–	–	2.7	1.3	2.08	–
Marsillea quadrifola	–	–	–	–	0.84	1	–
Sagittaria sagitlfolia	–	–	4.71	5.08	6.15	4.1	–
Phragmites australis	67.3	48.53	42.2	37.21	30.15	25.34	35.13
Sparganium ramosum	–	–	13.11	10.03	13.11	9.04	16.9
Typha angustata	33.82	27.15	24.1	22.1	18.13	12.1	24.3
Scirpus sp.	–	–	–	7.21	5.31	–	–
Carex sp.	–	–	–	–	–	–	–
Polygonum amphibium	–	–	6.33	–	3.71	–	–
Ranunculus scleratus	–	–	–	2.04	3.08	–	–
Menyanthese trifolata	–	12.21	–	–	–	–	–
Sium latijugam	–	–	9.15	6.27	8.21	5.03	–
Nasturtium officinale	–	–	–	2.03	4.31	5.41	–
Rorripa islandica	–	11.23	7.32	4.21	2.31	3	–
B. Rooted floating-leaf type							
Nelumbo nucifera	–	–	15.07	20.3	24.31	25.18	27.3
Nymphaea sp.	–	20.21	5.21	10.21	11.21	8.15	12.15
Nymphoides peltatum	–	20	21.15	30.31	42.08	45.17	39.57
Hydrocharis dubia		8.81	7.13	4.21	–	2.29	3.26
Potamogeton natans	27.23	–	23.06	15.1	9.21	10.11	18.16
C. Free floating							
Lemna sp.	–	–	3.07	10.21	16	30.13	23.06
Salvinia natans	–	–	2.18	9.21	10.21	17.6	13.5
D. Submerged							
Ceratophyllum demersum	54.73	50.62	47.13	31.15	23	20.11	15.2
Myriophyllum spicatum	18.21	20.26	10.21	–	3.23	4.07	3.71
Hydrilla verticillata	–	–	4.15	3.15	6.15	6.9	4.81
Potamogeton crispus	26.41	19.23	5	6.13	4.31	5.13	6.31
P.pectinatus	–	–	–	–	–	–	–
P. lucens	16.27	6	4.15	5.03	–	6.06	8.16
P. pucillus	–	–	–	–	–	–	–
Utricularia aurea	–	–	2.15	1.17	1.81	4.3	–
Najas sp.	–	–	1.1	0.82	–	–	–

Table 45 Importance Value Index (IVI) of macrophytes in the Dal lake at the floating gardens

Taxa	April	May	June	July	August	September	October
ALISMA PLANTAGO AQUATICA	–	–	–	10.31	12.13	8.17	13.09
Berula sp.	–	–	–	–	–	8.33	–
Bidens cernua	–	–	–	–	8.15	10.08	14.19
Bromus japonica	–	20.21	–	3.71	–	13.11	15.39
Cyperus glomeratus	–	15.13	23.71	26.56	19.08	20.73	17.47
Echinocloa crusgalli	–	–	17.37	19.8	23.18	9.53	–
Eliocharis palustris	–	–	–	–	16.32	17.2	13.09
Epilobium hersutum	–	–	–	–	7.21	10.11	13.03
Gallium sp.	–	–	8.8	18.23	20.33	–	10.13
Lathyrus salicaria	–	18.11	23.11	–	–	–	–
Lycopus europus	–	–	10.08	18.09	12.07	8.23	–
Mentha longifolia	–	–	–	–	9.66	9.56	10.23
Nasturtium officinale	–	–	–	–	11.11	14.67	15.37
Paspalam sp.	–	25.23	28.06	30.19	21.37	16.17	9.08
Phragmites australis	135.1	90.13	63.13	45.11	39.33	37.13	63.33
Polygonum amphibium	68.21	33.33	27.32	36.36	28.3	35.33	39.61
P. minus	–	16.13	21.13	15.8	–	9.15	–
Ranunculus lingua	–	17.11	13.41	11.93	16	12.92	–
Rorripa islandica	–	–	–	–	8.11	10.31	4.27
Sium latijugam	–	14.11	22.44	24.73	19.93	22.87	9.33
Typha angustata	96.33	50.17	41.42	39.09	27.09	36.03	52.11

lingua, Sagittaria sagittifolia, Scirpus sp., Sium latijugam, Sparganium ramosum and Typha angustata. Among emergents *Phragmites australis, Myriophyllum verticillatum* and *Typha angustata* recorded highest IVI in all the months while the lowest IVI was recorded for *Marsilea quadrifolia, Galium sp.* and *Echinocloa crusgalli, Eliocharis palustris, Epilobium hersutum, Gallum sp., Lathyrus salicaria, Mentha longifolia, Paspalum sp., Polygonum minus* and *Ranunculus lingua.* The profuse growth of tall growing emergents suggest high productive status of the lake (Zutshi et al. 1980; Kaul and Handoo 1980; Pandit 2002a, b).

The rooted floating-leaf type species were represented by 5 species viz, *Hydrocharis dubia, Nelumbo nucifera, Nymphaea sp., Nymphoides peltatum* and *Potamogeton natans.* The highest IVI value among rooted floating-leaf types was recorded for *Nymphoides peltatum* (45.17 in September) *Lemna sp.* and *Salvinia natans* were the only representatives of free-floating types with *Lemna sp.* depicting overall higher importance value index (IVI) than *Salvinia natans.*

The submergeds were represented by 8 species including *Ceratophyllum demersum, Hydrilla verticillata, Myriophyllum spicatum, Najas* sp. *Potamogeton crispus, P. lucens, P. pectinalis and Utricularia aurea.* The *Ceratophyllum demersum* showed overall highest IVI followed by *Potamogeton crispus* while the lowest IVI was recorded for *Najas sp.* Kaul (1984) correlated the development of *Lemna-Salvinia* weed complex with excessive eutrophication. The dominance of

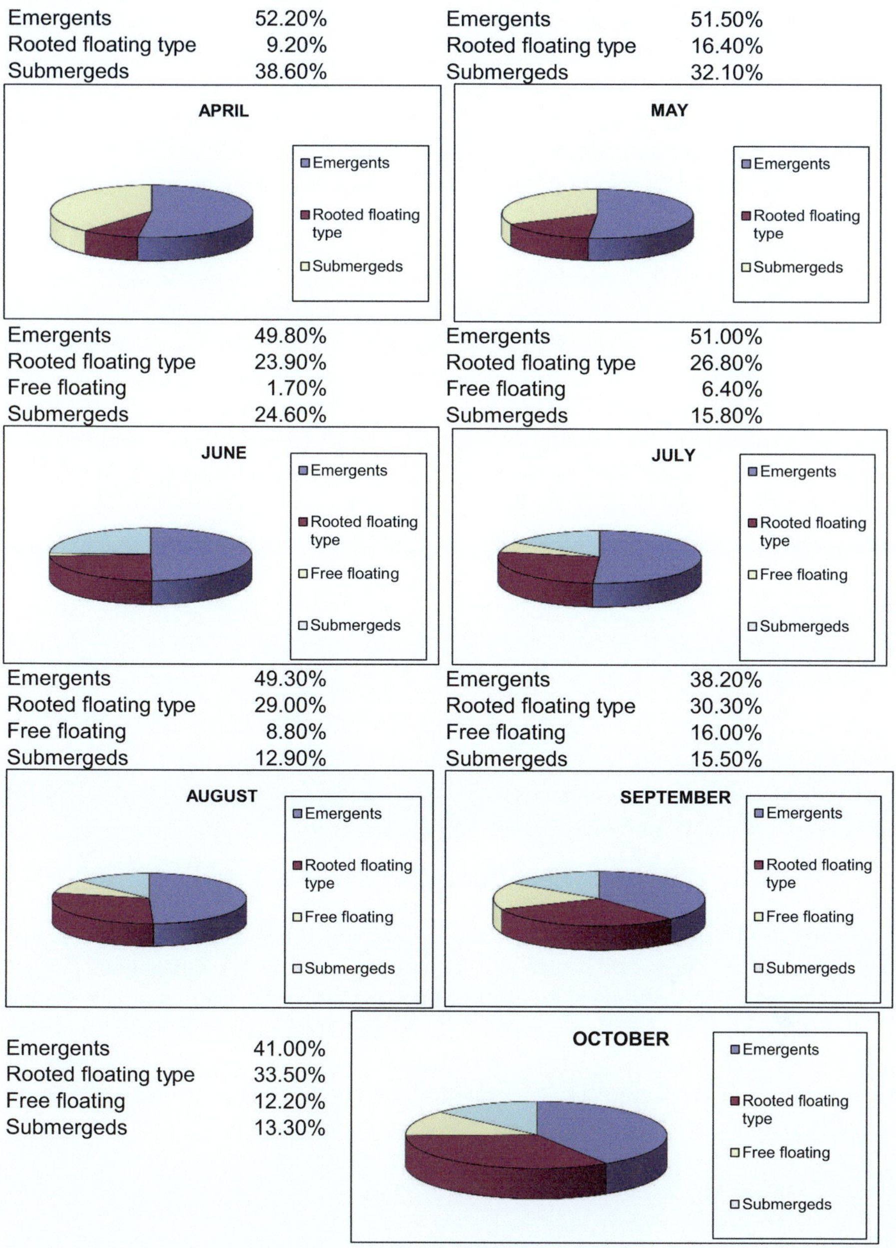

Fig. 34 Monthly percentage contribution of various life form classes in the Dal lake

Ceratophyllum demersum among submergeds may be attributed to nutrient enrichment of the lake waters as Goulder and Boatman (1971) opined that the species requires high inorganic nitrogen level in the medium. Kaul (1984) also reported that *Lemna-Salvinia* weed complex brings a shift in under water vegetation replacing *Potamogeton crispus* association by *C. demersum.*

The percentage contribution to IVI by various life-form classes in different months is depicted in.

In general emergents showed higher percentage contribution to IVI with the highest value of 52.2 % in April and lowest of 38.2 % in September. Rooted floating-leaf-types followed the emergents and recorded maximum value of 33.5 % in October and minimum of 9.2 % in April. Submergeds recorded maximum value in April (38.6) and minimum in August (12.9 %) while for free floating species the value ranged from (0) in April and May to 16 % in September. The higher percentage contribution to IVI by emergents and rooted floating-leaf types are indicative of productive status of the lake as they are main contributors of primary production (Westlake 1965; Kaul et al. 1978; Pandit 2002a, b).

Zooplankton

A total of 51 taxa of zooplankton belonging to Protozoa, Rotifer and Crustacea have been recovered in the Dal lake during 2 year of study. Out of the total, 11 taxa are contributed by Protozoa, 27 by Rotifers and 13 by Crustacea.

From the biological studies conducted during the entire study period, that the distribution of various groups indicate following trend.

Rotifers > Crustacea > Protozoa

Alikunhi (1957), George et al. (1966), Michael (1968), Pennak (1978), Singh and Sahai (1978), Khan and Zutshi (1979), Winner et al. (1979), Zutshi et al. (1980), Yousuf et al. (1986), Shukla et al. (1995), Kumar (1995a, b), Kaur et al. (1996a, b), Syal (1996), Sarwar and Parveen (1996), Unni and Naik (1997), Kaur et al. (1999) and Biswas and Konar (2000) have recorded the dominance rotifers while studying the zooplankton in different aquatic systems. Among the crustacea, cladocera dominated the copepoda.

Protozoa

The protozoa taxa distributed in the Dal lake are given in (table…). In all 11 protozoan taxa were recorded, they include : *Paraeuglypha sp., Diffulgia sp., D. oblonga,Centropyxis sp., Nebula sp., Coleps sp., Paraquadrula irregularis, Paramecium sp., Arcella mitrata, A. vulgaris,* and *Lesquereusia spiralis.* Out of

Table 46 List of Protozoan taxa recorded at the investigated sites of the Dal lake

S. No.	Protozoan taxa	Site-I	Site-II	Site-III	Site-IV
1	**ARCELLA VULGARIS *EHRENBERG***	–	–	+	+
2	*A. mitrata* Leidy	–	–	+	+
3	*Centropyxis sp.*	+	+	–	
4	*Coleps sp.*	–	+	–	
5	*Difflugia oblonga* Ehrenberg	–	+	+	+
6	*Difflugia sp.*	+	–	–	–
7	*Lesquereusia spiralis* Ehrenberg	–	–	+	+
8	*Nebula sp.*	+	+	+	+
9	*Paramecium sp.*	–	+	–	+
10	*Paraquadruda irregularis* Archer	–	+	–	
11	*Paraeuglypha sp.*	+	+	+	–

Table 47 Seasonal variations in the population density (Ind/L) of the Protozoa at the investigated sites of the Dal lake

Sites	Year	Autumn	Winter	Spring	Summer
Site-I	**2000–2001**	8	–	24	28
	2001–2002	–	–	14	40
	Average	4	0	19	34
Site-II	**2000–2001**	–	4	16	24
	2001–2002	4	4	16	32
	Average	2	4	16	28
Site-III	**2000–2001**	–	–	18	34
	2001–2002	4	–	28	44
	Average	2	0	23	39
Site-IV	**2000–2001**	16	1	20	26
	2001–2002	12	–	20	26
	Average	14	0.5	20	26

these, 4 taxa were recorded at site-I, 7 at site-II and 6 taxa each were recorded at site-III and site-IV (Table 46).

In the present study, maximum density of protozoan was observed during summer and minimum during winter at all the sites. Ramakrishnaiah and Sarkar (1982), and Kaur et al. (1997) also observed population density of protozoan to be highest during summer. Yousuf and Qadri (1985) while conducting studies on lake Manasbal, found population density of protozoan to be lowest in winter and then rising through early summer. Baruah and Das (2001) also observed lowest population density during winter but a rise during postmonsoon period. Patnaik (1973) and Sajeev (1999) observed three peaks with the major one in summer. Syal (1996), however, failed to find any definite seasonal trends in the population density of protozoa in running waters of Satluj river (Table 47, Figs. 35 and 36).

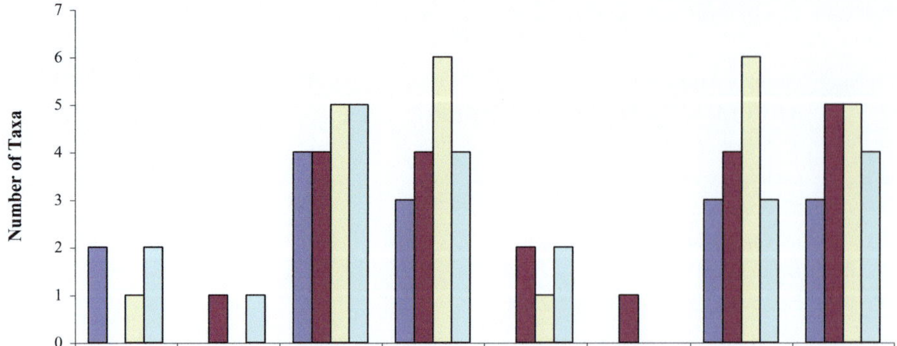

	Autumn	Winter	Spring	Summer	Autumn	Winter	Spring	Summer
Site-I	2		4	3			3	3
Site-II		1	4	4	2	1	4	5
Site-III	1		5	6	1		6	5
Site-IV	2	1	5	4	2		3	4

Season

Fig. 35 Seasonal variations in the species diversity of the Protozoa at the investigated sites of the Dal lake

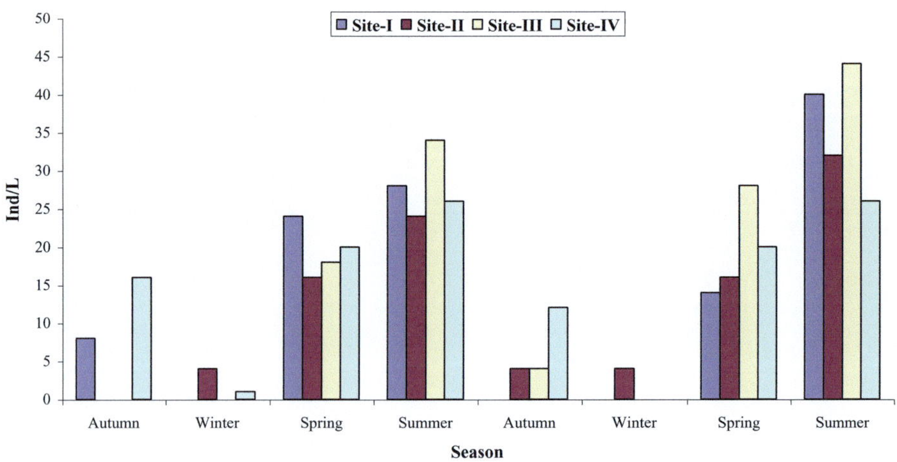

Fig. 36 Seasonal variations in the population density (Ind/L) of the Protozoa at the investigated sites of the Dal lake

The taxa recorded at site-I are *Paraeuglypha sp., Difflugia sp., Centropyxis sp., and Nebula sp.* This site is located in the open water area infested with submerged macrophytes and does not receive domestic or any other waste material. The water here is crystal clear and free from any human habitation and interference. Species diversity and density was observed to be lowest of all the sites. The low nutrient concentration at this site seems to be responsible for the low population density of protozoa. Zafar (1959a), Philipose (1960), Venkateshwarlu (1969) and Munawar (1970b) also observed similarly (Table 48).

Table 48 Monthly variations in the population density (Ind/L) of the Protozoa at the study site-I of the Dal lake

Taxa	Sep	Oct	Nov	Dec	Jan	Feb	Mar	Apr	May	Jun	Jul	Aug	Sep	Oct	Nov	Dec	Jan	Feb	Mar	Apr	May	Jun	Jul	Aug
Pareuglypha sp.	–	–	+	–	–	–	–	+	–	–	–	–	–	–	–	–		–	–	+	–	–	–	–
Difflugia sp.	–	–	–	–	–	–	–	+	+	+	+	+	–	–	–	–	–	–	+		+	+	+	+
Centropyxis sp.	+	–	–	–	–	–	–	+	+	+	+	–	–	–	–	–	–	–	–	–	–	+	+	+
Nebula sp.	–	–	–	–	–	–	+	–	+	+	+	+	–	–	–	–	–	–	–	–	+	+	–	–

0–5 Ind/L = +
6–10 Ind/L = ++
11–15 Ind/L = +++
and above = ++++

The protozoan taxa recorded at Site-II are *Paraeuglypha* sp., *Centropyxis sp., Nebula sp., Difflugia oblonga, Coleps sp., Paraquadrula irregularis, and Paramecium sp.* The population density was recorded to be more than site-I but less than site-III and site-IV. The species diversity was found to be highest of all the sites. This shallow water site is located in the lake where floating gardens cover a large area of the lake and macrophytes grow abundantly. Water here receives agricultural run-off from the floating gardens and dead organic matter by death and decay of macrophytes. The fertilizer rich water at this site provide nutrients required for growth of the protozoans. Kaur et al. (2003) observed that nutrient rich waters supporting rich vegetation favour the protozoan growth (Table 49).

The taxa recorded at Site-III include *Pareuglypha sp., Nebula sp., Difflugia oblonga, Arcella mitrata, A. vulgaris, and Lesquereusia spiralis* while at site-IV include *Nebula* sp., *Difflugia oblonga, Paramecium sp., Arcella mitrata, A. vulgaris, Lesquereusia spiralis.* Both these sites harbour rich growth of plants and receive domestic sewage discharges from the houseboats and hotels respectively which seem to have favoured the growth of protozoan as population density was found to be comparatively higher at both these sites, the maximum being at site-III. Philipose (1960); Venkateshwarlu (1969); Munawar (1970b), Syal (1996) and Sajeev (1999) have also observed that nutrients contributed by sewage discharges favour the growth of prozoans in water supporting rich macrophytic growth (Tables 50 and 51).

The only protozoan recorded from all the sites include *Nebula sp.* This species can, therefore, thrive well in unpolluted as well as in water polluted with agricultural run-off and sewage and hence can be termed as tolerant. This taxon has been found to be absent in Punjab rivers and reservoirs. (Bath 1996; Syal 1996; Sajeev 1999).

Among the protozoans, *Difflugia* sp. was the only one found to be exclusively present at site-I which is almost free from pollution source. Another species of *Difflugia, D.oblonga* was found at site-II, III and IV which are polluted by agricultural run-off and sewage. David and Ray (1966), Verma and Dalela (1975) and Syal (1996) reported this genus (sp. unidentified) in clean as well as polluted waters. Protozoan taxa which were exclusively found at site-II of the lake include, *Paraquadrula irregularis and Coleps sp.* The presence of these two taxa indicates that these can thrive well in waters which receive agricultural runoff and decomposed organic matter. *Paraquadrula irregularis* has also been found at all the sites of the Anchar lake in the present study. Both these taxa have not been recorded in water bodies of Punjab (Bath 1996; Syal 1996; Sajeev 1999) and. No taxon was found to be exclusively present at site-III and IV.

The genus *Arcella* has been recorded by various authors, however, its species has not been specified by anyone of them. In the present investigation, *A. vulgaris* and *A. mitrata* have been recorded. These species were present at site-III and site-IV only where lake water receives sewage from houseboats and hotels. Algarsamy et al. (1967) reported *Arcella* sp. in sewage waters. David and Ray (1966) and Bath (1996) found this genus in clean and polluted waters. *Lesquereusia spiralis* is the

Table 49 Monthly variations in the population density (Ind/L) of the Protozoa at the study site-II of the Dal lake

Taxa	Sep	Oct	Nov	Dec	Jan	Feb	Mar	Apr	May	Jun	Jul	Aug	Sep	Oct	Nov	Dec	Jan	Feb	Mar	Apr	May	Jun	Jul	Aug
Difflugia oblonga	–	–	–	–	–	+	+	–	–	–	++	–	+	–	–	–	–	+	+	–	–	–	++	+
Nebula sp.	–	–	–	–	–	–	–	+	–	–	–	–	–	–	–	–	–	–	–	+	–	–	–	–
Parauglypha sp.	–	–	–	–	–	–	–	+	–	–	–	–	–	–	–	–	–	–	–	+	–	++	–	–
Centropyxis sp.	–	–	–	–	–	–	+	–	–	–	+	–	–	–	–	–	–	–	+	–	–	–	–	–
Coleps sp.	–	–	–	–	–	–	–	–	–	–	–	–	–	+	–	–	–	–	–	–	–	+	++	–
Paraquadruda irregularis	–	–	–	–	–	–	–	–	–	–	–	++	–	–	–	–	–	–	–	–	–	–	–	+
Paramecuim sp.	–	–	–	–	–	–	–	–	–	–	+	–	–	–	–	–	–	–	–	–	–	–	+	–

0–5 Ind/L = +
6–10 Ind/L = ++
11–15 Ind/L = +++
and above = ++++

Table 50 Monthly variations in the population density (Ind/L) of the Protozoa at the study site-III of the Dal lake

Taxa	Sep	Oct	Nov	Dec	Jan	Feb	Mar	Apr	May	Jun	Jul	Aug	Sep	Oct	Nov	Dec	Jan	Feb	Mar	Apr	May	Jun	Jul	Aug
Difflugia oblonga	–	–	–	–	–	–	–	+	–	+	+	+	–	–	–	–	–	–	–	+	–	++	–	+
Nebula sp.	–	–	–	–	–	–	–	+	–	–	+	–	–	–	–	–	–	–	+	+	–	–	–	–
Paraeuglypha sp.	–	–	+	–	–	–	–	–	–	–	+	–	–	–	–	–	–	–	–	+	–	++	+	+
Arcella mitrata	–	–	–	–	–	–	+	–	–	–	+	–	–	–	–	–	–	–	–	+	–	–	++	–
A.vulgaris	–	–	–	–	–	–	+	+	–	–	+	–	–	+	–	–	–	–	+	–	–	–	–	+
Lesquereusia spiralis	–	–	–	–	–	–	–	+	–	+	+	–	–	–	–	–	–	–	+	+	–	+	+	–

0–5 Ind/L = +
6–10 Ind/L = ++
11–15 Ind/L = +++
and above = ++++

Table 51 Monthly variations in the population density (Ind/L) of the Protozoa at the study site-IV of the Dal lake

Taxa	Sep	Oct	Nov	Dec	Jan	Feb	Mar	Apr	May	Jun	Jul	Aug	Sep	Oct	Nov	Dec	Jan	Feb	Mar	Apr	May	Jun	Jul	Aug
Difflugia oblonga	+	+	–	–	–	–	–	+	–	–	–	–	+	+	–	–	–	–	–	+	–	–	–	–
Nebula sp.	–	–	–	–	–	–	–	–	–	+	–	+	–	–	–	–	–	–	–	–	–	–	+	+
Arcella mitrata	–	–	–	–	–	–	–	–	+	+	+	–	–	–	–	–	–	–	–	–	+	–	+	–
A.vulgaris	++	–	–	–	–	–	–	+	–	–	–	+	–	–	–	–	–	–	–	+	++	–	–	–
Lesquereusia spiralis	–	–	–	–	–	–	+	+	–	–	–	–	–	–	–	–	–	–	–	–	–	+	+	+
Paramecium sp.	–	–	–	–	+	–	+	–	–	+	+	+	+	+	–	–	–	–	–	–	–	+	–	–

0–5 Ind/L=+
6–10 Ind/L=++
11–15 Ind/L=+++
and above=++++

other species which occurred only at site-III and IV indicating its tolerance towards sewage. This species has been found at site-III of the Anchar lake, but absent in aquatic ecosystems of Punjab.

Centropyxis sp. in the present investigation was recorded from site-I and site-II only. David and Ray (1966), Bath (1996) and Syal (1996) found this at the sites with no direct source of pollution as well as at the site receiving sewage. Algarsamy et al. (1967) and Bilgrami and Dattamunshi (1979) found it in waters receiving sewage.

Paramecium sp. was found to be distributed at site-II and site-IV. Krishnamoorthi and Visvesvara (1966), Algarsamy et al. (1967), Venkateswarlu and Jayanti (1968) and Kaur et al. (1996a, b) found it in water contaminated with sewage. Kaur et al. (2003) reported its presence in water receiving fertilizer factory wastes.

Paraeuglypha sp.was reported at the sites-I, II and III only indicating it to be a tolerant taxon. In the Anchar lake this genus has been found at all the site

The Protozoa of the Dal lake have not been dealt with by earlier workers (Zutshi et al. 1980). The present study is therefore, the first one to report the data on Protozoa of the Dal lake.

Rotifera

The rotifers distributed at all the four sites in the Dal lake are *Anuraeopsis fissa, Asplanchna sp., Brachionus angularis, B.calycifloris, B. quadridentata, Cephalodella sp., Colurella adriatica, Epiphanes sp., Euchlanis dilatata, Filinia longiseta,, Gastropus sp., Keratella cochlearis, K. quadrata, K. valga, Lecane luna, L. ohiensis, Lepadella ovalis, Monostyla bulla, Mytilina sp., Philodina sp., Polyarthra vulgaris, Pompholyx sulcata, Squantinella mutica, Synchaeta sp., Trichocerca longiseta, Trichocerca sp., Trichotria tetractis.* In all 27 taxa were recorded of which 20 were recorded at site-I, 27 at site-II, 22 at site-III and 23 at site-IV (Table 52).

The rotifer fauna of the Dal lake shows single peak in population density recorded during summer at all the study sites. Sarwar and Parveen (1996) also recorded higher population density during summer in Khusalsar and Gilsar lakes of Kashmir. On the other hand Zuthsi et al. (1980) in their study on nine lakes of Kashmir including the Dal lake, found abundance of rotifers during both summer and winter in all the lakes excepting in Trigam lake where population peaks were observed during spring and autumn. Moore (1980), Nogrady (1983), Bath (1996), Kaur et al. (1996a, b) and Pulle and Khan (2003) also reported higher population during summer. Gabriel et al. (1975) recorded three peaks during May, July and November in a mesotrophic lake, a single peak during summer in eutrophic lake and two distinct peaks during spring and autumn in dystrophic lake. Biswas and Konar (2000) recorded dominance of rotifers in winter at least polluted site and in summer at polluted site. Three peaks in population density were recorded by Patnaik (1973) and

Table 52 List of Rotifer taxa recorded at the investigated sites of the Dal lake

S. No.	Rotifer taxa	Site-I	Site-II	Site-III	Site-IV
1	**ANURAEOPSIS FISSA *LAUTERBORN***	+	+	+	+
2	*Asplanchna sp.*	+	+	+	+
3	*Brachionus angularis* Pallus	–	+	–	+
4	*Brachionus calyciflоris* Pallus	–	+	+	+
5	*Brachionus quadridentata* Pallus	+	+	+	+
6	*Cephalodella sp.*	+	+	+	+
7	*Colurella adriatica* Bory de St. Vincent	+	+	+	+
8	*Epiphanes sp.*	+	+	–	–
9	*Euchlanis dilatata* Myers	+	+	+	+
10	*Filinia longiseta* Bory de St. Vincent	–	+	+	+
11	*Gastropus sp.*	+	+	–	–
12	*Keratella cochlearis* Bory de St. Vincent	+	+	+	+
13	*Keratella quadrata* Bory de St. Vincent	+	+	+	+
14	*Keratella valga* Bory de St. Vincent	+	+	+	+
15	*Lecane luna* Nitzsch	+	+	+	+
16	*Lecane ohiensis* Nitzsch	–	+	+	+
17	*Lepadella ovalis* Bory de St. Vincent	+	+	–	–
18	*Monostyla bulla* Ehrenberg	+	+	+	+
19	*Mytilina sp.*	+	+	+	+
20	*Philodina sp.*	+	+	+	+
21	*Polyarthra vulgaris* Ehrenberg	+	+	+	+
22	**POMPHOLYX SULCATA *GOSSE***	–	+	+	+
23	**SQUANTINELLA MUTICA BORY DE ST. VINCENT**	–	+	+	+
24	*Synchaeta sp.*	+	+	+	+
25	*Trichocerca longiseta* Lamarck	+	+	–	–
26	*Trichocerca sp.*	–	+	+	+
27	*Trichotria tetractis* Bory de St. Vincent	+	+	+	+

Sajeev (1999) in the Chilka lake and Ropar head works reservoir respectively. On the other hand, Vashist and Monga (1981) observed rotifer peak in January in Bhakra main line canal. Rotifer population was found to be maximum during monsoon by Sugunan (1995), Khatri (1985) and Vareethiah and Haniffa (1998), during postmonsoon by Thomas and Azis (1998) and during pre-monsoon and post-monsoon periods by Sabu and Aziz (1999). Kumar (1995a, b) recorded maximum population density during April (Table 53, Figs. 37 and 38).

The rotifer taxa recorded at site-I include *Anuraeopsis fissa, Asplanchna sp., Brachionus quadrata, Cephalodella sp., Colurella adriatica, Epiphanes sp., Euchlanis dilatata, Gastropus sp., Keratella cochlearis, K. quadrata, K. valga, Lecane luna, Lepadella ovalis, Monostyla bulla, Mytilina sp., Philodina sp.,*

Table 53 Seasonal variations in the population density (Ind/L) of the Rotifers at the investigated sites of the Dal lake

Sites	Year	Autumn	Winter	Spring	Summer
Site-I	**2000–2001**	144	60	128	280
	2001–2002	68	36	100	344
	Average	106	48	114	312
Site-II	**2000–2001**	196	72	160	524
	2001–2002	68	80	152	532
	Average	132	76	156	528
Site-III	**2000–2001**	92	28	153	412
	2001–2002	44	48	176	472
	Average	68	38	164.5	442
Site-IV	**2000–2001**	72	36	124	348
	2001–2002	52	44	156	312
	Average	62	40	140	330

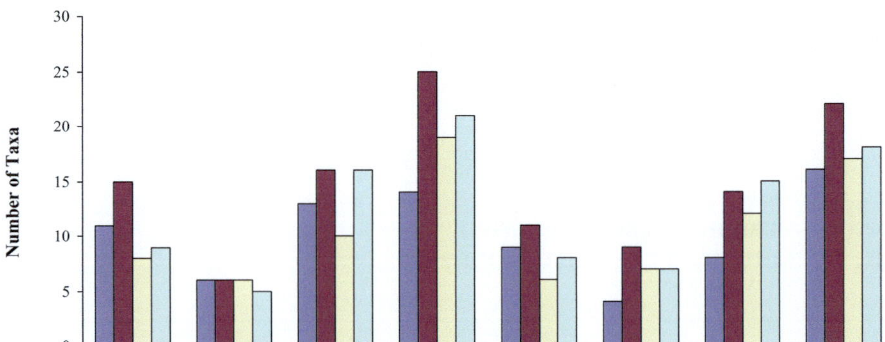

	Autumn	Winter	Spring	Summer	Autumn	Winter	Spring	Summer
Site-I	11	6	13	14	9	4	8	16
Site-II	15	6	16	25	11	9	14	22
Site-III	8	6	10	19	6	7	12	17
Site-IV	9	5	16	21	8	7	15	18

Fig. 37 Seasonal variations in the species diversity of the Rotifer at the investigated sites of the Dal lake

Polyarthra vulgaris, Synchaeta sp., Trichocerca longiseta, and Trichotria tetractis. This site is located in the open water area of the lake infested with submerged macrophytes and is least interfered by human influence and does not receive domestic waste. This site is also free of any human habitation. The species diversity was lowest at this site as compared to other sites. However, the population density was lower than site-I and site-II but slightly higher than site-IV (Table 54).

The taxa recorded at site-II include are *Anuraeopsis fissa, Asplanchna sp., Brachionus angularis, B. calycifloris, B. quadrata, Cephalodella sp., Colurella adriatica, Epiphanes sp., Euchlanis dilatata, Filinia longiseta,, Gastropus sp., Keratella cochlearis, K. quadrata, K. valga, Lecane luna, L. ohiensis, Lepadella*

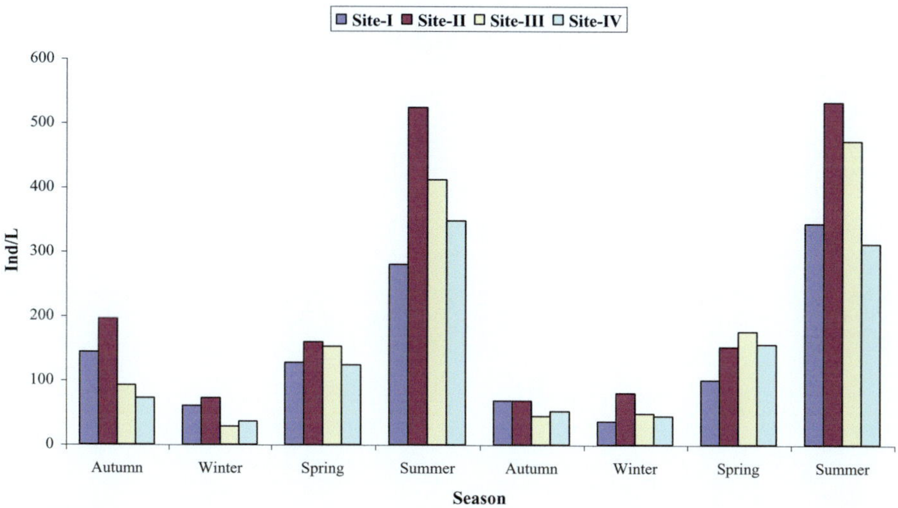

Fig. 38 Seasonal variations in population density (Ind/L) of the Rotifer at the investigated sites of the Dal lake

ovalis, Monostyla bulla, Mytilina sp., Philodina sp., Polyarthra vulgaris, Pompholyx sulcata, Squantinella mutica, Synchaeta sp., Trichocerca longiseta, Trichocerca sp., Trichotria tetractis. At this site species diversity and population density was highest of all the sites. Here water is shallow and is densely vegetated with macrophytes. Nutrients are added into water by the run-off from the floating gardens. The nutrient rich water and presence of macrophytes seem to be conducive for rotifers to flourish. Rich vegetation provides food and shelter to the growing rotifers as suggested by Vashisht and Sra (1979), Balkhi and Yousuf (1991), Pandit and Kaul (1981), Sexena and Mishra (1990), Kaur et al. (1996a, b, 2003), Bath and Kaur (1998) and Sajeev (1999). Syal (1996) observed higher population density in the Satluj river at sites receiving fertilizer factory wastes (Table 55).

The rotifer taxa recorded at site-III include *Anuraeopsis fissa, Asplanchna sp., Brachionus calycifloris, B. quadrata, Cephalodella sp., Colurella adriatica, Euchlanis dilatata, Filinia longiseta,, Keratella cochlearis, K.quadrata, K. valga, Lecane luna, L. ohiensis, Monostyla bulla, Mytilina sp., Philodina sp., Polyarthra vulgaris, Pompholyx sulcata, Rotifer eggs., Squantinella mutica, Synchaeta sp., Trichocerca sp., Trichotria tetractis.* At site-IV, in addition to all these species, *Brachionus angularis* is found. So species composition has been found to be almost similar at site-III and IV. Both these sites receive direct discharge of domestic sewage from houseboats and hotels. Macrophytic growth is also rich at both of these sites. Organic matter present in the sewage enrich the water with nutrients which favour the growth of rotifers. Furthermore, plants provide food and shelter to the rotifers. Sajeev (1999) also recorded that rotifers reproduce and grow well in receiving organic wastes. Kumar and Singh (1996), Haberman (1983) and Kumar (1997a) observed that rotifers prefer habitat rich in organic matter contributed by domestic sewage (Tables 56 and 57).

Table 54 Monthly variations in the population density (Ind/L) of the Rotifera at the study site-I of the Dal lake

Taxa	Sep	Oct	Nov	Dec	Jan	Feb	Mar	Apr	May	Jun	Jul	Aug	Sep	Oct	Nov	Dec	Jan	Feb	Mar	Apr	May	Jun	Jul	Aug
Anuraeopsis fissa	++++	+++	–	–	–	–	–	+	+++	++++	++++	–	+++	++	–	–	–	–	–	–	++++	++++	++++	+++
Asplanchna sp.	+	–	–	–	–	–	–	–	–	–	+	–	–	–	–	–	–	–	–	–	–	++	+	–
Brachionus quadridentata	–	–	–	–	–	–	–	–	–	+	++++	–	–	–	–	–	–	–	–	–	–	+++	–	–
Cephalodella sp.	–	–	–	–	–	–	–	+	++	+	++	–	–	–	–	–	–	–	–	–	–	++++	+++	+
Colurella adriatica	–	+	–	–	–	–	–	–	+	++	–	–	–	+	–	–	–	–	–	–	–	+	+	+
Epiphanes sp.	–	–	–	–	–	–	–	–	+	+	++	–	–	–	–	–	–	–	–	–	+	–	++	–
Euchlanis dilata	–	++++	–	–	–	–	–	–	–	–	++++	+	+	–	–	–	–	–	–	–	–	–	–	–
Gastropus sp.	–	–	–	–	–	–	–	–	–	–	+	–	–	–	–	–	–	–	–	–	+	–	–	+
Keratella cochlearis	++	+	+	–	+	+	–	–	–	–	++++	++++	+	–	–	+	+++	–	–	–	–	++++	++++	+++
Keratella quadrata	–	–	–	–	–	+	++	++++	+	–	–	–	–	–	–	–	–	++	++++	++++	–	–	–	–
Keratella valga	–	–	–	–	–	–	–	–	++++	+++	++++	–	–	–	–	–	–	–	–	–	–	++++	++++	+++
Lecane luna	–	+	–	++++	–	–	–	+	–	++	+	–	+	–	–	–	–	–	–	–	–	–	+++	–
Lepadella ovalis	+	+	–	++	–	–	–	+	–	++	+	–	–	–	–	–	–	+	–	–	–	–	–	+
Monostyla bulla	+++	–	+	++++	–	–	+	–	+++	++	++	–	–	–	+	–	–	–	–	+	++	–	++++	–
Mytilina sp.	++	–	–	–	–	–	–	–	–	–	–	–	–	+	–	–	–	–	–	–	–	–	–	–
Philodina sp.	–	+	–	–	–	–	–	+	–	–	–	–	+	–	–	–	–	–	–	–	–	–	–	–
Polyarthra vulgaris	–	–	–	–	–	+	+	–	–	+	–	–	–	–	–	–	–	–	–	++	–	–	–	+
Synchaeta sp.	–	–	–	–	–	–	–	+	+	–	–	–	–	–	–	–	–	–	–	–	–	+	+	+
Trichocerca longiseta	–	–	–	–	–	–	–	–	–	–	.	–	–	–	–	–	–	–	–	+	–	–	+	–
Trichotria tetractis	++	–	+++	–	–	–	–	–	+	–	–	–	–	–	+++	–	++	–	+	–	–	+	–	–

0–5 Ind/L = +
6–10 Ind/L = ++
11–15 Ind/L = +++
and above = ++++

Table 55 Monthly variations in the population density (Ind/L) of the Rotifera at the study site-II of the Dal lake

Taxa	Sep	Oct	Nov	Dec	Jan	Feb	Mar	Apr	May	Jun	Jul	Aug	Sep	Oct	Nov	Dec	Jan	Feb	Mar	Apr	May	Jun	Jul	Aug
Anuraeopsis fissa	++++	++++	–	–	–	–	–	–	+++	+++	++++	++++	++	–	–	–	–	–	–	–	+++	++++	++++	++++
Asplanchna sp.	–	–	–	–	–	–	–	–	–	–	–	–	–	–	–	–	+	–	–	–	–	–	–	–
Brachionius calyciflorus	–	–	–	–	++++	–	–	+++	+++	++++	++++	++++	–	–	–	–	+++	–	–	–	–	–	–	–
Brachionus quadridentata	+++	++	–	–	–	–	+++	–	–	–	+++	+++	+	–	–	–	–	++	–	–	–	–	+++	–
Brachnious angularis	–	–	–	–	–	–	–	+	+++	++++	+++	–	–	–	–	–	–	–	–	–	++	++++	++++	++++
Cephalodella sp.	–	–	–	–	–	–	–	–	–	+++	+	+	–	–	–	–	–	–	–	–	++	++++	++	–
Colurella adriatica	–	–	++	+	–	–	–	+	+	+	–	–	–	–	+	+	–	–	–	–	+++	+++	–	–
Epiphanes sp.	+	+	–	–	–	–	–	+	–	+	++	++	–	+	–	–	–	–	–	–	++	+	+	+
Euchlanis dilatata	–	++	–	–	–	–	–	–	–	–	–	–	++	+	–	–	–	–	–	–	–	–	–	–
Filinia longiseta	+	+	–	–	–	–	–	–	–	++	+	+	–	–	–	–	–	–	–	–	–	–	–	+
Gastropus sp.	+	+	–	–	–	–	–	–	–	–	+	+	+	–	–	–	–	–	–	–	–	–	–	++
Keratella cochlearis	+	+	+	–	–	+++	–	–	+++	++++	+	+++	++	–	–	–	++	+	–	–	+++	++++	++++	++++
Keratella quadrata	–	–	–	–	–	–	–	–	+	+	+++	+	–	–	–	–	–	–	–	–	–	–	+++	–
Keratella valga	–	–	–	–	–	–	–	+	–	++	+++	++++	–	–	–	–	–	–	++	–	+++	++++	++++	–
L.ohiensis	–	–	–	–	–	–	–	+	–	–	+	–	–	–	–	–	–	–	–	++	–	–	+	+

(continued)

Table 55 (continued)

Taxa	Sep	Oct	Nov	Dec	Jan	Feb	Mar	Apr	May	Jun	Jul	Aug	Sep	Oct	Nov	Dec	Jan	Feb	Mar	Apr	May	Jun	Jul	Aug
Lecane luna	++	+	–	+++	+	–	–	–	–	–	–	+	+	+	–	–	++	–	–	–	–	–	–	+
Lepadella ovalis	–	–	–	–	–	–	–	–	+	–	+	+	–	–	–	–	–	–	–	+	–	+	+	+
Monostyle bulla	+++	+	+	–	–	–	–	+	+	+	+	+	–	–	–	–	–	–	–	++	+++	++	+++	++++
Mytilina sp.	–	++	–	–	–	++	–	–	–	++	+	+	–	–	–	–	+	–	–	–	–	–	+++	+
Philodina sp.	+	+	+	–	–	–	–	–	–	+	+	+	–	–	–	–	–	–	–	–	–	–	–	–
Polyarthra vulgaris	–	–	–	–	–	–	+++	–	+	–	++	–	–	–	–	–	–	++	+	+	–	–	–	–
Pompholyx sulcata	–	–	–	–	–	–	–	–	–	–	+	–	–	–	–	–	–	–	–	–	–	+	–	–
Squantinella mutica	–	–	–	–	–	–	–	–	–	+	–	–	–	–	–	–	–	–	–	–	–	–	+	–
Synchaeta sp.	++	+	–	–	–	–	–	+	+	+	++	++	+	–	–	–	–	–	–	–	+	+	+	+
Trichocerca longiseta	–	–	–	–	–	–	–	–	+	–	+	–	–	–	–	–	–	–	–	–	++	–	+	–
Trichocerca sp.	–	–	+	–	–	–	–	–	–	–	+	+	+	–	–	–	–	–	–	–	–	+	+	–
Trichotria tetractis	+	+	–	–	–	++	–	–	+	++++	++++	++	+	–	–	–	–	++	–	++	–	++++	+++	++

0–5 Ind/L = +
6–10 Ind/L = ++
11–15 Ind/L = +++
and above = ++++

Table 56 Monthly variations in the population density (Ind/L) of the Rotifera at the study site-III of the Dal lake

Taxa	Sep	Oct	Nov	Dec	Jan	Feb	Mar	Apr	May	Jun	Jul	Aug	Sep	Oct	Nov	Dec	Jan	Feb	Mar	Apr	May	Jun	Jul	Aug
Anuraeopsis fissa	++	+	–	–	–	–	–	–	–	+	+++	++	–	–	–	–	–	–	–	–	–	+++	++++	+++
Asplanchna sp.	–	+	–	–	–	–	–	–	++	+	–	+++	++	–	–	–	–	–	–	+	–	–	–	–
Brachionus calyciflorus	–	–	–	–	–	–	–	–	–	+	+	+	–	–	–	–	–	–	–	–	–	++	+	–
Brachionus quadridentata	–	–	–	–	–	–	–	–	–	–	++	–	–	–	–	–	–	–	–	–	–	–	–	+++
Cephalodella sp.	–	–	–	–	–	–	–	–	–	–	–	–	–	–	–	–	–	–	–	–	–	+++	–	–
Colurella adriatica	–	–	–	–	–	–	–	–	–	–	–	–	–	–	–	–	–	–	+	–	++	+	+++	–
Euchlanis dilatata	–	–	++	++	–	–	–	–	–	+	+	+	–	+	–	–	–	–	–	–	–	+++	–	–
Filinia longiseta	–	–	–	–	–	–	–	–	–	–	+	–	–	–	–	–	–	–	–	–	+	+	+	–
Keratella cochlearis	–	+++	++	–	–	–	–	++++	+	+++	++++	++++	–	–	–	–	++	–	–	–	–	+++	++++	++++
Keratella quadrata	–	–	–	–	–	+	++	–	++++	+	–	–	–	–	–	–	–	++	+++	+++	+++	–	–	–
Keratella valga	++++	–	–	–	–	–	–	–	–	–	+++	–	+	–	–	–	–	–	–	–	–	+++	++++	–
Lecane ohiensis	+	+	–	–	–	–	–	–	+++	++	+	–	–	–	–	–	–	–	–	+	++	+++	–	–
Lecane luna	–	–	–	+	–	–	+++	–	++	++++	++++	+	+++	–	–	–	+	–	–	++	–	++++	++++	–
Monostyle bulla	–	–	–	–	–	–	–	+	+++	+++	++++	+++	+++	–	–	–	+	–	–	++	+	++++	++++	–
Mytilina sp.	–	+	–	+	–	–	+	–	+	+++	++++	–	–	–	–	–	–	++	+++	++	+++	+++	–	++++
Philodina sp.	+	–	–	–	–	–	–	–	–	+	–	–	–	–	–	–	+	–	–	+	–	–	–	–
Polyarthra vulgaris	–	–	–	–	–	+	–	–	–	+++	++	–	–	+	–	–	–	–	+++	+	+	–	–	–
Pompholyx sulcata	–	–	–	–	–	–	–	–	–	–	++	+	–	–	–	–	–	–	–	–	–	–	+	–
Squantinella mutica	–	–	–	–	–	–	–	–	–	–	+	–	–	–	–	–	–	–	–	–	–	+	+	–
Synchaeta sp.	–	–	–	–	–	–	–	–	++	++++	++	–	–	–	–	–	–	–	–	–	+	+++	+	+
Trichocerca sp.	–	–	–	–	–	–	–	–	++	+++	++	+	–	–	–	–	–	–	–	–	++	+	+	–
Trichotria tetractis	–	–	–	+	–	–	+	–	–	–	–	–	–	–	–	–	–	++	–	–	–	–	–	–

0–5 Ind/L=+
6–10 Ind/L=++
11–15 Ind/L=+++
and above=++++

Table 57 Monthly variations in the population density (Ind/L) of the Rotifera at the study site-IV of the Dal lake

Taxa	Sep	Oct	Nov	Dec	Jan	Feb	Mar	Apr	May	Jun	Jul	Aug	Sep	Oct	Nov	Dec	Jan	Feb	Mar	Apr	May	Jun	Jul	Aug
Anuraeopsis fissa	+	–	–	–	–	–	–	–	+	++	+++	+	+	–	–	–	–	–	–	–	+	++	+++	+
Asplanchna sp.	–	+	–	–	–	–	–	–	+	+	–	++	+	–	–	–	–	–	–	++	–	–	–	–
B. calyciflorus	–	–	–	–	–	–	–	–	+	+	++	+	–	–	–	–	–	–	–	–	–	++	+++	–
Brachionus quadridentata	–	–	–	–	–	–	–	–	–	++	+	+	–	–	–	–	–	–	–	–	–	–	+	+++
Brachnious angularis	–	–	–	–	–	–	–	–	–	++	+	–	–	–	–	–	–	–	–	–	–	+	+	–
Cephalodella sp.	–	–	–	–	–	–	–	–	+	–	–	++	+	–	–	–	–	–	–	–	–	+	+	+
Colurella adriatica	–	–	–	–	–	–	–	–	–	+	+	++	–	–	–	–	–	–	+	–	++	+	+++	–
Euchlanis dilatata	–	–	–	+	–	–	–	–	+	+	++	–	–	–	–	–	–	–	–	–	–	++	+	–
Filinia longiseta	–	–	–	–	–	–	–	–	+	+	+	–	–	–	–	–	–	–	–	–	+	–	–	+
Keratella cochlearis	+	–	–	–	–	–	–	–	+++	++	+	++	–	–	–	–	+	–	–	–	+	++	++++	+++
Keratella quadrata	–	–	+	–	+	++	+	–	+++	+	++	–	–	–	–	–	+	+++	++	+++	+	–	–	–
Keratella valga	++++	–	–	+	–	–	–	–	+	++	++++	+	+	–	–	–	–	–	–	–	+	++++	++++	–
Lecane ohiensis	+	+	–	–	–	–	–	–	++	+	+	–	–	–	–	–	–	–	+	+	++	++	–	–
Lecane luna	–	–	–	++	–	–	+++	–	+	++	+++	+	++	+	–	–	+	–	–	+	–	++	+++	–
Monostyle bulla	–	–	–	–	–	–	–	+	++	++	+++	++	++	–	–	–	+	–	–	+	++	++	++	–
Mytilina sp.	–	–	–	–	–	–	–	–	++	+++	++	–	+	–	–	–	–	+	+++	+	++	+++	+	++
Philodina sp.	+	–	–	–	–	–	–	–	–	–	–	–	–	–	–	–	+	–	–	+	–	–	–	–
Polyarthra vulgaris	–	–	–	–	–	+	–	–	–	+	++++	–	–	–	–	–	–	–	++	–	+	–	–	–
Pompholyx sulcata	++	–	+	–	–	–	–	–	–	+	–	–	–	–	–	–	–	–	–	–	–	+	–	–
Squantinella mutica	–	–	–	–	–	–	–	–	–	+	–	–	–	+	+	–	–	–	–	–	–	+	+	–
Synchaeta sp.	–	–	–	–	–	–	–	–	+	+++	+	+	–	–	–	–	–	–	–	–	++	+	+	–
Trichocerca sp.	–	–	–	–	–	–	–	–	++	+++	+	+	–	–	–	–	–	–	–	–	++	+	+	–
Trichotria tetractis	–	–	+	–	–	–	+	–	–	–	–	–	–	–	–	–	–	+	–	–	–	–	–	–

0–5 Ind/L=+
6–10 Ind/L=++
11–15 Ind/L=+++
and above=++++

In the present study, out of total 27 rotifer taxa,16 were encountered at all the sites. They include *Keratella quadrata; K. valga; K. cochlearis; Anuraeopsis fissa; Brachionus quadrata; Euchlanis dilatata; Mytilina sp.; Trichotria tetractis; Cephalodella sp.; Polyarthra vulgaris; Synchaeta sp.; Colurella adriatica; Monostyla bulla; Lecane luna; Asplanchna sp.* and *Philodina sp.*

The presence of *Keratella quadrata*; *K. valga* and *K. cochlearis* at all the sites indicates that they can thrive well in both clean waters and water contaminated with agricultural runoff and sewage. Arora (1966b), Sladecek (1983), Syal (1996) and Sajeev (1999) have also found *K. valga* in clean and sewage contaminated waters. However, Suduzuki et al. (1983) found it only in clean waters. Jayangoudar (1980) and Sajeev (1999) recorded *K. cochlearis* in clean as well as sewage contaminated waters while it was found to be present at sites receiving sewage and industrial effluents by Berzins (1949), Pejlar (1957), Hutchinson (1964); George et al. (1966); Radwan (1976); Hakkari (1978); Bilgrami and Datta Munshi (1979); Maemets (1983); Orcutt and Pace (1984); Sinha (1987) and Sinha and Prasad (1988). *Anuraeopsis fissa* was found to be distributed in moderately as well as highly polluted waters by Berzins (1949); Pejlar (1957); Hutchinson (1964); Radwan (1976); Hakkari (1978); Maemets (1983); Orcutt and Pace (1984); Balkhi and Yousuf (1991). However, Sladecek (1983) and Suduzuki et al. (1983) found this only in clean waters. The presence of *Brachionus quadridentata* in clean as well as water receiving sewage has also been observed by Syal (1996). Arora (1966b) found this in clean waters only while Sladecek (1983) noted it in moderately polluted waters.

Euchalanis dilatata was present at all the sites suggesting that it can survive both in clean as well as waters contaminated with sewage and agricultural run-off. David and Ray (1966); Sladecek (1983) and Syal (1996) also found it both in clean as well as waters polluted with sewage. Klimowicz (1973) found it in sludge from water purification plant.

Mytilina sp. was found at all the sites in the present investigation. Balkhi and Yousuf (1984) and Syal (1996), however, found this from highly polluted waters. Similarly *Polyarthra vulgaris* has been reported from highly polluted water by Spencer et al. (1983). *Colurella adriatica* was found at all the sites thereby indicating that it can tolerate water receiving sewage and agricultural run-off. Sladecek (1983), and Sajeev (1999) found this species in moderately and highly polluted waters. Vasisht and Sharma (1975) found it to be scanty in clean waters. However, it was found to be more or less evenly distributed at four sites of the Dal lake. *Monostyla bulla* was recorded by Syal (1996) from clean to highly polluted waters receiving sewage. The genus *Monostyla* has been recorded by other workers also but the species has not been specified by anyone of them. Goel and Chavan (1991) described it to be a pollution tolerant species. Sladecek (1983) discovered it in waters ranging from clean to moderately polluted, whereas George et al. (1966), Algarsamy et al. (1967), Bilgrami and Dattamunshi (1979) discovered it in waters contaminated with sewage and industrial wastes. *Lecane luna* was found to be present both in clean and moderately polluted waters also by Sladecek (1983) and Syal (1996). *Asplanchana* sp. was found in clean and sewage contaminated water by Pejler (1957); Hakkari (1978); Maemets (1983); Sladecek (1983); Suduzuki et al.

(1983); Sinha and Prasad (1988) and Yousuf and Qadri (1985). *Philodina* sp. was also found in clean and moderately polluted waters with sewage by George et al. (1966); Das and Pandey (1978), Vashisht and Sra (1979), Vasisht and Kapoor (1981) and Syal (1996).

The rotifer taxa absent from site-I and present at the site-II, III and IV include *B. calyciflorus, Filinia longiseta; Lecane ohiensis; Trichocerca sp., Pompholyx sulcata* and *Squantinella mutica* thereby indicating that these rotifers prefer nutrient rich waters. Among these rotifers, *Brachionus calyciflorus* was found to be distributed at the sites receiving fertilizer wastes and sewage by Pejler (1957), George et al. (1966), Radwan (1976), Hakkari (1978) and Syal (1996). Jyoti and Sehgal (1979) found it in polluted waters and described it as indicator of eutrophy.

Filinia longiseta was reported from moderately and highly polluted waters by Berzins (1949), De Smet and Evens (1972) reported it in moderately and highly polluted waters. However, Syal (1996) and Sladecek (1983) recorded its presence in both clean as well as moderately polluted waters. The presence of other taxa including *Lecane ohiensis*, *Trichocerca* sp., *Pompholyx sulcata* and *Squantinella mutica* at sites- II, III and IV indicate that they prefer waters enriched with agricultural run-off and sewage. These taxa have not been discussed in the literature available. *Gastropus* sp. and *Trichocerca longiseta* were found at site-I and site-II showing that these can thrive well in both clean as well as water receiving agricultural run-off. Maemets (1983), Sladecek (1983) and Sinha and Prasad (1988) described *Gastropus* sp. as clean water species while Syal (1996) found it to be present at sites receiving effluence from fertilizer factories in river Satluj of Punjab.

Zutshi et al. (1980) also reported good Rotifer taxa in the Dal lake. In the present investigation 27 taxa were reported. With increased human habitation in and around this lake, the enrichment of water with agricultural run-off and sewage had not made much impact on the rotifer diversity. The only taxon *Ascomorpha* sp. found to be present in their studies has not been recovered during the present study.

Crustacea

In all, 13 crustaceans belonging to Cladocera and Copepoda were discovered. Out of these, 12 were present at site-I and 13 each at all other sites of the Dal lake (Table 58). The crustacean fauna recorded at the study sites include, besides nauplius larvae, 8 cladocerans viz. *Acroperus harpae, Alona sp., Alonella exigua, Bosmina longirostris, Chydorus sp., Daphnia sp., Moina sp., and Moinadaphnia sp.* and 5 copepods viz. *Cyclops sp., Diaptomus sp., Eucyclops scutifer, Eucyclops sp.,* and *Sida sp.* The data shows dominance of cladocerans over copepods both in terms of species diversity and population density which is also observed by Sharma and Michael (1987) and Sarwar and Parveen (1996) in Indian lakes and in two inter connected lakes of Srinagar respectively. Zutshi et al. (1980), Yousuf and Qadri (1985) and Yousuf (1988) however, found copepods to be dominant over cladoceran in Kashmir lakes. Similar observations were made by Allan (1976), Singh et al. (1982), Ferguson (1982), Carmonze et al. (1983), Serruya and Pollingher (1983), Hecky

Table 58 List of Crustacean taxa recorded at the investigated sites of the Dal lake

S. No.	Crustacean taxa	Site-I	Site-II	Site-III	Site-IV
1	**ACROPERUS HARPAE *BAIRD***	+	+	+	+
2	*Alona sp.*	+	+	+	+
3	*Alonella exigua* Leydig	+	+	+	+
4	*Bosmina longirostris* Muller	+	+	+	+
5	*Chydorus sp.*	+	+	+	+
6	*Cyclops scutifer* Sars	+	+	+	+
7	*Cyclops sp.*	+	+	+	+
8	*Daphnia sp.*	+	+	+	+
9	*Diaptomas sp.*	+	+	+	+
10	*Eucyclops agills*	+	+	+	+
11	*Moinodaphnia sp.*	+	+	+	+
12	*Moina sp.*	+	+	+	+
13	*Sida sp.*	+	+	+	+

(1984), Sharma and Patnaik (1985), Drenner et al. (1987), Payne (1987), Mengastou and Fernando (1991) in other freshwater bodies.

Crustacean population in general increased during summer. A single peak was observed. George (1966), Battish (1968), Patalas (1972), Nassar (1975), Agrawal and Gautam (1985), Saha and Pandit (1988), Bath (1996); Sarwar and Parveen (1996); Biswas and Konar (2000) and Sharma et al. (2000) also observed maximum density of cladocera during summer. Literature reveals inconsistent observations with respect to seasonal distribution of cladocera and copepoda. Vasisht and Sharma (1975), O'Brien (1979), Kohli et al. (1982), Orcutt and Pace (1984) and Pulle and Khan (2003) recorded the maximum density of cladocera during winter while Syal (1996) failed to ascertain any definite seasonal trend in the Satluj river. Bath (1996) recorded two peaks of cladocerans in the Harike lake. Patnaik (1973) and Ramakrishnaiah and Sarkar (1982) noted that copepods were present throughout the year. Sukumaran et al. (1984) also observed copepods throughout the year in tanks of Maland region of Kataka but their population density was minimum in May and Maximum in November. On the other hand Sajeev (1999) recorded three peaks in copepod population in Ropar reservoir (Table 59, Figs. 39 and 40).

At site-I crustacean fauna include 7 cladocera; *Acroperus harpae, Alona sp., Alonella exigua, Bosmina longirostris, Chydorus sp., Daphnia sp.,* and *Moina sp.,* and 5 copepods viz. *Cyclops sp., Diaptomus sp., Eucyclops scutifer, Eucyclops sp.* and *Sida sp.* (Table 60). At sites II, III and IV in addition to all these taxa one more cladoceran, *Moinadaphnia* sp. is present (Tables 61, 62 and 63). Population density, however, was found to be higher at site II, III and IV as compared to the site-I. This may be due to nutrient rich water and presence of macrophytes at these sites which provide food and shelter to the crustaceans as has also been suggested by Dexter (1959), Holden and Green (1960), Macan (1963), Brooks and Dodson (1965), Krishnamoorthi and Visvesvara (1966), Jolly and Chapman (1966), Pahwa and Mehrotra (1966), Ray et al. (1966), Gideiri (1969), Hynes (1970), Kochsiek et al. (1971), Hall et al. (1976), Threlkeld (1979), O'Brien (1979), Tonopi (1980),

Table 59 Seasonal variations in the population density (Ind/L) of the Crustacea at the investigated sites of the Dal lake

Sites	Year	Autumn	Winter	Spring	Summer
Site-I	**2000–2001**	20	12	36	68
	2001–2002	20	4	24	68
	Average	20	8	30	68
Site-II	**2000–2001**	44	36	32	74
	2001–2002	52	32	44	124
	Average	48	34	38	99
Site-III	**2000–2001**	4	16	28	96
	2001–2002	8	32	28	112
	Average	6	24	28	104
Site-IV	**2000–2001**	4	12	32	92
	2001–2002	12	20	40	100
	Average	8	16	36	96

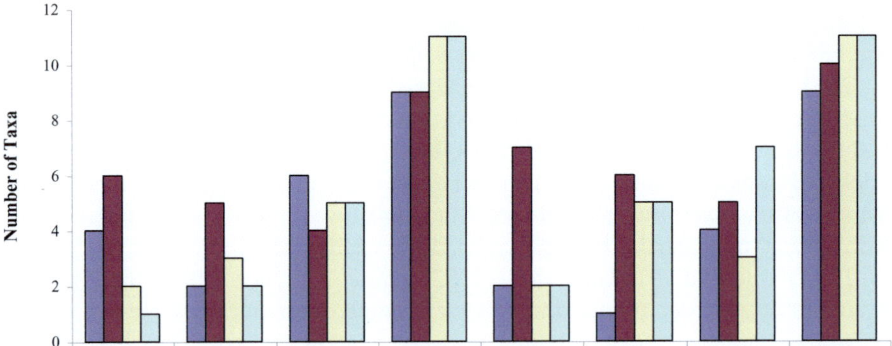

	Autumn	Winter	Spring	Summer	Autumn	Winter	Spring	Summer
Site-I	4	2	6	9	2	1	4	9
Site-II	6	5	4	9	7	6	5	10
Site-III	2	3	5	11	2	5	3	11
Site-IV	1	2	5	11	2	5	7	11

Fig. 39 Seasonal variations in species diversity of the Crustacea at the investigated sites of the Dal lake

Scullion et al. (1982), Singh (1982), Spencer *et al.* (1983), Orcutt and Pace (1984) Petts and Greenwood (1985), Saha and Pandit (1988), Bath (1996), Kaur et al. (1996a, b), Bath and Kaur (1997) and Sajeev (1999). The species composition of copepods was the same at all the sites. The absence of *Moinadaphnia* sp. at site-I indicates that this taxon prefer waters contaminated with agricultural run-off and sewage effluents. Its maximum population density was observed at site-II. Therefore, this taxon in the present study can be labelled as pollution indicator. However, no firm conclusion could be drawn as no relevant literature was found regarding its distribution and ecology. In Punjab aquatic ecosystems, it was found to be absent (Syal 1996; Bath 1996; Sajeev 1999).

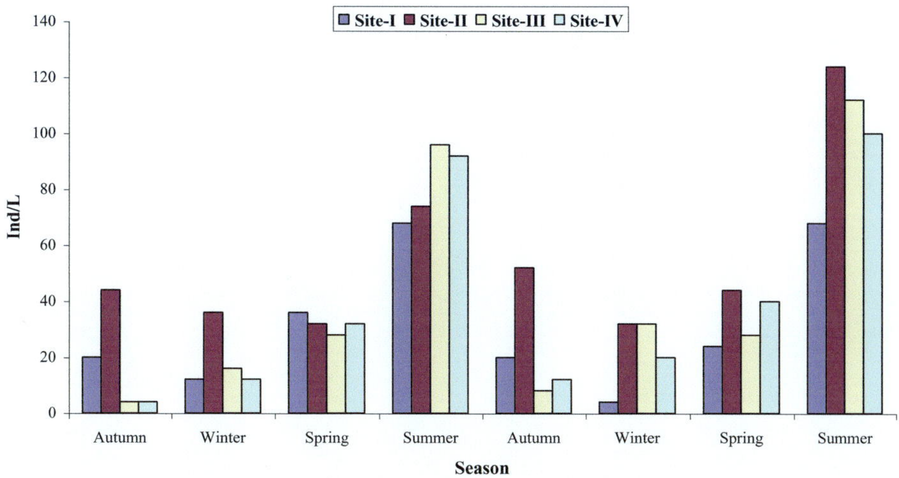

Fig. 40 Seasonal variations in the population density (Ind/L) of the Crustacea at the investigated sites of the Dal lake

Distribution of all other crustaceans at four sites of the Dal lake indicates that these are tolerant taxa. Literature reveals the presence of *Daphnia* sp. both in clean and polluted water (Das 1989; Syal 1996), in highly polluted waters (Sinha and Prasad 1988; Sharma et al. 1989; Sexena and Mishra 1990) and in clean waters only (Jolly and Chapman 1966). *Chydorus* sp. was found in clean as well as highly polluted waters by Khare et al. (1979) and Ponji et al. (1985). *Bosmina longirostris* was found in waters contaminated with sewage by Sharma et al. (1989). *Alona* sp. was found to be present in clean waters only by Verma and Shukla (1968) and Syal (1996). Besides, all these taxa, nauplius larvae were found to be present at all the sites. Das (1989) also found nauplius larvae both in clean and moderately polluted waters. Khare et al. (1979), Sukumaran et al. (1984) and Syal (1996) recorded nauplius larvae in polluted waters. *Diaptomus* sp. was found both in clean and highly polluted waters by Das (1989); in clean waters by Sukumaran et al. (1984), Jindal and Ghezta (1991) and Syal (1996); in highly polluted water by Krishnamoorthi and Visvesvara (1966) and Sexena and Mishra (1990). *Cyclops* sp. was found in both clean and highly polluted waters by Das (1989), in clean water by Sukumaran et al. (1984), Jindal and Ghezta (1991) and Syal (1996) and in highly polluted waters by David and Ray (1966) Another species *C. scutifer* was also recorded from all the studied sites. *Acroperus harpae, Alonella exigua, Eucyclops agilis* and *Sida* sp., too, were found at all the sites. Das (1989) and Syal (1996) reported *Sida crystallina* in recovery zone of paper mill effluents. In the present study, however, species of the genus *Sida* could not be identified. The relevant literature regarding distribution and ecology of rest of the species is not available.

Zutshi et al. (1980) reported 7 taxa of Crustacea in the Dal lake. It seems that with the enrichment of water with agricultural run-off and sewage, a number of new taxa have appeared in the Dal lake. *Thermocyclops* sp. and *Mesocyclops* sp. which

Table 60 Monthly variations in the population density (Ind/L) of the Crustacea at the study site-I of the Dal lake

Taxa	Sep	Oct	Nov	Dec	Jan	Feb	Mar	Apr	May	Jun	Jul	Aug	Sep	Oct	Nov	Dec	Jan	Feb	Mar	Apr	May	Jun	Jul	Aug
Acroperus harpae	–	–	–	+	–	+	–	–	+	–	–	–	–	–	–	–	–	–	–	–	–	+	–	–
Alona sp.	–	–	–	–	–	–	–	+	–	–	+	–	–	–	–	–	–	–	–	–	+	–	+	–
Alonella exigua	–	–	–	–	–	–	–	–	–	+	–	–	–	–	–	–	–	–	–	–	–	++	–	–
Bosmina longirostris	–	–	–	–	–	–	–	–	–	+	–	–	–	–	–	–	–	–	–	–	+	+	++	–
Chydorus sp.	–	–	–	–	–	–	–	–	+	+	+	–	–	–	–	–	–	–	–	–	–	–	+	–
Cyclops scutifer	–	–	–	–	–	–	–	–	–	+	+	–	+	–	+	+	–	–	–	–	–	+	–	–
Cyclops sp.	–	+	–	–	–	–	–	+	–	++	–	–	–	+	–	–	–	–	–	–	+	+	+	–
Daphnia sp.	–	–	+	+	–	–	–	+	–	–	–	–	–	–	–	–	–	–	–	+	+	–	–	–
Diaptomous sp.	–	–	–	–	–	–	–	–	–	+	+	–	–	–	–	–	–	–	–	–	–	+	++	–
Eucyclops sp.	–	–	–	–	–	–	–	–	–	–	–	–	–	–	–	–	–	–	–	–	–	–	–	+
Moina sp.	–	+	+	–	–	–	–	+	+	–	+	–	–	–	–	–	–	–	–	–	–	–	–	–
Sida sp.	+	–	–	–	–	–	–	–	–	+	+	–	–	–	–	–	–	–	–	–	–	–	–	–

0–5 Ind/L=+
6–10 Ind/L=++
11–15 Ind/L=+++
and above=++++

Table 61 Monthly variations in the population density (Ind/L) of the Crustacea at the study site-II of the Dal lake

Taxa	Sep	Oct	Nov	Dec	Jan	Feb	Mar	Apr	May	Jun	Jul	Aug	Sep	Oct	Nov	Dec	Jan	Feb	Mar	Apr	May	Jun	Jul	Aug
Daphnia sp.	–	–	+	+	–	+	+	+	+	+	–	–	–	+	+	+	+	+	+	+	+	++	+	–
Acroperus harpae	–	–	–	–	–	–	–	–	–	–	–	–	+	–	–	–	–	–	–	–	+	–	–	–
Alona sp.	–	–	–	–	–	–	–	–	–	–	–	–	–	–	–	–	–	–	–	–	–	+	–	–
Alonella exigua	–	–	–	–	–	–	–	–	++	+	+	–	–	–	–	–	–	–	–	–	++	+	+++	–
Bosmina longirostris	+	+	+	+	–	+	–	–	–	++	+	+	+	–	–	+	+	–	–	–	+	++	++	++
Chydorus sp.	–	–	–	–	–	–	–	–	–	+	+	–	–	–	–	–	–	–	–	–	–	–	+	–
Cyclops scutifer	–	–	–	–	–	–	–	–	–	–	++	–	+	–	–	–	–	–	–	–	–	+	–	–
Cyclops sp.	–	–	–	–	–	–	–	–	–	++	+	–	–	–	–	–	–	–	–	–	–	++	+++	+
Diaptomous sp.	–	–	–	–	–	++	+	+	–	–	+++	–	–	–	–	–	–	+	+	+	+	++	+++	–
Eucyclops sp.	+	+	–	–	–	–	–	–	–	–	–	+	++	+	–	–	–	–	–	–	–	–	–	+
Moinadaphnia sp.	+	+	+	+	+	–	–	–	+	–	–	+	+	+	+	–	+	–	–	–	–	–	–	+
Moina sp.	–	–	+	–	–	–	–	–	–	–	–	–	–	–	–	+	–	–	–	–	–	–	–	–
Sida sp.	–	–	+	+	–	–	–	–	–	–	–	–	–	+	+	–	+	–	–	–	–	–	–	–

0–5 Ind/L = +
6–10 Ind/L = ++
11–15 Ind/L = +++
and above = ++++

Table 62 Monthly variations in the population density (Ind/L) of the Crustacea at the study site-III of the Dal lake

Taxa	Sep	Oct	Nov	Dec	Jan	Feb	Mar	Apr	May	Jun	Jul	Aug	Sep	Oct	Nov	Dec	Jan	Feb	Mar	Apr	May	Jun	Jul	Aug
Daphnia sp.	–	–	–	–	–	–	+	–	+	–	++	–	+	–	–	–	+	–	–	+	–	++	–	–
Acroperus harpae	–	–	–	–	–	–	–	–	–	–	–	–	–	–	–	–	–	–	–	–	–	+	–	–
Alona sp.	–	–	–	–	–	–	–	–	–	–	+	–	–	–	+	–	–	–	–	–	–	+	+	–
Alonella exigua	–	–	–	–	–	–	–	–	–	+	–	–	–	–	–	–	–	–	–	–	–	+	+	–
Bosmina longirostris	–	–	–	–	–	–	–	–	–	++	+	–	–	–	–	–	–	–	–	–	+++	+	++	–
Chydorus sp.	–	–	–	–	–	–	–	–	–	–	+	+	–	–	–	–	–	–	–	–	–	+	–	–
Cyclops scutifer	–	–	–	–	–	–	–	–	+	–	+	+	–	–	–	–	+	–	–	–	–	–	+	–
Cyclops sp.	–	+	–	–	–	–	–	–	+	++	+	+++	–	–	–	–	–	–	–	–	–	+++	+	–
Diaptomous sp.	–	–	–	–	–	–	–	–	+	–	–	–	–	–	–	–	–	+	–	++	–	++	++++	–
Eucyclopssp.	–	–	–	–	–	–	–	–	–	+	–	–	–	–	–	–	–	–	–	–	–	–	+	+
Moinadaphnia sp.	–	–	–	+	–	–	+	–	–	–	++	+	–	–	–	–	+	–	–	–	–	–	–	–
Moina sp.	–	–	–	–	+	–	–	–	–	+	+	–	–	–	–	–	–	–		–	–	++	–	–
Sida sp.	–	–	+	+	+	–	–	–	–	–	+	–	–	–	–	+	+	+	–	–	–	–	–	–

0–5 Ind/L = +
6–10 Ind/L = ++
11–15 Ind/L = +++
and above = ++++

Table 63 Monthly variations in the population density (Ind/L) of the Crustacea at the study site-IV of the Dal lake

Taxa	Sep	Oct	Nov	Dec	Jan	Feb	Mar	Apr	May	Jun	Jul	Aug	Sep	Oct	Nov	Dec	Jan	Feb	Mar	Apr	May	Jun	Jul	Aug
Daphnia sp.	–	–	–	–	–	–	+	–	+	–	++	–	+	–	–	–	+	–	–	+	–	++	–	–
Acroperus harpae	–	–	–	–	–	–	+	–	–	–	–	–	–	–	–	–	–	–	+	+	–	–	–	–
Alona sp.	–	–	–	–	–	–	–	–	–	+	+	–	–	–	+	–	–	–	–	–	–	+	+	–
Alonella exigua	–	–	–	–	–	–	–	–	–	–	–	+	–	–	–	–	–	–	–	–	–	+	–	–
Bosmina longirostris	–	–	–	–	+	–	–	–	–	+	–	–	–	–	–	–	–	–	–	–	++	++	+	–
Chydorus sp.	–	–	–	–	–	–	–	–	–	–	+	+	–	–	–	–	–	–	–	–	–	+	–	–
Cyclops scutifer	–	–	–	–	–	–	–	–	+	–	+	+	–	–	–	–	+	–	–	+	–	–	+	–
Cyclops sp.	+	–	–	–	–	–	–	+	+	++	++	+	–	–	–	–	–	–	–	–	–	+++	+	+
Diaptomous sp.	–	–	–	–	–	–	–	–	–	–	+	+	–	–	–	–	–	+	–	+	–	+	++	–
Eucyclops sp.	–	–	–	–	–	–	–	–	–	+	–	–	–	–	–	–	–	–	–	–	–	–	+	+
Moinadaphnia sp.	–	–	–	–	–	–	+	–	–	–	+	–	–	–	–	–	–	+	+	–	–	–	–	–
Moina sp.	–	–	–	–	–	–	–	–	–	–		–	–	–	–	–	–	–		+		–	+	
Sida sp.	–	–	–	+	–	–	–	–	–	–	+	+	–	–	–	+	–	–	–	–	–	+	+	–

0–5 Ind/L = +
6–10 Ind/L = ++
11–15 Ind/L = +++
and above = ++++

were found to be present in their studies have not been recovered during the present investigation. *Mesocyclops* sp. has been found in Ropar reservoir at clean and slightly polluted water sites by Sajeev (1999).

Zoobenthos

Zoobenthic inverbrate fauna of the Dal lake revealed representatives belonging to only Annelida (100 %) which included *Placobdella* sp. and *Haemopis* sp. (Hirudinea); *Allonais* sp., *Tubifex* sp. and *Nais communis* (Oligochaetes). Out of these, 2 taxa were recorded at site-I, 4 at site-II and 3 each at site-III and site-IV. The insects were conspicuous by their absence at the study area of the Dal lake which is likely to be due to addition of enormous amount of sewage and agricultural runoff. Kumar (1996a, b) also observed a similar phenomenon in a pond receiving sewage. He found that oligochaetes contributed predominantly towards the benthos and attributed the dominance of oligochaetes to the direct sewage discharges which eliminate other intolerant species including some insects also. According to Odum (1971), oligochaetes are common inhabitants of water polluted with sewage. Scott (1966), Brusven and Prather (1974), McClelland and Brusven (1980) and Farrow et al. (1983) have opined that higher levels of suspension can alter and affect aquatic insects by altering biochemical conditions, food resources, respiratory diffusion, and habitat. Khan (1993) has also showed fewer fauna at the sites with pollution inputs. Cummins (1966) suggested that the sediment load influences the substrate composition the most. Another probable and attention-grabbing phenomenon in the present study was the impact of motorized water boats which are being frequently used by the security forces for patrolling in the Dal lake and by the tourists as water sport. The oil and petrol leakages enter the lake water and might have damaging effect on insect community. Cordery (1976) and Angino et al. (1992) have also opined that a hydrocarbon leakage stresses the insect fauna. Mason (1981) opined that when the environment becomes stressed under the impact of a particular pollutant, the species sensitive to that particular stress are eliminated. At the same time, certain other species may be favoured so that they either dominate or eliminate the intolerant species..

The zoobenthic fauna which was composed of annelids only showed a single peak in species diversity and population density observed during summer. Kumar (1997b) found oligochaetes to be the second most abundant group of benthic population in a river and observed their highest number in November. Mandal and Moitra (1975) also reported population peak in November. Dhillon et al. (1993a), Syal (1996) and Yousuf et al. (2002). Munawar (1970a, b), Das (1979), Rai and Datta Munshi (1979), Chowdhary (1984), Sharma (1986) and Kaushik et al. (1991), on the other hand, failed to record any seasonal trend in the population density of zoobenthic organisms. Jindal and Singh (2005) recorded maximum peak in monsoon. Chauhan (1988) reported the maximum abundance in winter and minimum in summer (Figs. 41 and 42).

The zoobenthic taxa recorded at site-I include two annelids represented by *Allonais* sp. and *Tubifex* sp. This site is located in the lake where water is least inter-

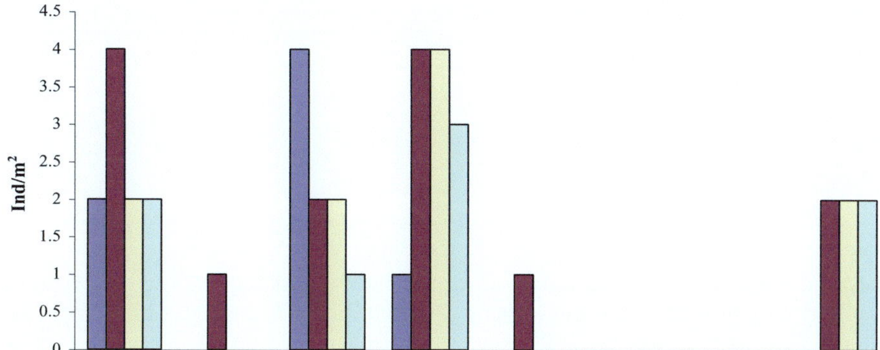

	Autumn	Winter	Spring	Summer	Autumn	Winter	Spring	Summer
Site-I	2		4	1				
Site-II	4	1	2	4	1			2
Site-III	2		2	4				2
Site-IV	2		1	3				2

Season

Fig. 41 Seasonal variations in species diversity of Zoobenthos at the investigated sites of the Dal lake

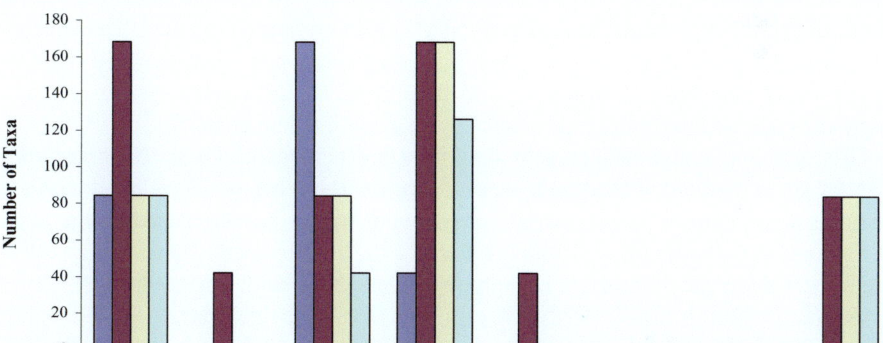

	Autumn	Winter	Spring	Summer	Autumn	Winter	Spring	Summer
Site-I	84		168	42				
Site-II	168	42	84	168	42			84
Site-III	84		84	168				84
Site-IV	84		42	126				84

Season

Fig. 42 Seasonal variations in population density (number of organisms/m^2) of the Zoobenthos at the investigated sites of the Dal lake

fered by human habitation and does not directly receive domestic sewage or any other type of waste. There are thick mats of only submerged type of vegetation and the water is crystal clear.

The zoobenthic taxa recorded at site-II include four annelids represented by *Placobdella* sp. *Haemopis* sp. (Hirudinea); *Nais communis* and *Tubifex* sp. Oligochaeta). This site is located in the lake where water is occupied by floating

Table 64 List of Zoobenthic Taxa recorded at the investigated sites of the Dal Lake

TAXA	SITE-I	SITE-II	SITE-III	SITE-IV
Allonais sp.	+	–	–	–
Tubifex sp.	+	+	+	+
Nais communis	–	+	+	+
Placobdella sp.	–	+	+	+
Haemopis sp.	–	+	–	–

gardens. The site receives agricultural run-off, harbours rich growth of macrophytes and thus is rich in nutrients and organic deposits. Jindal and Singh (2005) recorded oligochaetes more in number at Sangal Sohal pond and opined that it might be due to its depth and availability of decomposed matter. The species diversity and density of zoobenthic fauna was maximum at this site. Kumar (1996a, b, 1997b) and Chukwu and Nwankwo (2004) have observed the presence of annelids in higher density in organically rich waters. Krecker (1939), Andrews and Hasler (1943), Tonapi (1980), Sharma (1988), Kaushik et al. (1991) have opined that macrophytes provide protection to the benthic fauna and they thrive well under nutrient rich conditions.

The zoobenthic taxa recorded at site-III include three annelids namely *Placobdella* sp. *Tubifex* sp. and *Nais communis.* The same annelid composition is observed at site-IV. These two sites receive sewage from houseboats and hotels. The water at both these sites is more organically rich as compared to site-I. The species diversity and population density was higher than site-I but lower than site-II. The taxa present here have been reported to inhabit such places where water receives organic waste and harbours rich macrophytic growth (Pennak 1978).

The only zoobenthic taxa present at all the sites include *Tubifex* sp. (Oligochaeta). *Tubifex* sp. is reported to feed on bacteria down to 10 cm below the sediment surface (Brinkurst and Jamison 1971). It can tolerate severe organic pollution and thus has been recognized as pollution tolerant annelid (Purdy 1922 and Hellawell 1986). Lang (1989) regarded *Tubifex sp.* an indicator of eutrophy in lake Neuchatel (Switzerland).

Kumar (1996a, b, 1997, 1998) also reported *Tubifex* sp. in sewage polluted water. Bath (1996) recorded this genus at region of the Harike lake where water was slow and rich in macrophytes. *Allonais* sp. was reported only at site-I.

The zoobenthic taxa present at site-II, III and IV include *Placobdella* sp. belonging to Hirudinea and *Nais communis* belonging to oligochaeta. Here the lake water is rich in nutrients, lake bed is muddy, and the water current is serene. These conditions seem favourable for their growth. Both these taxa have been found to inhabit water which is rich in organic matter and harbours aquatic macrophytes (Bath 1996). Pennak (1978) found *Placobdella* sp. at the places which are protected by macrophytes and reported that *Placobdella* sp. can withstand some degree of pollution. *Nais communis* is reported to tolerate severe organic pollution (Hellawell 1986). Kumar (1997b) has reported this from a shallow eutrophic pond in Bihar (Table 64).

Bioindicators From the studies of the physico-chemical parameters, it is concluded that there is an increase in calcium, magnesium, chloride, total alkalinity, total hardness, sodium, ammomical nitrogen and total phosphate phosphorus at site-II, III and IV due to addition of agricultural run-off from floating gardens and sewage effluents from house-boats and hotels. The dissolved oxygen has been recorded to decrease at site-III and IV. Due to high ionic concentration of these salts and presence of rich growth of macrophytes, water at sites receiving agricultural waste and sewage effluents is more turbid, less transparent and shows high TDS values. On the basis of total phosphate phosphorus and nitrate nitrogen values, the lake has been categorized as grossly eutrophicated and highly productive at site-II, III and IV. The studies on the phytoplankton, zooplankton and zoobenthos also indicate that site-II, III and IV are more productive as compared to site-I.

Based on distribution of phytoplankton, zooplankton and zoobenthos at different sites, certain bioindicators of clean water and eutrophicated water have been identified which are listed as follows:

Group	Clean water taxa (Site-I)	Polluted water taxa (Site-II, III and IV)
Bacillariophycae	*Aphanocapsa* sp., *Tabellara* sp.	–
Chlorophyceae	*Pediastrum ovatum and P. spinosum*	–
Cyanophyceae	*Merismopedia glauca*	*Oscillatoria princeps*
Euglenophyceae	–	–
Dinophyceae	–	–
Chrysophyceae	–	*Dinobryon* sp.
Protozoa	–	*Arcella vulgaris,*
		Difflugia oblonga, Paraguadrula irregularis, Paramecium sp.
Rotifera	–	*Brachiouns calyciflori, Filinia longiseta,*
		Lecane ohiensis,
		Pompholyx sulcata, Trichocerca sp.
Crustacea	–	*Moinadaphnia* sp.
Zoobenthos	*Allonais* sp.	–

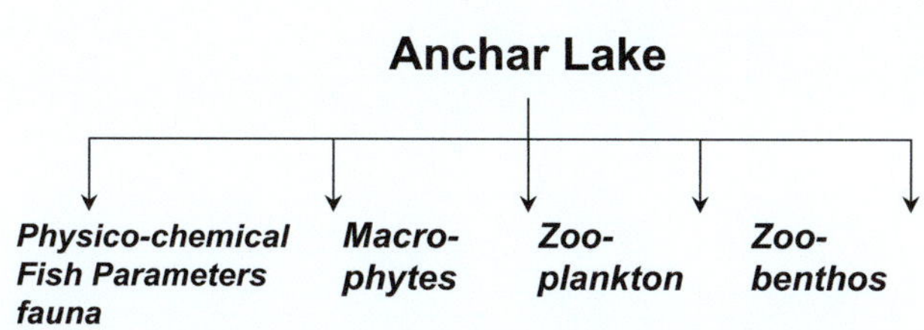

Physico-Chemical Parameters

Water Temperature (°C)

In the Anchar lake, water temperature ranged from 3 to 28 °C over a period of 2 year study (Table 65 and Fig. 43).

Table 65 Monthly fluctuations in water temperature (°C) at the investigated sites of the Anchar lake

	Months																			
2000–2001	**Sep**	**Oct**	**Nov**	**Avg.**	**± S.D.**	**Dec**	**Jan**	**Feb**	**Avg.**	**± S.D.**	**Mar**	**April**	**May**	**Avg.**	**± S.D.**	**June**	**July**	**Aug**	**Avg.**	**± S.D.**
SITE-I	–	12.5	10.0	7.5	1.3	9.0	3.0	4.5	5.5	0.8	8.5	13.0	18.0	13.2	2.5	18.0	16.0	19.0	17.7	1.5
SITE-II	12.0	14.0	9.5	11.8	2.3	9.1	5.0	5.0	6.4	0.0	9.5	14.0	19.0	14.2	2.5	18.5	16.0	19.0	17.8	1.5
SITE-III	13.0	12.0	9.8	11.6	1.1	9.0	5.0	4.5	6.2	0.3	9.5	14.0	19.0	14.2	2.5	17.0	19.0	19.0	18.3	0.0
2001–2002	**Sep**	**Oct**	**Nov**	**Avg.**	**± S.D.**	**Dec**	**Jan**	**Feb**	**Avg.**	**± S.D.**	**Mar**	**April**	**May**	**Avg.**	**± S.D.**	**June**	**July**	**Aug**	**Avg.**	**± S.D.**
SITE-I	16.0	16.0	7.0	13.0	4.5	6.0	2.0	5.0	4.3	1.5	6.0	13.0	19.0	12.7	3.0	28.0	27.0	25.0	26.7	1.0
SITE-II	18.0	18.0	8.0	14.7	5.0	8.0	3.0	5.0	5.3	1.0	7.4	13.0	17.5	12.6	2.3	22.4	18.0	19.0	19.8	0.5
SITE-III	18.0	16.0	8.0	14.0	4.0	6.0	3.0	6.8	5.3	1.9	10.0	16.0	19.5	15.2	1.8	23.0	20.0	25.0	22.7	2.5

Avg.: Average
S.D.: Standard deviation
– : No sampling

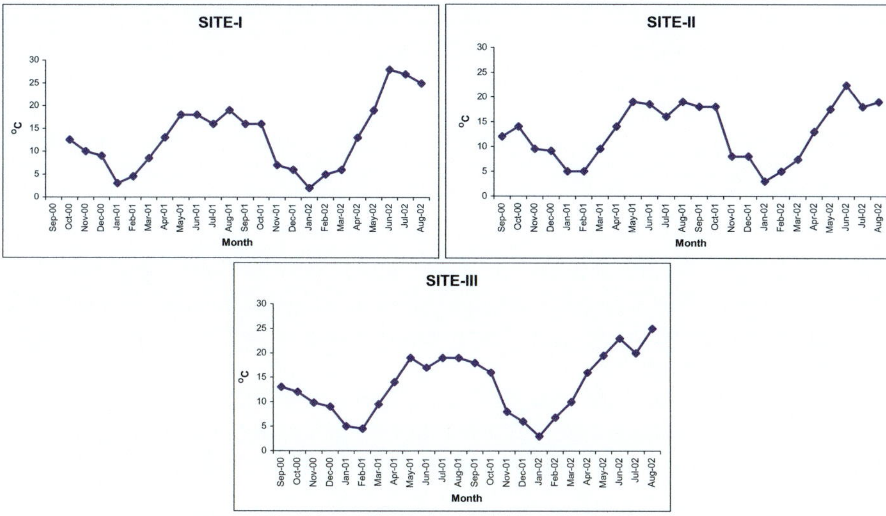

Fig. 43 Monthly fluctuations in water temperature (°C) at the Investigated Sites of the Anchar lake

The maximum water temperature was recorded in the month of June, followed by gradual fall to its minimum in January and then an increase to its maximum in June. In overall terms, the temperature do follow set pattern dictated by seasonal cycle. Similar seasonal variations in temperature has been observed in Kashmir water bodies by Yousuf and Qadri (1986), Sarwar (1987) and Shah (1988) and in other water bodies by Adholia et al. (1991), Welch (1952), Rao (1955), Zafar (1955), Munawar (1970a, b), Hannan and Young (1974), Swarup and Singh (1979), Harshey et al. (1982), Patra and Nayak (1983), Bagde and Verma (1985a, b), Goel et al. (1980), Palharya and Malviya (1988), Shyamsunder (1988), Jindal and Kumar (1993), Bisht (1993), Prakash (1994), Kumar (1995a, b), Bath (1996), Kumar and Singh (1996), Syal (1996), Batcha (1997), Bath and Kaur (1998), Kaur et al. (1999), Roy (2000), and Khabade et al. (2003).

When the temperature variations at various sites of Anchar lake were compared, it was found that site-I showed slightly lower water temperature as compared to other sites. It may be because water at this site is fast flowing while at other sites water is standing.

Site-II which is deep and infested with rich plant growth and receives water from site-III, temperature was slightly higher than the site-I. Jamil (1990) and Dhillon et al. (1993a, b) observed that the macrophytic plants tend to raise the water temperature of a water body by preventing the contact of air with water. At site-III temperature was found to be the highest. This site is also rich in macrophytic growth and receives effluents from human habitation as well as from SKIMS hospital. Jolly and Chapman (1966), Venkateshwarlu and Jayanti (1968), Munawar (1970a, b), Sharma et al. (1981), Mahadevan and Krishnaswamy (1983), Palharya and Malviya (1988), Bath (1996), Syal (1996) and Kaur et al. (2003) recorded an increase in water temperature with addition of effluents. James et al. (1990) have also shown an increase in water temperature due to hospital waste while studying impact of hospital wastes in Tirucherapalii, Tamil Nadu.

Depth (m)

The Anchar lake is situated in the low lying area (1583–1600 m) and in the flood plains of the river Sind which is the major source of water of the lake. Water here is maintained mainly by snow melt. Local precipitation is only a weak contributor towards the water level.

Overall depth in the lake ranges from 0.1 to 1.8 m. No seasonal trend was observed. At site-I, it ranges from 0.1 to 1.8 m and at site-II, from 0.8 to 1.8 m and at site-III, from 0.3 to 1.4 m. Minimum depth was recorded at site-III (Table 66 and Fig. 44).

Site-I is located in the lake where lake is fed by river Sind. The flowing water of river Sind carries along its course silt which gets accumulated at this site resulting in decrease in the lake bed. The gentle slope at this site also helps in the accumulation of silt.

The site-III is located near the point where hospital complex of SKIMS discharges much of its effluents into the lake. Run-off from surrounding agricultural fields and sewage from the adjoining human settlements are also added at this site. The site is characterised by thick growth of macrophytes. Accumulation of decomposed vegetation and silt coming from the surrounding areas have resulted in raising of the bed of the lake at this site.

Site-II is located in the centre of the lake and is the deepest portion of the lake. Water from site-I and site-III mixes in this region of the lake. The site also harbours rich growth of macrophytes.

Earlier studies of Zutshi et al. (1980) recorded a depth of 2.75 m. In the present investigation maximum depth is recorded to be 1.8 m.A comparison clearly indicates the raising of the Anchar lake bed owing to the factors described above.

Water Transparency (m)

Transparency values in the Anchar lake fluctuated from 0.1 to 1.8 m. From the studies no seasonal trend was observed. The transparency was found to be ranging from 0.1 to 1.8 m at site-I, from 0.8 to 1.8 m at site-II and from 0.1 to 1.3 m at site-III (Table 67 and Fig. 45).

The higher values of transparency at site-I is due to absence of any macrophytic growth and less anthropogenic activities. At site-II and III the transparency was found to be decreased. The reduction in water transparency at site-II and III is due to the silt loaded inflow from the river Sind and dense macrophytic growth. These

Table 66 Monthly fluctuations in water depth (m) at the investigated sites of the Anchar lake

	Months																		
2000–2001	**Sep**	**Oct**	**Nov**	**Avg.**	**± S.D.**	**Dec**	**Jan**	**Feb**	**Avg.**	**± S.D.**	**Mar**	**April**	**May**	**Avg.**	**± S.D.**	**June**	**July**	**Aug**	**Avg.**	**± S.D.**
SITE-I	–	1.7	1.8	1.2	0.1	1.7	0.3	0.9	1.0	0.3	1.1	1.0	0.4	0.8	0.3	0.4	0.4	0.5	0.5	0.1
SITE-II	1.7	1.7	1.7	1.7	0.0	1.7	1.3	1.6	1.5	0.1	1.7	1.9	1.1	1.5	0.4	1.0	1.0	1.1	1.0	0.1
SITE-III	0.6	0.6	0.7	0.6	0.1	0.6	0.3	0.6	0.5	0.1	0.7	0.7	0.8	0.7	0.0	0.7	0.7	1.1	0.8	0.2
2001–2002	**Sep**	**Oct**	**Nov**	**Avg.**	**± S.D.**	**Dec**	**Jan**	**Feb**	**Avg.**	**± S.D.**	**Mar**	**April**	**May**	**Avg.**	**± S.D.**	**June**	**July**	**Aug**	**Avg.**	**± S.D.**
SITE-I	0.3	0.1	0.2	0.2	0.0	0.2	0.3	0.8	0.5	0.3	1.0	0.8	0.7	0.8	0.1	0.6	0.5	0.4	0.5	0.1
SITE-II	1.3	0.9	0.8	1.0	0.0	1.0	0.9	1.7	1.2	0.4	1.7	1.8	1.7	1.8	0.0	1.3	1.2	1.2	1.2	0.0
SITE-III	0.5	0.4	0.4	0.4	0.0	0.5	0.4	0.7	0.5	0.1	1.4	0.7	0.7	0.9	0.0	0.7	0.7	0.8	0.8	0.0

Avg.: Average
S.D.: Standard deviation
– : No sampling

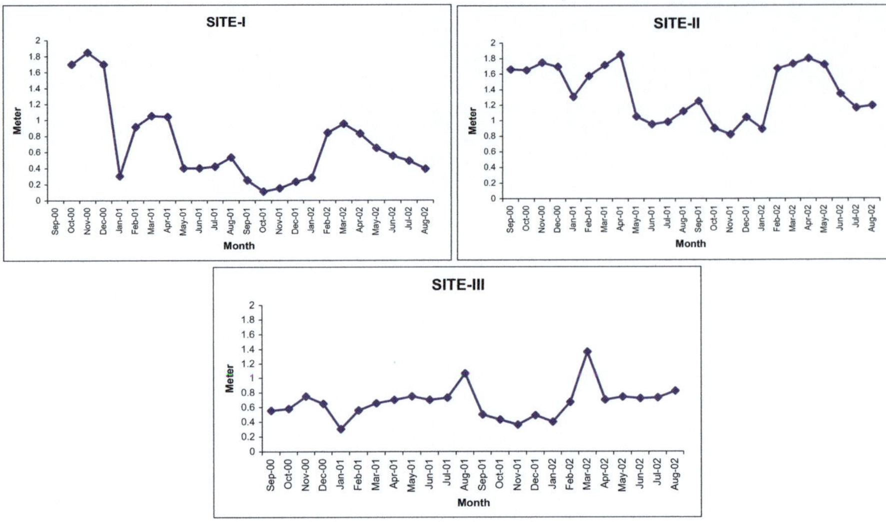

Fig. 44 Monthly fluctuations in water depth (m) at the investigated sites of the Anchar lake

two factors combined with anthropogenic activities further decrease the transparency. Zutshi et al. (1980) attributed the decrease in light penetration (transparency) in Anchar lake due to silt coming from the river Sind. Khan and Zutshi (1980) also indicated the turbid nature of water mainly due to the suspended material brought in by the feeding channel for Nilnag lake.

According to Yoshimura (1933), the lakes showing transparency values of 1.50 m or less exhibit a tendency towards eutrophication. The transparency values recorded in the investigated lake indicate its eutrophic nature.

Earlier Zutshi et al. (1980) recorded the transparency of the lake ranging from 0.25 to 1.75 m. A comparison clearly indicates that the transparency has decreased over these years owing to the factors described above.

Turbidity (NTU)

The turbidity values ranged from 2 to 32 NTU in the Anchar lake water over a period of 2 year study. No seasonal trend in turbidity of water is recorded. At site-I, the values ranged from 2NTU to 6NTU, at site-II from 5NTU to 17 NTU and at Site-III, from 13 to 32 NTU (Table 68 and Fig. 46).

A comparison of values at three selected sites of the lake reveals the following trend.

Site III > Site II > Site I

The maximum turbidity is found at site-III which receives hospital waste and sewage from the surroundings areas. This site also harbours rich growth of macro-

Table 67 Monthly fluctuations in water transparency (m) at the investigated sites of the Anchar lake

	Months																		
2000–2001	**Sep**	**Oct**	**Nov**	**Avg.**	**± S.D.**	**Dec**	**Jan**	***Feb***	**Avg.**	**± S.D.**	**Mar**	**April**	**May**	**Avg.**	**± S.D.**	**June**	**July**	**Aug**	**Avg.**	**± S.D.**
SITE-I	–	1.7	1.8	1.2	0.0	1.7	0.3	0.9	1.0	0.3	1.0	1.0	0.4	0.8	0.3	0.4	0.4	0.5	0.4	0.1
SITE-II	1.6	1.6	1.7	1.6	0.1	1.6	1.3	1.5	1.5	0.1	1.7	1.8	1.0	1.5	0.4	0.9	0.9	1.1	1.0	0.1
SITE-III	0.5	0.5	0.7	0.6	0.1	0.6	0.3	0.5	0.5	0.1	0.6	0.7	0.7	0.7	0.0	0.7	0.7	1.0	0.8	0.2
2001–2002	**Sep**	**Oct**	**Nov**	**Avg.**	**± S.D.**	**Dec**	**Jan**	**Feb**	**Avg.**	**± S.D.**	**Mar**	**April**	**May**	**Avg.**	**± S.D.**	**June**	**July**	**Aug**	**Avg.**	**± S.D.**
SITE-I	0.2	0.1	0.1	0.1	0.0	0.2	0.2	0.8	0.4	0.3	0.9	0.8	0.6	0.8	0.1	0.5	0.4	0.3	0.4	0.1
SITE-II	1.2	0.9	0.8	1.0	0.0	1.0	0.8	1.6	1.1	0.4	1.7	1.8	1.7	1.7	0.0	1.3	1.1	1.1	1.2	0.0
SITE-III	0.5	0.4	0.3	0.4	0.1	0.4	0.4	0.6	0.5	0.1	1.3	0.7	0.7	0.9	0.0	0.7	0.7	0.8	0.7	0.1

Avg.: Average
S.D.: Standard deviation
– : No sampling

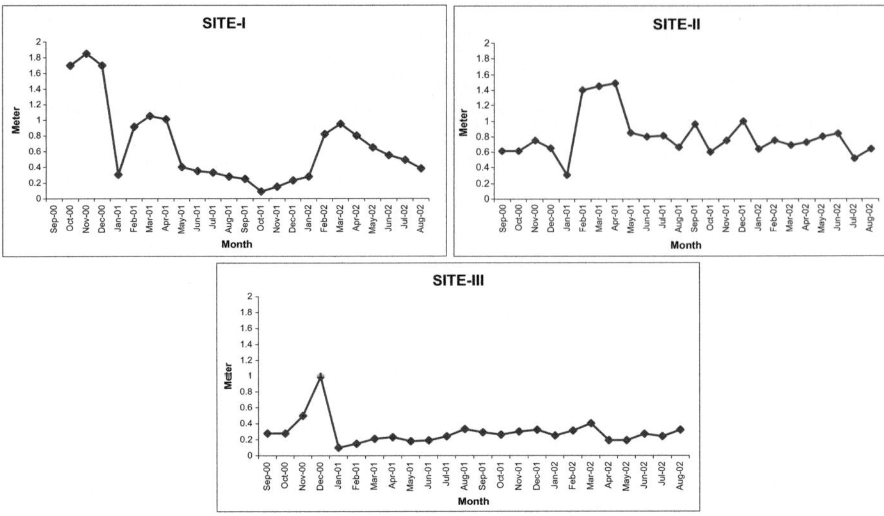

Fig. 45 Monthly fluctuations in water transparency (m) at the investigated sites of the Anchar lake

phytes. Both these factors add organic matter to water. David (1956), Sengar and Sing (1986), Pagioro et al. (1997), Maruthi et al. (2000), Thorat and Sultana (2000), Das et al. (2003) and Sharma and Verma (2003) also recorded increase in turbidity due to sewage and organic matter. Moreover, water at this site is frequently disturbed due to plying of boats and other agricultural activities which result in the mixing of mud with water and thus increase the turbidity. Pagioro et al. (1997) and Thorat and Sultana (2000) observed increase in turbidity in shallow water bodies due to frequent disturbances leading to re-suspension of solids. Khan and Zutshi (1980) and Wanganeo et al. (1996) attributed high turbid nature of water to suspended material brought in by the feeding channel.

ISI (1986) has suggested the maximum turbidity value for drinking/potable water as 10NTU. Based on this, the water at site-I, remained well within the permissible limit, at site-II (5–17), it exceeded the limit occasionally while at site-III, it was found to cross the limit for most of the period of study.

pH

The pH values in the Anchar lake ranged from 6.9 to 8.6.

A slight increase in pH value was observed in spring and summer at site-I but such trend was not evident at site-II and site-III. However, works of Zutshi and Vass (1978), Khan and Zutshi (1980) and Wanganeo et al. (1996) for Kashmir lakes and Mohanty (1975), Sharma and Dheneshwar (1986), Mishra (1988), Bandopadhyay and Gopal (1991), Jindal and Kumar (1993), Khalique and Afser (1995), Islam (1996) and Narain and Chauhan (2000) for other lakes reported high pH values dur-

Table 68 Monthly fluctuations in water turbidity (N.T.U.) at the investigated sites of the Anchar lake

	Months																			
2000–2001	**Sep**	**Oct**	**Nov**	**Avg.**	**± S.D.**	**Dec**	**Jan**	**Feb**	**Avg.**	**± S.D.**	**Mar**	**April**	**May**	**Avg.**	**± S.D.**	**June**	**July**	**Aug**	**Avg.**	**± S.D.**
SITE-I		2.0	4.0	2.0	1.0	4.0	2.0	2.0	2.7	0.0	2.0	2.0	2.0	2.0	0.0	3.0	3.0	4.0	3.3	0.5
SITE-II	5.0	5.0	10.0	6.7	2.5	10.0	10.0	8.0	9.3	1.0	5.0	5.0	5.0	5.0	0.0	5.0	9.0	11.0	8.3	1.0
SITE-III	14.0	13.0	15.0	14.0	1.0	15.0	20.0	20.0	18.3	0.0	16.0	29.0	25.0	23.3	2.0	25.0	25.0	26.0	25.3	0.5
2001–2002	**Sep**	**Oct**	**Nov**	**Avg.**	**± S.D.**	**Dec**	**Jan**	**Feb**	**Avg.**	**± S.D.**	**Mar**	**April**	**May**	**Avg.**	**± S.D.**	**June**	**July**	**Aug**	**Avg.**	**± S.D.**
SITE-I	6.0	5.0	5.0	5.3	0.0	5.0	5.0	4.0	4.7	0.5	4.0	4.0	5.0	4.3	0.5	4.0	4.0	4.0	4.0	0.0
SITE-II	10.0	17.0	14.0	13.7	1.5	13.0	12.0	7.0	10.7	2.5	5.0	6.0	5.0	5.3	0.5	6.0	8.0	12.0	8.7	2.0
SITE-III	28.0	32.0	25.0	28.3	3.5	22.0	20.0	25.0	22.3	2.5	21.0	26.0	28.0	25.0	1.0	25.0	25.0	27.0	25.7	1.0

Avg.: Average
S.D.: Standard deviation
– : No sampling

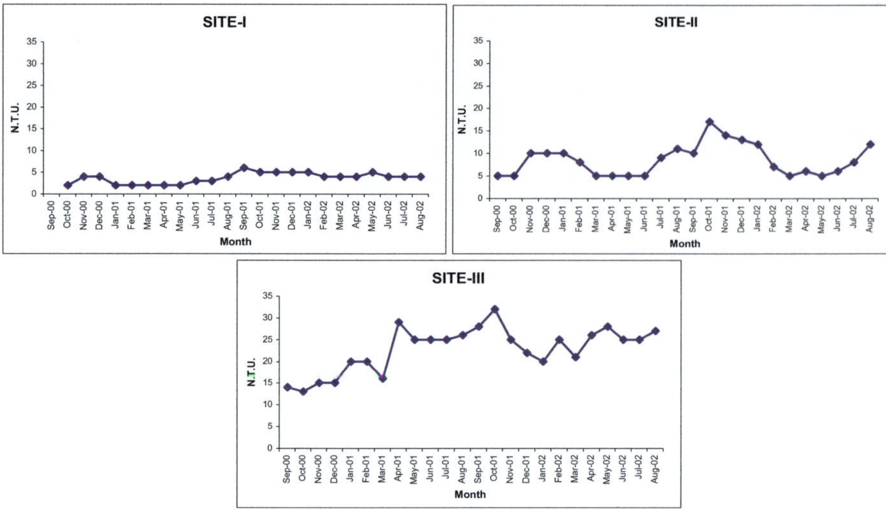

Fig. 46 Monthly fluctuations in water turbidity (N.T.U.) at the investigated sites of the Anchar lake

ing the summer months. On the other hand, Sarwar (1999) for Kashmir lakes and Bath (1996), Syal (1996) and Valarmathi et al. (2002) for other lakes failed to record any seasonal trend in pH values.

The pH values of water in the Anchar lake show highest value of 8.6 at site-III in November, 2000 and the lowest of 6.9 at the site-III in October 2001. It ranged from 7.1 to 8.5 both at site-I and site-II and 6.9 to 8.6 at site-III (Table 69 and Fig. 47).

The pH values indicate water to be on alkaline side in the Anchar lake. The alkaline nature of water is attributed to the macrophytic infestation at several regions of the lake. Qadri and Yousuf (1978), Khan (1979), Sarwar (1987, 1999) and Wanganeo et al. (1996) in various Kashmir lakes related alkaline pH of water to rich and intense macrophytes vegetation. Sarwar and Majid (1997) worked on the Wular lake of Kashmir and attributed the alkaline nature of the lake to the presence of calcium rich rocks in the catchments area. Sarwar (1999) attributed it to higher silicate content in the rocky beds and sediments due to the production of silicic acid by hydrolysis of the silicate. The alkaline nature of other water bodies have also been reported by Bist (1986), Prakash (1994), Bath (1996) and Khabade et al. (2003).

Dissolved Oxygen (mg/L)

Dissolved Oxygen values fluctuated from 0.5 to 9.5 mg/L in the Anchar lake. A look at the data of the 2 year study reveals that dissolved oxygen contents of water at all the selected sites of the lake were higher in winter and lower in summer. Lower

Table 69 Monthly fluctuations in pH at the investigated sites of the Anchar lake

	Months																		
2000–2001	**Sep**	**Oct**	**Nov**	**Avg.**	**± S.D.**	**Dec**	**Jan**	**Feb**	**Avg.**	**± S.D.**	**Mar**	**April**	**May**	**Avg.**	**± S.D.**	**June**	**July**	**Aug**	**Avg.**	**± S.D.**
SITE-I		7.5	7.1	4.9	0.2	7.4	7.1	7.5	7.3	0.2	8.5	8.5	8.5	8.5	0.0	8.3	7.9	7.4	7.9	0.3
SITE-II	8.0	8.1	8.1	8.1	0.0	8.5	7.4	7.2	7.7	0.1	8.1	8.1	8.1	8.1	0.0	8.1	7.5	7.4	7.7	0.0
SITE-III	7.3	7.2	8.6	7.7	0.7	7.9	7.2	7.5	7.5	0.2	7.6	7.3	7.5	7.5	0.1	7.5	7.3	7.1	7.3	0.1
2000–2001	**Sep**	**Oct**	**Nov**	**Avg.**	**± S.D.**	**Dec**	**Jan**	**Feb**	**Avg.**	**± S.D.**	**Mar**	**April**	**May**	**Avg.**	**± S.D.**	**June**	**July**	**Aug**	**Avg.**	**± S.D.**
SITE-I	7.5	7.1	7.4	7.3	0.2	7.1	7.5	7.4	7.3	0.0	7.5	7.5	7.1	7.4	0.2	7.9	7.6	7.7	7.7	0.0
SITE-II	7.5	7.1	7.6	7.4	0.3	7.6	7.6	7.9	7.7	0.2	7.8	7.9	7.8	7.8	0.1	7.8	7.9	7.7	7.8	0.1
SITE-III	7.3	6.9	7.2	7.1	0.2	7.5	7.4	8.1	7.7	0.4	7.7	7.4	7.6	7.6	0.1	7.9	7.8	7.5	7.7	0.1

Avg.: Average
S.D.: Standard deviation
– : No sampling

Fig. 47 Monthly fluctuations in pH at the investigated sites of the Anchar lake

contents of dissolved oxygen during summer is due to lower oxygen holding capacity of water and higher rate of decomposition of organic matter at high temperature. In winter, oxygen holding capacity of water gets increased due to low temperature and thus the dissolved oxygen contents are increased. Furthermore rate of decomposition of organic matter by micro-organisms is lowered during winter. Hutchinson (1957), Chakrabarty et al. (1959), Reid (1961), Pahwa and Mehrotra (1966), Ray et al. (1966), Vasisht and Sharma (1975), Badola and Singh (1981), Bhowmick and Singh (1985), Bandopadhyay and Gopal (1991), Datta Munshi and Singh (1991), Shastree et al. (1991) and Esmaeili and Johal (2005) have also given similar explanations for lower dissolved oxygen values during summer and higher during winter. A sharp increase in dissolved oxygen in the spring of 2001 may be due to rainfall during these months as is apparent from the data on precipitation obtained from office of the Director Meteorological Centre Rambagh, Srinagar. Verma et al. (1978), Mishra (1988) and Shastree et al. (1991) also observed an increase is dissolved oxygen values during the rainy months. Mishra (1988) contended that it was due to the inflow of highly oxygenated flood waters during the rainy period.

Dissolved oxygen contents at site-I ranged from 5.8 to 9.5 mg/L, at Site-II from 2.8 to 8.9 mg/L and at Site-III from 0.5 to 4.8 mg/L (Table 70 and Fig. 48).

The data reveals the following trend.

$$\text{Site III} < \text{Site II} < \text{Site I}$$

At site-I, water holds good amount of dissolved oxygen. This sampling site is located where the Anchar lake is fed by the Sind river. The site is characterised by shallow and flowing water and is least interfered by human habitation and does not receive domestic sewage or any other type of waste. The water here is crystal clear,

Table 70 Monthly fluctuations in dissolved oxygen (mg/L) at the investigated sites of the Anchar lake

	Months																		
2000–2001	**Sep**	**Oct**	**Nov**	**Avg.**	**± S.D.**	**Dec**	**Jan**	**Feb**	**Avg.**	**± S.D.**	**Mar**	**April**	**May**	**Avg.**	**± S.D.**	**June**	**July**	**Aug**	**Avg.**	**± S.D.**
SITE-I	–	5.8	6.3	4.0	0.3	7.9	8.4	7.9	8.1	0.3	8.9	9.5	8.5	9.0	0.5	6.5	5.9	5.9	6.1	0.0
SITE-II	8.4	8.9	6.4	7.9	1.3	7.5	7.2	7.0	7.2	0.1	3.5	8.5	5.5	5.8	1.5	4.0	3.5	1.5	3.0	1.0
SITE-III	2.6	3.0	1.5	2.4	0.7	2.5	2.5	2.3	2.4	0.1	2.9	2.1	1.9	2.3	0.1	1.6	1.5	1.9	1.7	0.2
2001–2002	**Sep**	**Oct**	**Nov**	**Avg.**	**± S.D.**	**Dec**	**Jan**	**Feb**	**Avg.**	**± S.D.**	**Mar**	**April**	**May**	**Avg.**	**± S.D.**	**June**	**July**	**Aug**	**Avg.**	**± S.D.**
SITE-I	5.9	6.8	7.8	6.8	0.5	8.9	8.5	7.2	8.2	0.7	7.0	5.8	6.0	6.3	0.1	6.3	5.8	5.8	6.0	0.0
SITE-II	5.4	2.7	5.8	4.6	1.6	8.5	8.5	4.5	7.2	2.0	3.3	5.4	5.9	4.9	0.3	3.8	4.2	2.8	3.6	0.7
SITE-III	2.3	0.5	2.5	1.8	1.0	2.8	2.8	2.7	2.8	0.0	5.0	4.5	4.8	4.8	0.1	3.3	2.5	3.2	3.0	0.4

Avg.: Average
S.D.: Standard deviation
– : No sampling

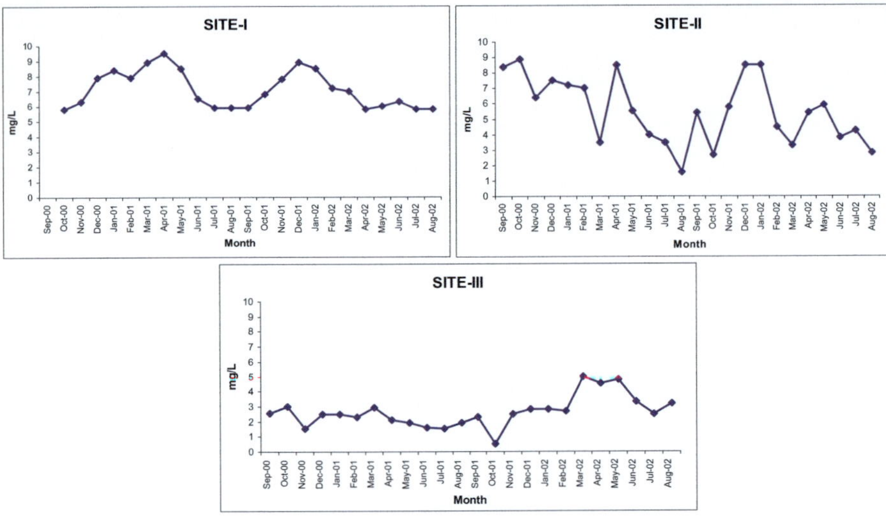

Fig. 48 Monthly fluctuations in dissolved oxygen (mg/L) at the investigated sites of the Anchar lake

and is free from any type of macrophytic vegetation and thus is well aerated. The addition of atmospheric oxygen through aeration to the water is also suggested by Roy (2000) and Khatri and Dhankhar (2003).

Site-III is characterized by standing water with rich growth of emergent and floating type of macrophytes. The surface cover by *Salvinia* sp. and *Lemna* sp. lower the dissolved oxygen content of water at this site by restricting the contact of air with water. Reduction in the oxygen content of Kashmir lake waters due to *Salvinia* cover has also been reported by Zutshi and Vass (1971). Furthermore this site receives sewage from the adjoining inhabited areas, agricultural runoff from the catchment area and hospital effluents from S.K. Institute of medical Sciences. This results in enrichment of water with organic matter. The decomposition of organic matter consumes much of dissolved oxygen, leading to a sharp decline in dissolved oxygen of water. Gonzalves and Joshi (1946), Vasisht and Sharma (1975) have also suggested that presence of organic matter lowers the dissolved oxygen contents of water. Pahwa and Mehrotra (1966), Badola and Singh (1981), Bhatt et al. (1988), Palharya and Malviya (1988), Rana and Palria (1988) and Goel and Chavan (1991) reported the lowering of dissolved oxygen content of water contaminated with sewage discharges. The another reason for reduced dissolved oxygen at this site may be due to high turbidity as Bhatt et al. (1988) while inter-relating the turbidity, photosynthetic activity and dissolved oxygen suggested that in turbid water, photosynthetic activity is retarded leading to decline in the release of oxygen into water.

Site-II showed comparatively lower dissolved oxygen values than site-I but more than site-III. This site as such does not receive any type of agricultural or sewage effluents but the water from site-I and site-III mixes at this site. Here water is comparatively deeper, and harbours intense growth of macrophytes.

Earlier Sarwar (1999) recorded dissolved oxygen ranging from 3.2 to 13.6 mg/L during his study in 1983. Sarwar and Zutshi (1989) recorded average dissolved oxygen to be 9.8.mg/L. If compared with present studies, a sharp decline in dissolved oxygen in the Anchar lake is recorded.

Total Alkalinity (mg/L)

Total alkalinity values varied from 59 to 315 mg/L at three sites of the Anchar lake.

Seasonal trend is not indicated at site-I. At site-II and III, however, total alkalinity was found to be higher during spring followed by a sharp decline during summer. The high values of alkalinity at Site-II and site-III during spring may be due to leaching of fertilizers from the adjoining cultivation fields. A sharp decline in summer at these two sites may be due to absorption of nutrients by the macrophytes which grow to their maximum during this season. Kaul and Handoo (1980) recorded lower values of total alkalinity from May to September and attributed this to the consumption of dissolved carbonates as a carbon source during rapid assimilation by autotrops during summer.

Total alkalinity ranged from 59 to 165 mg/L at site-I, from 102 to 301 mg/L at site-II and from 137 to 315 mg/L at site-III (Table 71 and Fig. 49).

Overall trend observed in the alkalinity values is as under

Site III > Site II > Site I

Alkalinity was observed to be highest at site-III, moderate at site-II and lowest at site-I. At site-I with least human interference, total alkalinity present is mainly contributed by rocks rich in carbonaceous salts. Gibbs (1970), Sreenivasan (1970), Zutshi et al. (1980), Patra and Nayak (1982), Weiler (1988), Chapman and Kimstach (1992), Meybeek et al. (1992) and Syal (1996) also suggested that alkalinity of water is primarily due to presence of carbonates and bicarbonates which come from weathering of rocks.

Alkalinity values were highest at the site-III where waste from the hospital, agricultural run-off from surrounding fields and sewage from the catchment are added. Water (1957), Sarwar and Zutshi (1987a, b), Sarwar (1991a, b), Sarwar et al. (1996), Valarmathi et al. (2002), Kumar et al. (2003) and Prasannakumari et al. (2003) also observed increase in alkalinity due to addition of sewage. At site-II alkalinity was found to be moderate because water from site-I and site-III mixes up and gets diluted.

Freiser and Fernando (1966) and Qadri and Yousuf (1980a, b) have suggested that H_2CO_3 predominate in water when pH is 6.0 or below 6.0, HCO_3 predominate when the pH is between 6.0 and 10 and CO_3 predominate when pH is above 10. The pH values were found in the range of 6.9–8.5 during present investigation suggesting that the alkalinity at all the sites of the lake was primarily due to bicarbonates. Zutshi et al. (1980) made similar observations for Kashmir lakes.

Table 71 Monthly fluctuations in total alkalinity (mg/L) at the investigated sites of the Anchar lake

	Months																		
2000–2001	**Sep**	**Oct**	**Nov**	**Avg.**	**± S.D.**	**Dec**	**Jan**	**Feb**	**Avg.**	**± S.D.**	**Mar**	**April**	**May**	**Avg.**	**± S.D.**	**June**	**July**	**Aug**	**Avg.**	**± S.D.**
SITE-I	–	115.0	112.0	75.7	1.5	112.0	64.0	59.0	78.3	2.5	65.0	105.0	165.0	111.7	30.0	101.0	127.0	101.0	109.7	13.0
SITE-II	109.0	105.0	120.0	111.3	7.5	102.0	107.0	110.0	106.3	1.5	120.0	195.0	210.0	175.0	7.5	120.0	124.0	160.0	134.7	18.0
SITE-III	210.0	205.0	215.0	210.0	5.0	203.0	217.0	212.0	210.7	2.5	220.0	210.0	315.0	248.3	52.5	175.0	154.0	152.0	160.3	1.0
2001–2002	**Sep**	**Oct**	**Nov**	**Avg.**	**± S.D.**	**Dec**	**Jan**	**Feb**	**Avg.**	**± S.D.**	**Mar**	**April**	**May**	**Avg.**	**± S.D.**	**June**	**July**	**Aug**	**Avg.**	**± S.D.**
SITE-I	120.0	115.0	125.0	120.0	5.0	120.0	105.0	95.0	106.7	5.0	105.0	115.0	125.0	115.0	5.0	127.0	109.0	125.0	120.3	8.0
SITE-II	135.0	140.0	129.0	134.7	5.5	143.0	142.0	241.0	175.3	49.5	301.0	245.0	205.0	250.3	20.0	138.0	110.0	160.0	136.0	25.0
SITE-III	175.0	170.0	160.0	168.3	5.0	180.0	175.0	234.0	196.3	29.5	223.0	240.0	184.0	215.7	28.0	137.0	140.0	149.0	142.0	4.5

Avg. : Average
S. D.: Standard deviation
– : No sampling

Fig. 49 Monthly fluctuations in total alkalinity (mg/L) at the investigated sites of the Anchar lake

Moyle (1946) while classifying lake waters on the basis of the total alkalinity labelled waters having alkalinity upto 40 mg/L as soft, with 40–90 mg/L as medium and above 90 mg/L as hard. On the basis of this categorization, the Anchar lake can be considered as medium to hard water lake with values ranging from 59 to 315 mg/L. Sarwar and Rifat (1991), on the other hand, suggested that total alkalinity level of 40 mg/L provides a dividing line between soft and hard water. Philipose (1960) and Spence (1964) classified the lake as nutrient rich if the bicarbonate alkalinity is more than 60 mg/L. Based on this classification, the Anchar lake water can be considered nutrient rich and productive.

Earlier Sarwar (1999) recorded alkalinity of 58–275 mg/L in the Anchar lake in a study conducted in 1983. Sarwar and Zutshi (1989) recorded average alkalinity to be 149 mg/L. Present study shows the alkalinity values to rise over these years due to various anthropogenic activities.

ICMR (report no. 44) recommended the maximum permissible limits of total alkalinity in water as 120 mg/L. On the basis of these recommendations it can be concluded that total alkalinity of water at all the sites exceeds the limit.

Total Hardness (mg/L)

Total hardness values in the Anchar lake ranged from 73 to 377 mg/L. It varied from 73 to 275 mg/L at site-I, from 81 to 335 mg/L at site-II and from 74 to 377 mg/L at site-III (Table 72 and Fig. 50).

Table 72 Monthly fluctuations in total hardness (mg/L) at the investigated sites of the Anchar lake

	Months																		
2000–2001	**Sep**	**Oct**	**Nov**	**Avg.**	**± S.D.**	**Dec**	**Jan**	**Feb**	**Avg.**	**± S.D.**	**Mar**	**April**	**May**	**Avg.**	**± S.D.**	**June**	**July**	**Aug**	**Avg.**	**± S.D.**
SITE-I	–	74.0	102.0	58.7	14.0	119.0	181.0	92.0	130.7	44.5	130.0	211.0	275.0	205.3	32.0	84.0	167.0	73.0	108.0	47.0
SITE-II	194.0	206.0	335.0	245.0	64.5	260.0	199.0	195.0	218.0	2.0	92.0	227.0	276.0	198.3	24.5	87.0	81.0	98.0	88.7	8.5
SITE-III	186.0	169.0	92.0	149.0	38.5	74.0	377.0	200.0	217.0	88.5	196.0	210.0	253.0	219.7	21.5	213.0	160.0	216.0	196.3	28.0
2001–2002	**Sep**	**Oct**	**Nov**	**Avg.**	**± S.D.**	**Dec**	**Jan**	**Feb**	**Avg.**	**± S.D.**	**Mar**	**April**	**May**	**Avg.**	**± S.D.**	**June**	**July**	**Aug**	**Avg.**	**± S.D.**
SITE-I	110.0	124.0	143.0	125.7	9.5	184.0	185.0	115.0	161.3	35.0	110.0	115.0	163.0	129.3	24.0	131.0	149.0	132.0	137.3	8.5
SITE-II	199.0	188.0	288.0	225.0	50.0	255.0	299.0	176.0	243.3	61.5	191.0	223.0	213.0	209.0	5.0	178.0	159.0	160.0	165.7	0.5
SITE-III	215.0	198.0	186.0	199.7	6.0	187.0	240.0	221.0	216.0	9.5	200.0	218.0	194.0	204.0	12.0	199.0	196.0	204.0	199.7	4.0

Avg.: Average
S.D.: Standard deviation
– : No sampling

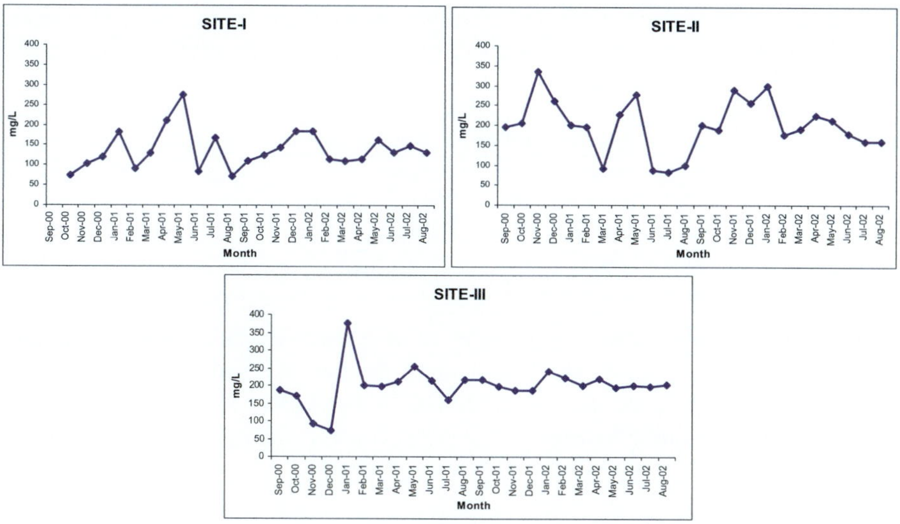

Fig. 50 Monthly fluctuations in total hardness (mg/L) at the investigated sites of the Anchar lake

Overall trend observed in the total hardness values is as under

$$\text{Site III} > \text{Site II} > \text{Site I}$$

At site-I, minimum values of hardness were recorded. Here water is least disturbed by human habitation and whatever hardness present here is because of calcium salts added by the calcium rich rocks. Zutshi (1968), Vass (1973), Zutshi et al. (1980), Kundanger and Zutshi (1985), Sarwar (1987, 1991a, b), Sarwar and Wazir (1988), Sarwar and Rifat (1991) and Sarwar et al. (1996) who worked on the Dal and various other lakes of Kashmir also suggested that calcium is one of the component of hardness contributed from the rocks. Thomson (1952), Bhuyan (1970), Upadhyaya and Roy (1982), Trivedy and Goel (1986), Chapman and Kimstach (1992) and Meybeck et al. (1992) also suggested that calcium get dissolved in water from the rocks along its course.

At site-III, water receives sewage and agricultural run-off from the catchment area and wastes from the hospital. All these factors are responsible for the increase in hardness of water at this site. Singh et al. (1969), Bagde and Verma (1985a, b), Trivedy and Goel (1986), Sinha (1987), Sinha and Prasad (1988), Gopal (1990), Sinha et al. (1991), Syal (1996) and Kaur et al. (2003) observed an increase in hardness due to addition of sewage. James et al. (1990) also observed an increase in hardness in a channel of Tamilnadu receiving hospital wastes.

At site-II, where water from site-I and site-III mixes shows moderate value of hardness. This is because of dilution of effluents added at site-III. James et al. (1990) and Syal (1996) observed fall in hardness with dilution of water.

There are no records of earlier studies of total hardness in this lake.

Naturally occurring waters are usually classified into hard and soft waters. Kannan (1991) has provided the following chart depicting the nature of water on the basis of total hardness value.

Hardness (PPM)	Water quality
0–60	Soft
61–120	Moderately hard
121–180	Hard
>180	Very hard

Based on the above categorization, the 2 year study reveals that water is quite hard in the Anchar lake. At site-I, in spite of the fact that water is free from any human interference, water is very hard. This is because of rich calcareous rocks conceded by water along its way.

Calcium (mg/L)

The calcium values in the Anchar lake ranged from 9.0 to 43.0 mg/L. It has been found to be maximum during spring at all the sites. Summer and autumn record a comparatively lower calcium levels. This is followed by an increase in winter to reach to its maximum in spring. Sarwar (1987) and Shyamsunder (1988) observed rise of calcium ions during autumn and winter months and attributed it to its great solubility at lower temperature. Zafar (1964), Hussainy (1965), Boznaik and Kennedy (1968), Munawar (1970a, b), Zutshi and Vass (1978), Qadri and Yousuf (1980a, b), Chapman and Kimstach (1992) and Saradhamani and Siwakumar (1995) observed no change in concentration of calcium ions with seasonal variations. Its utilization by the phytoplankton has been observed by Swarup and Singh (1979).

The calcium concentration ranged from 9.0 to 40.5 mg/L at site-I, 10.5 to 43.0 mg/L at site-II and 10.7 to 40.5 mg/L at site-III respectively (Table 73 and Fig. 51) showing the following trend.

Site III > Site II > Site I

At site-I the calcium contents were found to be lowest of all other sites. The site is almost free from polluting elements. Here the only source of calcium are rocks from which the calcium salts dissolve in water while flowing over them. This has been suggested by Thomson (1952), Zutshi (1968), Bhuyan (1970), Vass (1973), Zutshi et al. (1980), Upadhyaya and Roy (1982), Kundangar and Zutshi (1985), Trivedy and Goel (1986), Sarwar (1987, 1991a, b), Sarwar and Wazir (1988), Sarwar and Rifat (1991), Chapman and Kimstach (1992) and Meybeek et al. (1992).

At site-III calcium contents were found to be maximum. At this site, sewage from the adjoining human habitation, wastes from the hospital and fertilizer run-off from the agricultural fields are added. Singh et al. (1969), Bagde and Verma (1985a, b),

Table 73 Monthly fluctuations in calcium (mg/L) at the investigated sites of the Anchar lake

	Months																		
2000–2001	**Sep**	**Oct**	**Nov**	**Avg.**	**± S.D.**	**Dec**	**Jan**	**Feb**	**Avg.**	**± S.D.**	**Mar**	**April**	**May**	**Avg.**	**± S.D.**	**June**	**July**	**Aug**	**Avg.**	**± S.D.**
SITE-I	–	9.0	12.3	7.1	1.7	13.5	20.2	15.5	16.4	2.3	19.2	28.0	40.5	29.2	6.3	14.3	17.2	12.9	14.8	2.1
SITE-II	24.5	25.4	42.2	30.7	8.4	35.3	23.7	20.5	26.5	1.6	15.0	29.0	39.5	27.8	5.3	10.5	12.4	18.6	13.8	3.1
SITE-III	21.1	19.3	10.7	17.0	4.3	12.5	40.5	32.5	28.5	4.0	29.5	30.5	38.1	32.7	3.8	26.5	21.1	24.9	24.2	1.9
2001–2002	**Sep**	**Oct**	**Nov**	**Avg.**	**± S.D.**	**Dec**	**Jan**	**Feb**	**Avg.**	**± S.D.**	**Mar**	**April**	**May**	**Avg.**	**± S.D.**	**June**	**July**	**Aug**	**Avg.**	**± S.D.**
SITE-I	12.5	14.3	13.9	13.6	0.2	18.5	16.2	14.5	16.4	0.8	15.2	14.3	22.3	17.3	4.0	18.2	16.3	15.2	16.6	0.6
SITE-II	23.7	24.3	28.9	25.6	2.3	27.5	29.3	35.2	30.7	3.0	43.0	42.0	35.8	40.3	3.1	33.2	30.1	29.0	30.8	0.6
SITE-III	25.5	22.5	21.7	23.2	0.4	19.5	25.2	39.5	28.1	7.2	33.4	35.8	34.0	34.4	0.9	33.5	28.0	27.0	29.5	0.5

Avg.: Average
S.D.: Standard deviation
– : No sampling

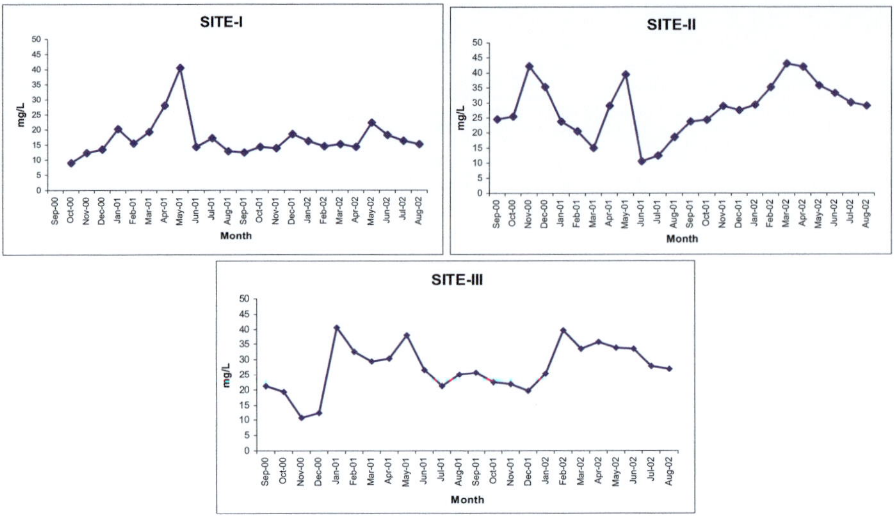

Fig. 51 Monthly fluctuations in calcium (mg/L) at the investigated sites of the Anchar lake

Trivedy and Goel (1986), Sinha (1987), Sinha and Prasad (1988), Gopal (1990), Sinha et al. (1991) Syal (1996) and Kaur et al. (2003) observed an increase in calcium contents due to addition of sewage.

At site-II, calcium contents were found to be moderate. The effluents which were added at site-III get diluted when water from site-I mixes here.

As per Ohle (1934), water of Anchar lake range from medium to nutrient rich.

Earlier Sarwar (1999) recorded calcium ranging from 23.2 to 58.0 mg/L in 1983. While Sarwar and Zutshi (1989) recorded average calcium concentration in this lake to be 38.1 mg/L. During the present investigation, calcium concentration has been found to vary from 9.0 to 43.0 mg/L, thus showing not much variation in calcium level in water.

ISI (1986) has prescribed maximum permissible limit of calcium in water for drinking to be 75 mg/L. The present investigation revealed that calcium concentration at selected sites remained between 9 and 43.0 mg/L which is well within the permissible limits.

Magnesium (mg/L)

The magnesium values in the Anchar lake ranged from 4.3 to 28.5 mg/L.

No definite seasonal trend is indicated during the present study. Zafar (1964), Zutshi and Vass (1978), Zutshi et al. (1980), Trivedy and Goel (1986), Sarwar and Rifat (1991) and Saradhamani and Siwakumar (1995) also failed to observe any seasonal change in concentration of magnesium in their studies.

The magnesium concentration ranged from 4.3 to 18.1 at Site-I, 5.3 to 23.8 at Site-II and 4.5 to 28.5 at Site-III (Table 74 and Fig. 52) showing the following trend.

Table 74 Monthly fluctuations in magnesium (mg/L) at the investigated sites of the Anchar lake

	Months																			
2000–2001	**Sep**	**Oct**	**Nov**	**Avg.**	**± S.D.**	**Dec**	**Jan**	**Feb**	**Avg.**	**± S.D.**	**Mar**	**April**	**May**	**Avg.**	**± S.D.**	**June**	**July**	**Aug**	**Avg.**	**± S.D.**
SITE-I	–	5.4	7.4	4.3	1.0	8.9	13.5	5.6	9.3	4.0	8.6	15.0	18.1	13.9	1.6	5.1	12.8	4.3	7.4	4.3
SITE-II	13.8	14.8	23.8	17.5	4.5	17.9	14.5	14.9	15.8	0.2	5.7	16.0	18.5	13.4	1.3	6.3	5.3	5.5	5.7	0.1
SITE-III	13.8	12.5	6.8	11.0	2.9	4.5	28.5	12.4	15.1	8.1	12.8	14.0	16.5	14.4	1.3	15.2	11.2	15.9	14.1	2.4
2001–2002	**Sep**	**Oct**	**Nov**	**Avg.**	**± S.D.**	**Dec**	**Jan**	**Feb**	**Avg.**	**± S.D.**	**Mar**	**April**	**May**	**Avg.**	**± S.D.**	**June**	**July**	**Aug**	**Avg.**	**± S.D.**
SITE-I	8.2	9.2	11.2	9.5	1.0	14.2	14.9	8.2	12.4	3.4	7.5	8.2	11.2	9.0	1.5	8.9	11.2	9.8	10.0	0.7
SITE-II	14.5	13.2	16.2	14.6	1.5	19.3	16.2	9.4	15.0	3.4	9.0	12.5	13.0	11.5	0.3	10.0	8.9	9.2	9.4	0.1
SITE-III	15.7	14.7	13.7	14.7	0.5	14.3	18.3	12.9	15.2	2.7	12.2	13.5	11.5	12.4	1.0	12.1	13.1	14.2	13.1	0.6

Avg.: Average
S.D.: Standard deviation
– : No sampling

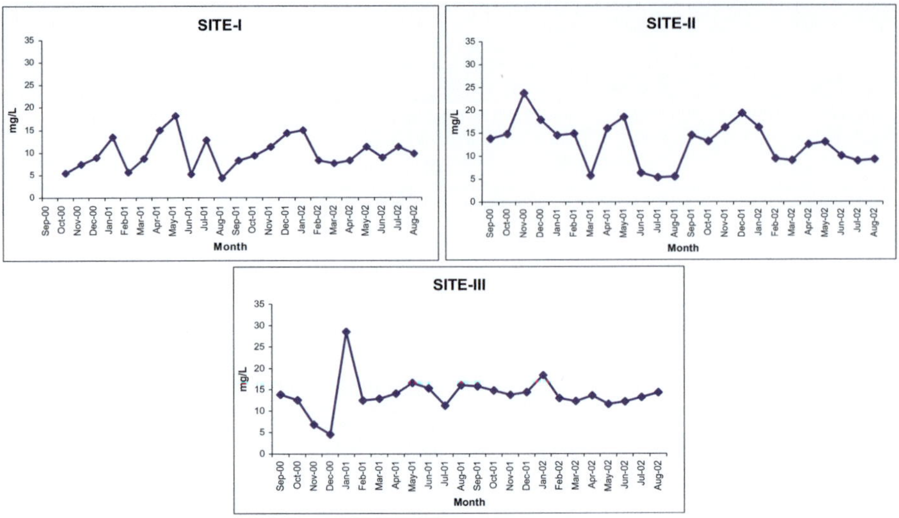

Fig. 52 Monthly fluctuations in magnesium (mg/L) at the investigated sites of the Anchar lake

Site III > Site II > Site I

At site-I magnesium contents were found to be lower than other sites. This site is free of human interference. The only source of magnesium here seems to be the rocks from which magnesium salts dissolve in water while flowing over them. This has also been suggested by Zutshi et al. (1980) for Kashmir lakes and Bhuyan (1970), Trivedy and Goel (1986),Chapman and Kimstach (1992) and Meybeek et al. (1992).

At site-III magnesium contents were found to be the highest. The increased magnesium contents seem to be due to addition of sewage, agricultural run-off and hospital wastes which are added at this site. The contribution of sewage to magnesium contents of water has also been reported by Singh et al. (1969), Munawar (1970a, b), Bagde and Verma (1985a, b) and Sinha et al. (1991), Syal (1996) and Kaur et al. (2003).

At site-II magnesium contents were found to be slightly higher than site-I and lower than site-III as water from site-I and III mixes at this site.

Earlier Sarwar (1999) reported the magnesium contents ranging from 1.9 to 21.8 mg/L in his studies carried out in 1983 and Sarwar and Zutshi (1989) recorded an average value of 4.1 mg/L. A comparison with the present studies show that magnesium concentration has increased considerably during these years.

ISI (1986) has prescribed maximum permissible limit of magnesium in water to be 30 mg/L. The present studies have revealed the magnesium concentration range to be 4.3–28.5 mg/L. which is well within the permissible limits.

Iron (μg/L)

Iron concentration at three selected sites of the Anchar lake ranged from 100 to 600 μg/L.

No definite seasonal pattern with respect to iron concentration of water at all the study sites was recorded. Sajeev (1999) also failed to observe any seasonal trend in iron concentration in Ropar reservoir of Punjab. However, Hasan and Pandy (1985) observed lower values of iron both in winter and summer and Wanganeo et al. (1996) and Suvarna and Somashekar (2000) observed higher values in Summer.

The iron concentration ranged from 100 to 600 μg/L at site-I, 200 to 600 μg/L at site-II and 200 to 600 μg/L at site-III (Table 75 and Fig. 53).

At site-I, iron concentration ranged between 100 and 600 μg/L, the values remaining lowest of all the sites. This site is located where no direct pollution source is seen. Sajeev (1999) also observed lower iron values in Ropar wetland at sites devoid of any pollution.

At site-III iron concentration ranged from 200 to 600 μg/L. Here the water is contaminated with sewage discharges, agricultural run-off and hospital wastes. Trivedy and Goel (1986) and Israili (1991) have asserted that wastes contaminated with sewage contribute towards the iron content of natural waters. Sajeev (1999) also observed higher iron contents in the Ropar reservoir at sites which received organic wastes.

At site-II also, iron concentration ranged from 200 to 600 but values were mostly found to be lower than site-III. This is due to absence of any pollution source and dilution of sewage coming from the site-III.

Earlier Sarwar (1999) recorded iron concentration of Anchar lake (worked in 1983) to vary from 60 to 400 μg/L while Sarwar and Zutshi (1989) recorded an average iron concentration to be 144 μg/L. During the present investigation iron concentration was found to vary from 100 to 600 μg/L thus clearly depicting a considerable rise in iron contents of lake water during these years.

I.S.I. (1986) has prescribed a maximum permissible limit of iron in drinking water as 300 μg/L. The concentration of iron in the investigated lake remained between 100 and 600 μg/L thus exceeding the limits prescribed for drinking water.

Sodium (mg/L)

The level of sodium ranged between 3.0 and 14 mg/L in the waters of the Anchar lake.

No seasonal trend is indicated from the 2 year data.

Sodium concentration ranged from 3 to 6 mg/L at site-I, from 3 to13mg/L at site-II and from 6 to 14 mg/L at site-III (Table 76 and Fig. 54) showing the following trend.

Table 75 Monthly fluctuations in iron (μg/L) at the investigated sites of the Anchar lake

	Months																		
2000–2001	**Sep**	**Oct**	**Nov**	**Avg.**	**± S.D.**	**Dec**	**Jan**	**Feb**	**Avg.**	**± S.D.**	**Mar**	**April**	**May**	**Avg.**	**± S.D.**	**June**	**July**	**Aug**	**Avg.**	**± S.D.**
SITE-I	–	300.0	400.0	233.3	50.0	500.0	200.0	100.0	266.7	50.0	100.0	600.0	100.0	266.7	250.0	200.0	100.0	371.0	223.7	135.5
SITE-II	300.0	400.0	200.0	300.0	100.0	600.0	200.0	300.0	366.7	50.0	280.0	500.0	550.0	443.3	25.0	550.0	290.0	381.0	407.0	45.5
SITE-III	600.0	600.0	400.0	533.3	100.0	300.0	400.0	600.0	433.3	100.0	600.0	500.0	600.0	566.7	50.0	300.0	200.0	360.0	286.7	80.0
2001–2002	**Sep**	**Oct**	**Nov**	**Avg.**	**± S.D.**	**Dec**	**Jan**	**Feb**	**Avg.**	**± S.D.**	**Mar**	**April**	**May**	**Avg.**	**± S.D.**	**June**	**July**	**Aug**	**Avg.**	**± S.D.**
SITE-I	295.0	405.0	455.0	385.0	25.0	403.0	305.0	285.0	331.0	10.0	295.0	265.0	205.0	255.0	30.0	165.0	210.0	218.0	197.7	4.0
SITE-II	315.0	388.0	350.0	351.0	19.0	382.0	395.0	350.0	375.7	22.5	384.0	475.0	437.0	432.0	19.0	475.0	390.0	369.0	411.3	10.5
SITE-III	405.0	555.0	514.0	491.3	20.5	449.0	430.0	421.0	433.3	4.5	520.0	495.0	515.0	510.0	10.0	445.0	395.0	385.0	408.3	5.0

Avg.: Average
S.D.: Standard deviation
– : No Sampling

Fig. 53 Monthly fluctuations in iron (µg/L) at the investigated sites of the Anchar lake

Site III > Site II > Site I

At site-I, water is least interfered by human activities and whatever sodium present is due to the weathering of rocks.

At site-III water receives sewage, agricultural run-off from adjoining fields, sewage from surrounding habitation and effluents from SKIMS hospital. High concentration of sodium shows that these wastes add sodium to water at this site. Some sodium may also be added to the water by the decomposition of plants which are present in abundance at this site. According to Gonzalves and Joshi (1946), Singh (1960), Munawar (1970a), Trivedy and Goel (1986) and Sharma et al. (2000) the domestic sewage are rich in sodium and increases the concentration in waters after disposal.

Sodium concentration gets decreased at site-II where no direct source of pollution is present and water from site-I and site-III mixes up resulting in dilution of sodium concentration. A similar dilution has been observed in the concentration. of calcium and magnesium.

Concentration of sodium has been found to be lower than calcium and magnesium Zutshi et al. (1980) and Sarwar and Zutshi (1989) also found sodium to be lower than calcium in most of the lakes of Kashmir except in Trigam lake where sodium is dominant over calcium.

Earlier Sarwar (1999) recorded sodium to be ranging from 1.3 to 14.4 mg/L during his study conducted in 1983 while Sarwar and Zutshi (1989) recorded and average concentration of sodium as 5.3 mg/L. Not much difference in sodium concentration has been observed during the present investigation.

Table 76 Monthly fluctuations in sodium (mg/L) at the investigated sites of the Anchar lake

	Months																		
2000–2001	**Sep**	**Oct**	**Nov**	**Avg.**	**± S.D.**	**Dec**	**Jan**	**Feb**	**Avg.**	**± S.D.**	**Mar**	**April**	**May**	**Avg.**	**± S.D.**	**June**	**July**	**Aug**	**Avg.**	**± S.D.**
SITE-I	–	3.0	3.0	2.0	0.0	5.0	5.0	3.0	4.3	1.0	3.0	4.0	5.0	4.0	0.5	3.0	3.0	5.0	3.7	1.0
SITE-II	7.0	6.0	4.0	5.7	1.0	12.0	13.0	3.0	9.3	5.0	3.0	8.0	8.0	6.3	0.0	10.0	3.0	5.0	6.0	1.0
SITE-III	7.0	7.0	8.0	7.3	0.5	14.0	14.0	8.0	12.0	3.0	7.0	13.0	9.0	9.7	2.0	9.0	8.0	10.0	9.0	1.0
2001–2002	**Sep**	**Oct**	**Nov**	**Avg.**	**± S.D.**	**Dec**	**Jan**	**Feb**	**Avg.**	**± S.D.**	**Mar**	**April**	**May**	**Avg.**	**± S.D.**	**June**	**July**	**Aug**	**Avg.**	**± S.D.**
SITE-I	5.0	5.0	5.0	5.0	0.0	5.0	5.0	6.0	5.3	0.5	6.0	4.0	5.0	5.0	0.5	5.0	5.0	5.0	5.0	0.0
SITE-II	9.0	8.0	7.0	8.0	0.5	7.0	8.0	9.0	8.0	0.5	8.0	7.0	7.0	7.3	0.0	6.0	7.0	8.0	7.0	0.5
SITE-III	9.0	8.0	7.0	8.0	0.5	7.0	8.0	10.0	8.3	1.0	7.0	8.0	7.0	7.3	0.5	6.0	6.0	6.0	6.0	0.0

Avg.: Average
S.D.: Standard deviation
– : No Sampling

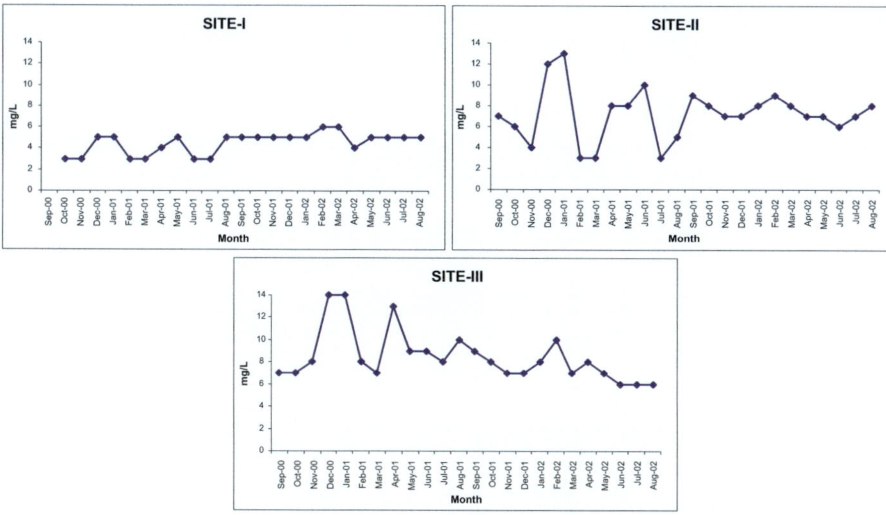

Fig. 54 Monthly fluctuations in sodium (mg/L) at the investigated sites of the Anchar lake

Potassium (mg/L)

The potassium concentration in the Anchar lake ranged form 2 to 9 mg/L in the Anchar lake over a period of 2 years.

No seasonal trend is indicated in the potassium values in the present study.

Potassium concentration varied from 2 to 4 mg/L at site-I, 2 to 9 mg/L at site-II and 2 to 8 mg/L at site-III (Table 77 and Fig. 55) showing the overall pattern as follows

$$\text{Site III} > \text{Site II} > \text{Site I}$$

Minimum values at site-I is attributed to the absence of any pollution source. Increase in potassium concentration at site-II and III seems to be due to addition of sewage, agricultural run-off and hospital waste. A comparison with the other cations show the following trend.

$$Ca^2+ > Mg^2+ > Na+ > K+$$

A similar trend has been observed by Zutshi et al. (1980), Trivedy and Goel (1986), Sarwar and Zutshi (1989) and Sarwar (1999).

Sarwar (1999) recorded potassium to range from 1 to 10.6 mg/L during his work done in 1983, while Sarwar and Zutshi (1989) recorded an average value of 4.0 mg/L. Range of potassium recorded during the present investigation show that there is no change in potassium concentration in water of the lake.

Table 77 Monthly fluctuations in potassium (mg/L) at the investigated sites of the Anchar lake

	Months																		
2000–2001	**Sep**	**Oct**	**Nov**	**Avg.**	**± S.D.**	**Dec**	**Jan**	**Feb**	**Avg.**	**± S.D.**	**Mar**	**April**	**May**	**Avg.**	**± S.D.**	**June**	**July**	**Aug**	**Avg.**	**± S.D.**
SITE-I	–	2.0	3.0	1.7	0.5	3.0	3.0	3.0	3.0	0.0	3.0	4.0	3.0	3.3	0.5	2.0	4.0	4.0	3.3	0.0
SITE-II	5.0	5.0	3.0	4.3	1.0	9.0	8.0	3.0	6.7	2.5	3.0	4.0	2.0	3.0	1.0	6.0	3.0	4.0	4.3	0.5
SITE-III	4.0	4.0	5.0	4.3	0.5	8.0	8.0	2.0	6.0	3.0	3.0	5.0	5.0	4.3	0.0	4.0	5.0	7.0	5.3	1.0
2001–2002	**Sep**	**Oct**	**Nov**	**Avg.**	**± S.D.**	**Dec**	**Jan**	**Feb**	**Avg.**	**± S.D.**	**Mar**	**April**	**May**	**Avg.**	**± S.D.**	**June**	**July**	**Aug**	**Avg.**	**± S.D.**
SITE-I	3.0	3.0	4.0	3.3	0.5	4.0	4.0	4.0	4.0	0.0	3.0	2.0	4.0	3.0	1.0	3.0	3.0	3.0	3.0	0.0
SITE-II	8.0	8.0	6.0	7.3	1.0	6.0	8.0	3.0	5.7	2.5	4.0	3.0	3.0	3.3	0.0	4.0	4.0	6.0	4.7	1.0
SITE-III	8.0	8.0	6.0	7.3	1.0	6.0	8.0	3.4	5.8	2.3	4.0	3.0	4.0	3.7	0.5	3.0	4.0	4.0	3.7	0.0

Avg.: Average
S.D.: Standard deviation
– : No sampling

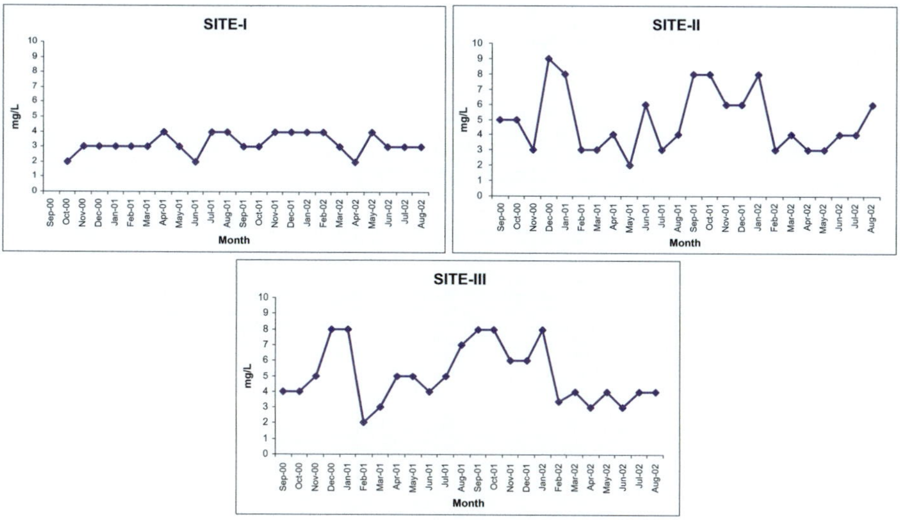

Fig. 55 Monthly fluctuations in potassium (mg/L) at the investigated sites of the Anchar lake

Chloride (mg/L)

The Chloride concentration in the Anchar lake ranged from a maximum of 85 mg/L to a minimum of 9 mg/L (Table 78 and Fig. 56).

From the 2 year study of Anchar lake, any definite seasonal pattern of chloride concentration could not be detected. Shastree et al. (1991), Bath (1996) and Sajeev (1999) also failed to observe any definite seasonal pattern with respect to chloride values. Zutshi and Vass (1978) however, recorded lower chloride contents in autumn in Dal lake of Kashmir. Ownbey and Kee (1967) and Ajmal et al. (1985) recorded higher chloride values during the winter months and and lower during summer.

At site-I the chloride values were observed to be lowest as the water at this site is least interfered by any human interference. This site is not contaminated by any kind of waste or sewage discharges. Prasannakumari et al. (2000) also observed low chloride contents in water devoid of sewage contamination.

Site-III shows the highest contents of chloride concentration in the water. This site is located where human and animal excreta, agricultural run-off, sewage discharges and hospital waste are added. Drischel (1940), Thresh et al. (1944), Gonzalves and Joshi (1946), Blum (1957), Sawyer (1960), Zafar (1964), Lakshminarayan (1965), Singh (1965), Ownbey and Kee (1967), Venkateshwarlu (1969), Munawar (1970a, b), Cole (1975), Goel et al. (1980), Trivedy and Goel (1986), Das (1989), APHA (1989) and Shastree et al. (1991), Bath (1996) and Sarwar (1999) suggested that addition of sewage in water increases chloride concentration. Hospital wastes are also instrumental in increasing chloride contents of water as has also been observed by James et al. (1990).

Table 78 Monthly fluctuations in chloride (mg/L) at the investigated sites of the Anchar lake

	Months																		
2000–2001	**Sep**	**Oct**	**Nov**	**Avg.**	**± S.D.**	**Dec**	**Jan**	**Feb**	**Avg.**	**± S.D.**	**Mar**	**April**	**May**	**Avg.**	**± S.D.**	**June**	**July**	**Aug**	**Avg.**	**± S.D.**
SITE-I	–	25.0	21.0	15.3	2.0	31.0	22.0	15.0	22.7	3.5	20.0	19.0	35.0	24.7	8.0	9.0	16.0	61.0	28.7	22.5
SITE-II	39.0	36.0	31.0	35.3	2.5	45.0	23.0	27.0	31.7	2.0	29.0	25.0	25.0	26.3	0.0	35.0	31.0	35.0	33.7	2.0
SITE-III	65.0	60.0	75.0	66.7	7.5	69.0	49.0	65.0	61.0	8.0	75.0	95.0	69.0	79.7	13.0	34.0	29.0	22.0	28.3	3.5
2001–2002	**Sep**	**Oct**	**Nov**	**Avg.**	**± S.D.**	**Dec**	**Jan**	**Feb**	**Avg.**	**± S.D.**	**Mar**	**April**	**May**	**Avg.**	**± S.D.**	**June**	**July**	**Aug**	**Avg.**	**± S.D.**
SITE-I	25.0	22.0	21.0	22.7	0.5	25.0	30.0	12.0	22.3	9.0	13.0	14.0	19.0	15.3	2.5	16.0	19.0	22.0	19.0	1.5
SITE-II	55.0	75.0	62.0	64.0	6.5	60.0	64.0	41.0	55.0	11.5	28.0	35.0	28.0	30.3	3.5	31.0	19.0	25.0	25.0	3.0
SITE-III	50.0	85.0	82.0	72.3	1.5	80.0	75.0	24.0	59.7	25.5	40.0	55.0	50.0	48.3	2.5	40.6	49.0	47.0	45.5	1.0

Avg.: Average
S.D.: Standard deviation
– : No sampling

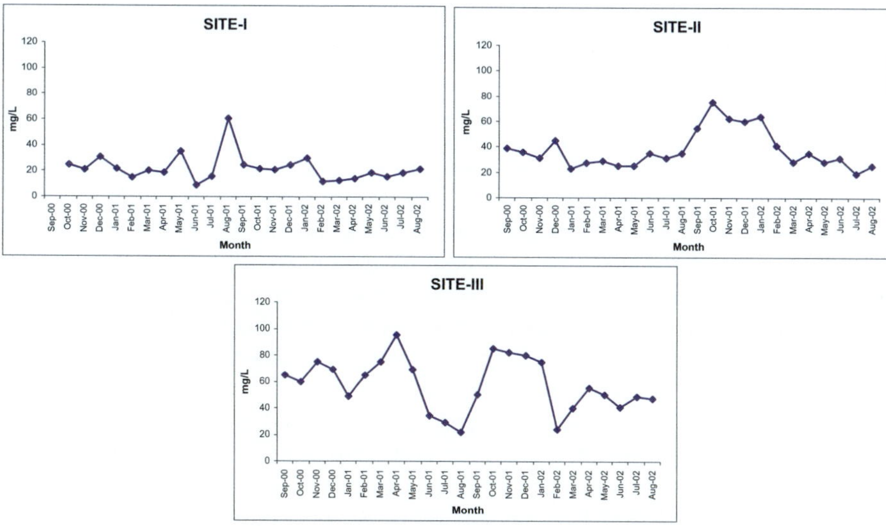

Fig. 56 Monthly fluctuations in chloride (mg/L) at the investigated sites of the Anchar lake

At site-II the range of chloride contents falls between 19 and 75 mg/L being higher than site-I and lower than site-III. At this site no direct source of pollution is present and pollutants added at site-III get diluted at this site by the water coming from the site-I.

Earlier Sarwar (1999) recorded chloride contents ranging from 11.0 to 38.0 mg/L during a study conducted in 1983 while Sarwar and Zutshi (1989) recorded average value of chloride to be 24.0 mg/L. A comparison with the present study shows a considerable increase in chloride contents in the Anchar lake which is mainly due to the use of lake as dumping ground for waste disposal by hospital and surrounding habitants.

During the entire period of study, the chloride values were observed to remain within the maximum permissible limits which is 250 mg/L as suggested by ISI (1986).

Ammonical Nitrogen (μg/L)

Ammonical nitrogen in the Anchar lake, varied from 285 to 705 μg/L.

Concentration of ammonical nitrogen is found to be minimum in summer and higher in winter and autumn. Harold (1934), Adoni and Joshi (1987) and Bandopadhyay and Gopal (1991) observed lower values during the summer. On the otherhand, Zutshi and Vass (1978) and Ramakrishnaiah and Sarkar (1986) failed to record any seasonal pattern.

A study of the Anchar lake revealed much difference in the values of ammonical nitrogen at the three sites. The concentration of ammonical nitrogen varied from 285 to 490 g/L at site-I, from 310 to 610 g/L at site-II and from 305 to 705 g/L at site-III (Table 79 and Fig. 57) showing the overall trend as follows

$$\text{Site III} > \text{Site II} > \text{Site I}$$

At site-I, ammonical nitrogen concentration was found to be lowest. Here whatever ammonical nitrogen is present, besides the natural source, comes by microbial decomposition and excretory products of the biota present at this site. Any external source of nitrogenous compounds is absent. Wetzel and Likens (2000) and Bhatt et al. (1999) also suggested that ammonical nitrogen is added by microbial decomposition and excreta of biota present in water.

High contents of ammonical nitrogen were observed at site-III. This site is located close to the point where sewage is added from the catchment area bringing in household effluents and residues of detergents. Besides, agricultural run-off and effluents from the SKIMS hospital are added at this site. Thresh et al. (1944), Seenayya (1971), King (1981), Trivedy and Goel (1986) and Das (1989), Sarwar (1991a, b), Bhatt et al. (1999) and Narain and Chauhan (2000) have also observed high concentration of ammonical nitrogen in waters contaminated with sewage. Hospital waste also add ammonical nitrogen to water as suggested by James et al. (1990).

At site-II, ammonical nitrogen was comparatively lower than site-III and higher than site-I. At this site no source of pollution is present and pollutants added at site-III get diluted at this site by the water coming from site-I. As there is no direct source of pollution at this site, ammonical nitrogen present here is likely to be due to decomposition of abundantly growing macrophytes and excretory products of the biota present at this site as also suggested by Wetzel (1975) and Bhatt et al. (1999).

Earlier Sarwar and Zutshi (1989) recorded average value of 16 μg/L which is much lower than the values recorded during the present study, thereby showing concentration of ammonical nitrogen on the rise due to sewage and agricultural run-off.

ISI. (1986) has recommended maximum permissible limit of this form of nitrogen to be 500 μg/L. The present data reveals that ammonical nitrogen has crossed the permissible limits at site-II and III respectively.

Nitrate-Nitrogen (μg/L)

The Nitrate-nitrogen values ranged from 245 to 611 μg/L at three sites of the Anchar lake.

Nitrate-nitrogen was found to be minimum in summer and high in autumn and winter at all the sites. The minimum value during summer season may be attributed to its uptake and utilization by phytoplanktons and macrophytic which grow pro-

Table 79 Monthly fluctuations in ammonical nitrogen (μg/L) at the investigated sites of the Anchar lake

	Months																		
2000–2001	**Sep**	**Oct**	**Nov**	**Avg.**	**± S.D.**	**Dec**	**Jan**	**Feb**	**Avg.**	**± S.D.**	**Mar**	**April**	**May**	**Avg.**	**± S.D.**	**June**	**July**	**Aug**	**Avg.**	**± S.D.**
SITE-I	–	390.0	420.0	270.0	15.0	480.0	490.0	450.0	473.3	20.0	450.0	480.0	380.0	436.7	50.0	400.0	410.0	352.0	387.3	29.0
SITE-II	610.0	590.0	592.0	597.3	1.0	590.0	610.0	406.0	535.3	102.0	400.0	410.0	450.0	420.0	20.0	310.0	395.0	435.0	380.0	20.0
SITE-III	499.0	458.0	691.0	549.3	116.5	693.0	692.0	680.0	688.3	6.0	365.0	431.0	329.0	375.0	51.0	389.0	350.0	308.0	349.0	21.0
2001–2002	**Sep**	**Oct**	**Nov**	**Avg.**	**± S.D.**	**Dec**	**Jan**	**Feb**	**Avg.**	**± S.D.**	**Mar**	**April**	**May**	**Avg.**	**± S.D.**	**June**	**July**	**Aug**	**Avg.**	**± S.D.**
SITE-I	310.0	385.0	351.0	348.7	17.0	325.0	395.0	302.0	340.7	46.5	285.0	290.0	335.0	303.3	22.5	390.0	309.0	315.0	338.0	3.0
SITE-II	495.0	520.0	478.0	497.7	21.0	490.0	510.0	410.0	470.0	50.0	390.0	340.0	350.0	360.0	5.0	330.0	337.0	340.0	335.7	1.5
SITE-III	705.0	659.0	651.0	671.7	4.0	639.0	620.0	505.0	588.0	57.5	465.0	485.0	305.0	418.3	90.0	382.0	345.0	365.0	364.0	10.0

Avg.: Average
S.D.: Standard deviation
– : No sampling

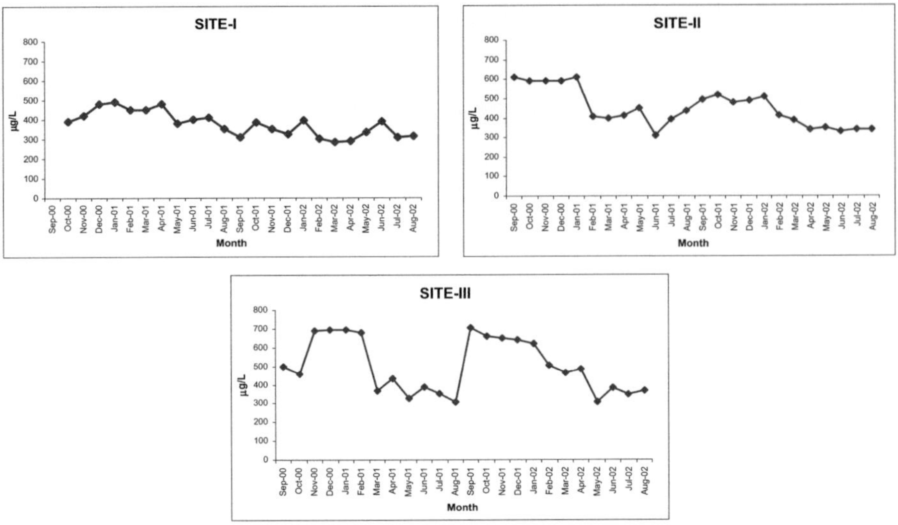

Fig. 57 Monthly fluctuations in ammonical nitrogen (µg/L) at the investigated sites of the Anchar lake

fusely during summer. Harold (1934), Harvey (1940), Vasisht and Sharma (1975), Ajmal et al. (1985) Adoni and Joshi (1987) and Bandopadhyay and Gopal (1991) also recorded lower values of nitrate-nitrogen during the summer and contended that phytoplanktons and macrophytes flourish during summer and pickup nutrients from their aquatic environment. As a result, the concentration of nitrates gets decreased. Zutshi and Vass (1971) and Ramakrishnaiah and Sarkar (1986), Sarwar and Majid (1997) and Sajeev (1999), however, failed to record well marked seasonal variations in nitrate nitrogen concentration while Willem et al. (1972), King (1981), Shyamsunder (1988), Shah (1988) and Shastree et al. (1991) observed maximum nitrate values during the winters.

The nitrate nitrogen values ranged from 245 to 495 µg/L at site-I, from 315 to 548 µg/L at site-II and from 335 to 611 µg/L at site-III (Table 80 and Fig. 58) showing the overall pattern as follows

$$\text{Site III} > \text{Site II} > \text{Site I}$$

Site-I is not polluted by any sewage effluence. The macrophytic vegetation at this site is also scarce. Nitrate-nitrogen concentration here is mainly due to natural weathering of rocks. Lower values of Nitrate-nitrogen here may be attributed to absence of any polluting source. Jhingran (1991a, b) also opined that natural waters get their nitrate through weathering of rocks.

Site-III where the nitrate concentration was highest, seems to be due to multiple types of wastes added at this site. Sewage is added in abundance from the adjoining habitations which bring in house hold effluents and faecal matter. Agricultural run-off and discharges from SKIMS hospital are also added at this site. Therefore,

Table 80 Monthly fluctuations in nitrate nitrogen (μg/L) at the investigated sites of the Anchar lake

	Months																		
2000–2001	**Sep**	**Oct**	**Nov**	**Avg.**	**± S.D.**	**Dec**	**Jan**	**Feb**	**Avg.**	**± S.D.**	**Mar**	**April**	**May**	**Avg.**	**± S.D.**	**June**	**July**	**Aug**	**Avg.**	**± S.D.**
SITE-I	–	490.0	460.0	316.7	15.0	495.0	396.0	395.0	428.7	0.5	415.0	430.0	360.0	401.7	35.0	390.0	396.0	341.0	375.7	27.5
SITE-II	500.0	495.0	420.0	471.7	37.5	420.0	400.0	521.0	447.0	60.5	453.0	431.0	474.0	452.7	21.5	430.0	475.0	420.0	441.7	27.5
SITE-III	520.0	500.0	520.0	513.3	10.0	540.0	520.0	580.0	546.7	30.0	392.0	490.0	435.0	439.0	27.5	335.0	410.0	345.0	363.3	32.5
2001–2002	**Sep**	**Oct**	**Nov**	**Avg.**	**± S.D.**	**Dec**	**Jan**	**Feb**	**Avg.**	**± S.D.**	**Mar**	**April**	**May**	**Avg.**	**± S.D.**	**June**	**July**	**Aug**	**Avg.**	**± S.D.**
SITE-I	280.0	292.0	265.0	279.0	13.5	320.0	325.0	360.0	335.0	17.5	325.0	343.0	292.0	320.0	25.5	275.0	263.0	245.0	261.0	9.0
SITE-II	538.0	548.0	513.0	533.0	17.5	520.0	529.0	504.0	517.7	12.5	453.0	364.0	315.0	377.3	24.5	395.0	340.0	392.0	375.7	26.0
SITE-III	550.0	605.0	602.0	585.7	1.5	544.0	550.0	580.0	558.0	15.0	611.0	480.0	385.0	492.0	47.5	430.0	355.0	395.0	393.3	20.0

Avg.: Average
S.D.: Standard deviation
– : No sampling

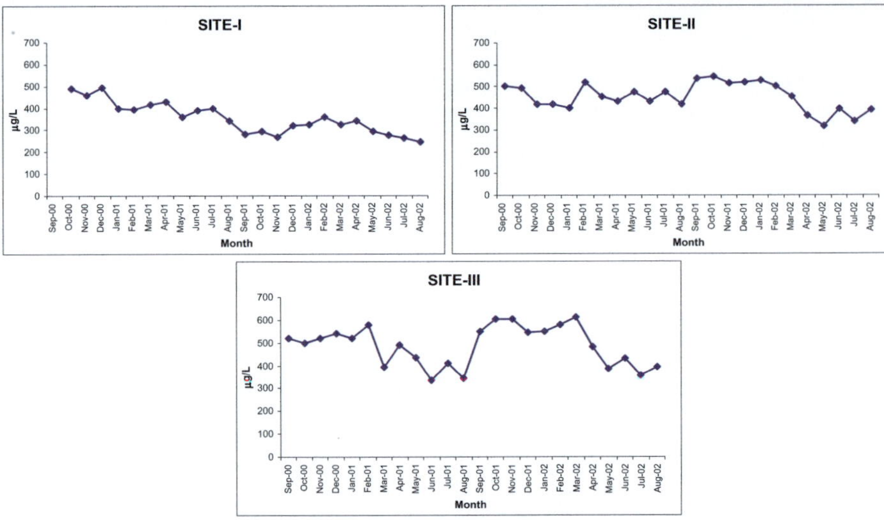

Fig. 58 Monthly fluctuations in nitrate nitrogen (µg/L) at the investigated sites of the Anchar lake

the water remains rich in nitrates. Thresh et al. (1944), Sylvester (1961), Jolly and Chapman (1966), Munawar (1970a), Seenayya (1971), King (1981), Trivedy and Goel (1986), Das (1989) and Suvarna and Somashekar (2000), have also observed high nitrate contents in water bodies contaminated with Sewage. James et al. (1990) reported an increase in nitrate concentration in water as result of hospital waste contamination. Inspite of rich growth of macrophytes, which pickup nitrates for their growth, the nitrate concentration remained very high which is attributed to rich input of this nutrient by multiple sources of pollution at this site.

Site-II, where nitrate-nitrogen concentration is lower than site-III but higher than site-I. Here direct source of pollution is absent, water from site-III and site-I mixes up and pollutants added at site-III gets diluted. Furthermore, macrophytes also pick up nitrate-nitrogen for their growth thereby decreasing the nitrate in water as has also been suggested by Harold (1934), Harvey (1940), Vasisht and Sharma (1975), Ajmal et al. (1985) Adoni and Joshi (1987) and Bandopadhyay and Gopal (1991).

It is evident from the aforesaid discussion that site-II and site-III are rich in organic pollution due to intense anthropogenic pressure which is manifested in the prolific growth of obnoxious weeds like *Salvinia natans* and *Lemna* sp. at these sites of the lake. Ellis et al. (1946) reported growth of these macrophytes in organically rich waters.

According to Alikunhi (1957) presence of 60–100 µg/L of nitrate-nitrogen indicate water to be highly productive. Nitrate-nitrogen is considered to be the primary factor for enrichment of waters. In view of these statements, the Anchar lake may be regarded as a highly productive and nutriently rich water body.

Earlier Sarwar (1999) recorded nitrate-nitrogen concentration ranging from 20 to 360 during the study carried out in 1983 while Sarwar and Zutshi (1989) recorded an average value of 91 μg/L. A comparison with the present study clearly depicts that nitrate concentration is on the rise in the lake due to various activities.

ISI (1986) has recommended maximum permissible limit of this form of nitrogen to be 4500 μg/L. The present data reveals that nitrate- nitrogen is within the permissible limits in the Anchar lake water.

Total Phosphate-Phosphorus (μg/L)

In the Anchar lake, total phosphate phosphorus values at the three sites ranged from 200 to 505 μg/L.

The data reveals minimum total phosphate phosphorus values were present at all the studied sites in summer and higher values in colder months. During the summers, aquatic plants are found to grow vigorously. *Typha* sp. *Ceratophyllum* sp. are seen to grow abundantly in the lake during summer. Due to high growth rate, the phosphates which are one of the major constituents needed for growth are picked up thereby decreasing its concentration in water. The decline in phosphate contents of water during summer has also been reported by Welch (1952), Rutter (1953), Hutchinson (1957), Michael (1969), Kaul (1977), Zutshi and Vass (1978), Harshey et al. (1982), Bath (1996), Bhatt et al. (1999), Kumar (2003) and Kumar et al. (2003). These workers have suggested that phosphates and nitrates are utilized by macrophytes such as *Typha* sp. and *Ceratophyllum* sp.

The concentration of total phosphate phosphorus ranged from 205 to 460 μg/L at site-I, 200 to 350 μg/L at site-II and 300 to 505 μg/L at site-III (Table 81 and Fig. 59) showing the overall trend as follows

$$\text{Site III} > \text{Site I} > \text{Site II}$$

At site-I direct source of pollution is absent. Welch (1952) and Rutter (1953) have also reported smaller amounts of phosphorus in waters free of pollution impregnation.

Total phosphate phosphorus level was found to be very high at site-III where a substantial amount of agricultural run-off, domestic wastes and hospital wastes are added. All these wastes are likely to contain large amount of phosphates and thus increase the concentration of phosphates in water. An increase in the phosphate contents due to the addition of sewerage wastes is also reported by Welch (1952), Rutter (1953), Blum (1956, 1957), Hutchinson (1957), Michael (1969), Zutshi and Vass (1978), Clarke (1924), Heron (1961), Trivedy and Goel (1986), Ghose and Sharma (1988), Mishra (1988), Rana and Palria (1988), Das (1989), Kannan (1991), Mishra and Saxena (1993), Sarwar (1991a, b), Bandopadhyay and Gopal (1991), Bath (1996), Khabade et al. (2003) and Prasannakumari et al. (2003).

At site-II total phosphate phosphorus concentration was found to be lower than site-I as well as site-III. Here direct source of pollution is absent. total phosphate phosphorus here are those which come from the site-III and site-I as water mixes up

Table 81 Monthly fluctuations in total phosphate phosphorus (μg/L) at the investigated sites of the Anchar lake

	Months																			
2000–2001	**Sep**	**Oct**	**Nov**	**Avg.**	**± S.D.**	**Dec**	**Jan**	**Feb**	**Avg.**	**± S.D.**	**Mar**	**April**	**May**	**Avg.**	**± S.D.**	**June**	**July**	**Aug**	**Avg.**	**± S.D.**
SITE-I	–	370.0	315.0	228.3	27.5	359.0	395.0	460.0	404.7	32.5	420.0	350.0	353.0	374.3	1.5	351.0	360.0	280.0	330.3	40.0
SITE-II	281.0	285.0	298.0	288.0	6.5	213.0	200.0	295.0	236.0	47.5	240.0	250.0	200.0	230.0	25.0	295.0	215.0	290.0	266.7	37.5
SITE-III	480.0	479.0	475.0	478.0	2.0	410.0	420.0	391.0	407.0	14.5	395.0	455.0	300.0	383.3	77.5	430.0	412.0	350.0	397.3	31.0
2001–2002	**Sep**	**Oct**	**Nov**	**Avg.**	**± S.D.**	**Dec**	**Jan**	**Feb**	**Avg.**	**± S.D.**	**Mar**	**April**	**May**	**Avg.**	**± S.D.**	**June**	**July**	**Aug**	**Avg.**	**± S.D.**
SITE-I	285.0	345.0	295.0	308.3	25.0	280.0	274.0	245.0	266.3	14.5	275.0	285.0	265.0	275.0	10.0	275.0	285.0	205.0	255.0	40.0
SITE-II	308.0	320.0	350.0	326.0	15.0	335.0	305.0	302.0	314.0	1.5	325.0	325.0	315.0	321.7	5.0	295.0	315.0	285.0	298.3	15.0
SITE-III	505.0	409.0	309.0	407.7	50.0	375.0	379.0	457.0	403.7	39.0	424.0	435.0	428.0	429.0	3.5	410.0	352.0	420.0	394.0	34.0

Avg.: Average
S.D.: Standard deviation
– : No sampling

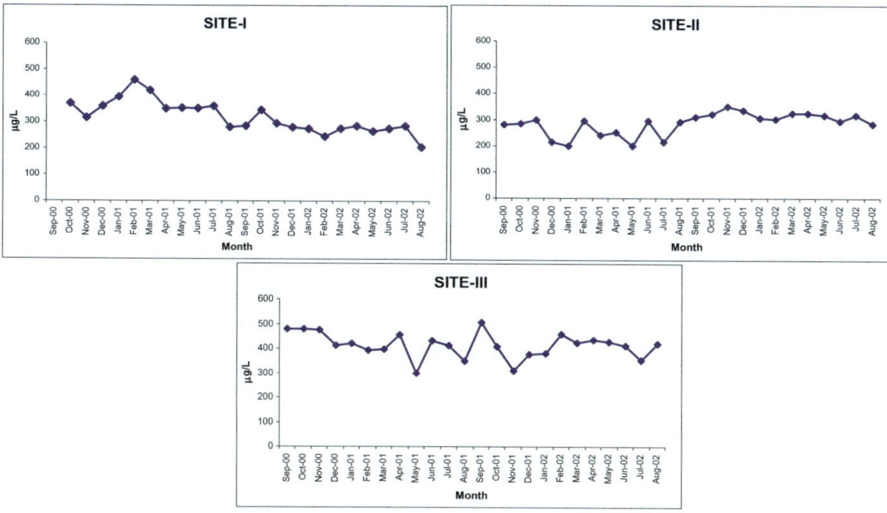

Fig. 59 Monthly fluctuations in total phosphate phosphorus (µg/L) at the investigated sites of the Anchar lake

here. The lowering of total phosphate phosphorus content may be due to its uptake by extensively growing macrophytes at this site. Bath (1996), Trivedy and Goel (1986) and Das (1989) also recorded low total phosphate phosphorus in water due to its uptake by plants.

Sawyer (1947) reported that beyond 300 g/L total phosphate phosphorus, algal blooms indicative of eutrophicated status can be expected in fresh water bodies. Vollenweider (1972) regarded phosphorus as a key element in the process of eutrophication. The 2 year monitoring of the studied lake indicates that site-III attained the status of advanced eutrophication as total phosphate phosphorus values here ranged between 300 and 500 µg/L.At site-I and site-II also total phosphate phosphorus values have been found to be more than 300 µg/L during several observations thereby indicating that water in the lake is grossly eutrophicated. The US Department of interior Division of Technical Support (1969) reported that to prevent biological nuisance, the total phosphate phosphorus concentration in the lake should not exceed 50 µg/L. The total phosphate phosphorus concentration in the Anchar lake remained above the threshold limit and hence really a nuisance for the lake.

Earlier Sarwar (1999) recorded total phosphate phosphorus value to be between 100 and 250 µg/L during his study conducted in 1983, while Sarwar and Zutshi (1989) recorded an average value of 91 g/L in the Dal lake. During the present investigation, total phosphate phosphorus values have been observed to cross the threshold limit at all the sites showing addition of ample amount of total phosphate phosphorus in the lake due to various anthropogenic activities.

ISI (1986) has recommended maximum permissible limit of this form of total phosphate phosphorus to be not more than 500 µg/L. The present data reveals that values are slightly higher than the permissible limits in the Anchar lake water particularly at site-III.

Silicate (mg/L)

The silicate values in the Anchar lake ranged from 0.8 to 8.1 mg/L.

No definite seasonal pattern in silicate concentration has been revealed from the 2 year data. Sarwar (1999) has also not revealed any seasonal trend in silicate values of the Anchar lake. Wanganeo et al. (1996) recorded maximum amount of silicate in bottom water during summer in the Nilnag whereas Kumar et al. (2003) observed high silicate contents during monsoon.

Concentration of silicate varied from 0.9 to 8.1 mg/L at site-I, 0.8 to 7.5 mg/L at site-II and from 1.4 to 5.4 mg/L at site-III µg/L (Table 82 and Fig. 60) showing the site-wise pattern shows the following trend.

Site I > Site II > Site III

At site-I, water is least disturbed by human activities and any external source of silicate is absent. Trivedy and Goel (1986) reported that natural waters have silicate from 1 to 30 mg/L. The concentration of silicate shows a decline at site-II and site-III. These sites receive waters from agricultural run-off, sewage and hospital wastes from the surrounding area and also harbours rich growth of macrophytes. Pearsall (1932) and Welch (1952) reported that silicon is most important element for the development of diatom frustules. In the present study, diatoms have not been investigated, but a study by Sarwar (1999) reveals rich diatom fauna in the Anchar lake. It is therefore possible that concentration of silicates in water gets decreased because of its uptake and utilization bu diatoms present in the lake as is suggested by Stockner and Benson (1971). However, present study reveals that there is no contribution of sewage, agricultural run-off and hospital effluents towards the silicate concentration of water in the Anchar lake.

Earlier Sarwar and Zutshi (1989) reported average silicate concentration in the Anchar lake to be 1.96 mg/L which is lower to the level (0.8–8.1 mg/L) observed during the present investigation.

Total Dissolved Solids (mg/L)

The total dissolved solids in the Anchar lake ranged between 153 and 435 mg/L. The total dissolved solids ranged from 153 to 337 mg/L at site-I, from 228 to 363 mg/L at site-II and from 207 to 435 mg/L at site-III (Table 83 and Fig. 61) showing the overall trend as follows

Site III > Site II > Site I

Total dissolved solids at all the three sites indicate the values are highest at site-III as compared to site-I.

At site-I, total dissolved solids values were found to be lower as this site is free of human interference. The only contribution of total dissolved solids at this site might be from the natural waters, as in natural waters, total dissolved solids are

Table 82 Monthly fluctuations in silicate (mg/L) at the investigated sites of the Anchar lake

	Months																		
2000–2001	**Sep**	**Oct**	**Nov**	**Avg.**	**± S.D.**	**Dec**	**Jan**	**Feb**	**Avg.**	**± S.D.**	**Mar**	**April**	**May**	**Avg.**	**± S.D.**	**June**	**July**	**Aug**	**Avg.**	**± S.D.**
SITE-I	–	8.1	3.9	4.0	2.1	4.0	5.4	4.5	4.6	0.5	4.1	7.0	4.9	5.3	1.1	6.9	4.1	0.9	4.0	1.6
SITE-II	5.5	5.4	4.0	5.0	0.7	3.9	5.0	7.5	5.5	1.3	5.0	5.0	5.0	5.0	0.0	4.9	4.5	0.8	3.4	1.9
SITE-III	1.4	1.7	4.0	2.4	1.2	3.9	4.0	5.4	4.4	0.7	4.6	4.1	4.9	4.5	0.4	4.9	4.6	1.5	3.6	1.5
2001–2002	**Sep**	**Oct**	**Nov**	**Avg.**	**± S.D.**	**Dec**	**Jan**	**Feb**	**Avg.**	**± S.D.**	**Mar**	**April**	**May**	**Avg.**	**± S.D.**	**June**	**July**	**Aug**	**Avg.**	**± S.D.**
SITE-I	1.2	1.2	1.5	1.3	0.1	1.5	1.5	1.4	1.5	0.0	1.4	1.1	1.1	1.2	0.0	1.1	1.2	1.2	1.2	0.0
SITE-II	1.4	1.5	1.5	1.5	0.0	1.5	1.8	2.8	2.0	0.5	3.2	2.5	2.4	2.7	0.0	3.0	3.1	3.2	3.1	0.0
SITE-III	1.5	1.5	1.5	1.5	0.0	1.4	1.7	3.2	2.1	0.8	2.6	2.8	3.0	2.8	0.1	3.2	2.5	2.8	2.8	0.2

Avg.: Average
S.D.: Standard deviation
– : No sampling

Fig. 60 Monthly fluctuations in silicates (mg/L) at the investigated sites of the Anchar lake

represented by carbonates, bicarbonates, chlorides, sulphates, phosphates and nitrates of calcium, magnesium, sodium, potassium and iron. (Trivedy and Goel 1986).

At site- III increased concentration seem to be due to addition of sewage, agricultural run-off and hospital wastes which are added at this site.

At Site-II the increased concentration seem to be because, at this site water from site-I and III mixes.

There are no records of earlier study in this lake.

Conductivity (μS/cm)

Conductivity values in the Anchar lake ranged from 205 to 580 μScm^{-1}. The maximum conductance (μScm^{-1}) was obtained at site-III in autumn and the minimum value (205 μScm^{-1}) at site-I during spring.

The conductivity at three sites of the Anchar lake over a period of 2 years, showed lowest values during summer which might be due to the uptake of ions by macrophytes whose growth was seen to be maximum during this period. Conductivity was observed to rise in autumn might be because of release of nutrients into water as a result of death and decomposition of aquatic plants during this period.

The site-wise variations in conductivity values of the lake show an overall trend as under:

$$\text{Site III} > \text{Site II} > \text{Site I}$$

Conductivity was found to be lowest ranging from 205 to 450 μScm^{-1} at site-I This site is almost free from human habitation and thus water is least contaminated.

Table 83 Monthly fluctuations in total dissolved solids (mg/L) at the investigated sites of the Anchar lake

	Months																			
2000–2001	**Sep**	**Oct**	**Nov**	**Avg.**	**± S.D.**	**Dec**	**Jan**	**Feb**	**Avg.**	**± S.D.**	**Mar**	**April**	**May**	**Avg.**	**± S.D.**	**June**	**July**	**Aug**	**Avg.**	**± S.D.**
SITE-I	–	330.0	337.0	222.3	3.5	311.0	292.0	217.0	273.3	37.5	231.0	221.0	153.0	201.7	34.0	165.0	178.0	217.0	186.7	19.5
SITE-II	363.0	360.0	350.0	357.7	5.0	336.0	277.0	228.0	280.3	24.5	255.0	232.0	263.0	250.0	15.5	270.0	263.0	262.0	265.0	0.5
SITE-III	435.0	432.0	408.0	425.0	12.0	373.0	375.0	378.0	375.3	1.5	420.0	299.0	296.0	338.3	1.5	303.0	265.0	207.0	258.3	29.0
2001–2002	**Sep**	**Oct**	**Nov**	**Avg.**	**± S.D.**	**Dec**	**Jan**	**Feb**	**Avg.**	**± S.D.**	**Mar**	**April**	**May**	**Avg.**	**± S.D.**	**June**	**July**	**Aug**	**Avg.**	**± S.D.**
SITE-I	333.0	309.0	310.0	317.3	0.5	280.0	262.0	225.0	255.7	18.5	229.0	220.0	187.0	212.0	16.5	171.0	175.0	251.0	199.0	38.0
SITE-II	262.0	348.0	325.0	311.7	11.5	339.0	338.0	281.0	319.3	28.5	306.0	228.0	263.0	265.7	17.5	260.0	252.0	256.0	256.0	2.0
SITE-III	276.0	343.0	348.0	322.3	2.5	361.0	337.0	234.0	310.7	51.5	432.0	373.0	386.0	397.0	6.5	348.0	240.0	221.0	269.7	9.5

Avg.: Average
S.D.: Standard deviation
– : No sampling

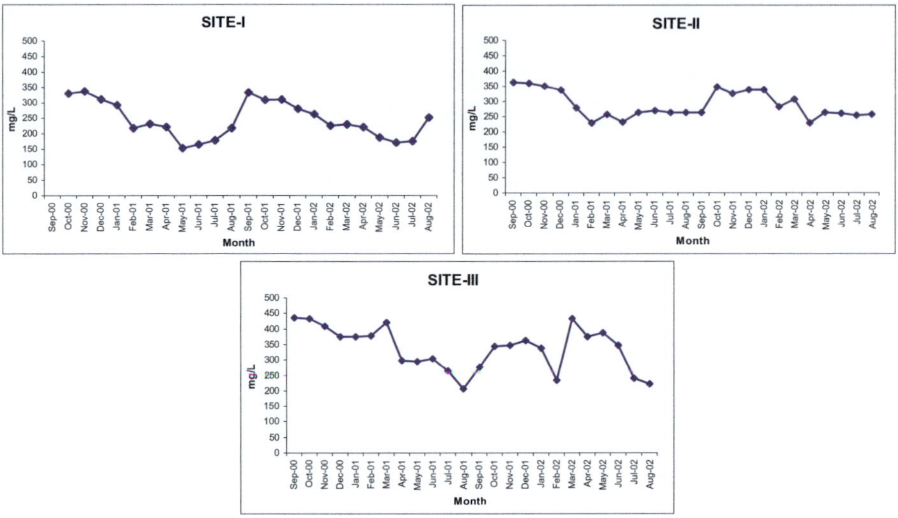

Fig. 61 Monthly fluctuations in total dissolved solids (mg/L) at the investigated sites of the Anchar lake

At site-II, conductivity was comparatively higher than at site-I, ranging from 305 to 485 μScm^{-1} (Table 84 and Fig. 62).

At site-III, conductivity was found to be highest ranging from 277 to 580 μScm^{-1}. This site receives wastes from agricultural fields, sewage from catchment area and the effluents from SKIMS hospital thereby increasing the organic and inorganic load in water. Decomposition of organic matter further release free ions into water thereby increasing its conductivity values. Bath (1996) and Sajeev (1999) have also suggested that decomposition of organic matter release free ions into water and increase the conductivity values..

This site lies in the centre of the lake and received no direct effluent discharge but indirectly via water coming from the site-III.

Anchar lake is a drainage type of lake. The overall conductivity values of this lake depict high ionic concentration. As per Wilcox (1995) categorization of water for irrigation purpose, the water of the Anchar lake is classified as good for irrigation on the basis of conductivity values. Oslen (1950) classified water with conductivity values upto 200 μS/cm as the Oligo-mesotrophic and from 200 to 500 μS/cm as the ß-mesotrophic.

On this basis, water of the lake remained mesotrophic at site-I and site-II but crosses the limit at site-III. However, as per Dunn (1962), water of the lake falls in the category of eutrophic level at all the sites.

Earlier Sarwar (1999) worked on the Anchar lake in 1983 and recorded the conductivity values ranging from 122 to 526 μS/cm. In a later study by Sarwar and Zutshi (1989) they recorded an average conductivity of 316 μS/cm in the Anchar

Table 84 Monthly fluctuations in conductivity (μ Scm^{-1}) at the investigated sites of the Anchar lake

	Months																			
2000–2001	**Sep**	**Oct**	**Nov**	**Avg.**	**± S.D.**	**Dec**	**Jan**	**Feb**	**Avg.**	**± S.D.**	**Mar**	**April**	**May**	**Avg.**	**± S.D.**	**June**	**July**	**Aug**	**Avg.**	**± S.D.**
SITE-I	–	440.0	450.0	296.7	5.0	415.0	390.0	290.0	365.0	50.0	309.0	295.0	205.0	269.7	45.0	220.0	238.0	290.0	249.3	26.0
SITE-II	485.0	480.0	440.0	468.3	20.0	403.0	370.0	305.0	359.3	32.5	340.0	310.0	351.0	333.7	20.5	361.0	351.0	352.0	354.7	0.5
SITE-III	580.0	577.0	544.0	567.0	16.5	498.0	500.0	505.0	501.0	2.5	560.0	399.0	395.0	451.3	2.0	405.0	354.0	277.0	345.3	38.5
2001–2002	**Sep**	**Oct**	**Nov**	**Avg.**	**± S.D.**	**Dec**	**Jan**	**Feb**	**Avg.**	**± S.D.**	**Mar**	**April**	**May**	**Avg.**	**± S.D.**	**June**	**July**	**Aug**	**Avg.**	**± S.D.**
SITE-I	445.0	413.0	415.0	424.3	1.0	374.0	350.0	300.0	341.3	25.0	315.0	295.0	250.0	286.7	22.5	228.0	239.0	335.0	267.3	48.0
SITE-II	350.0	464.0	434.0	416.0	15.0	452.0	450.0	375.0	425.7	37.5	408.0	305.0	351.0	354.7	23.0	350.0	336.0	342.0	342.7	3.0
SITE-III	368.0	458.0	465.0	430.3	3.5	482.0	450.0	312.0	414.7	69.0	570.0	498.0	515.0	527.7	8.5	465.0	320.0	295.0	360.0	12.5

Avg.: Average
S.D.: Standard deviation
–: No sampling

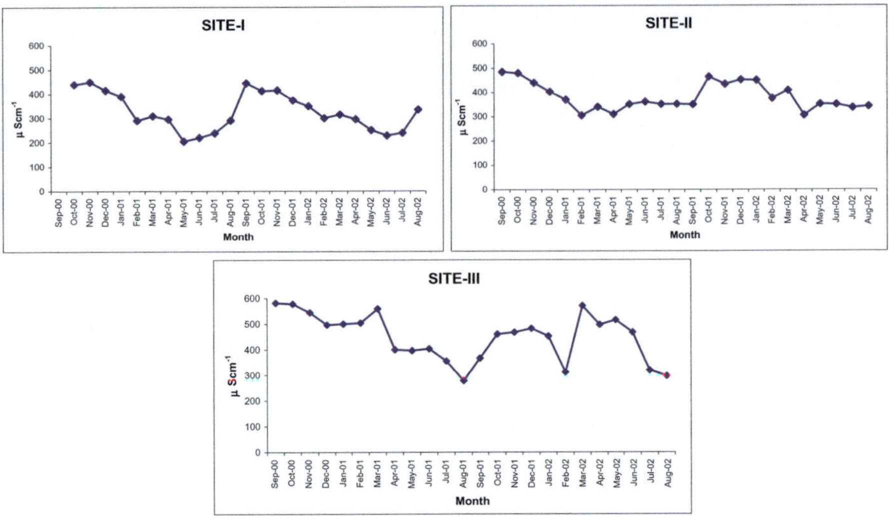

Fig. 62 Monthly fluctuations in conductivity (µ Scm^{-1}) at the investigated sites of the Anchar lake

lake. The range of conductivity recorded in the present study reveals that ionic load is increasing in lake due to various anthropogenic pressures.

ISI (1986) have suggested the maximum permissible limit of 500 µS/cm. On the basis of these recommendations it can be concluded that the water at all the sites exceed the limits.

Macrophytes

In Anchar lake, the macrophytes show all the four life forms viz-emergents, rooted floating-leaf type, free floating type and submergeds. The lake littorals are colonized mostly by the emergent vegetation and the open waters are dominated by other three life form classes besides the emergents.

Regarding species composition, 36 species of macrophytes were recorded in the Anchar lake during the course of present investigation. These included 21 species of emergents, 5 rooted floating-leaf type species, 2 free floating species and 8 species of submergeds. However, monthly fluctuations in the number of species in this lake were quite high. The macrophytes recovered include *Alisma plantago aquatica, Bidens cernua, Carex sp., Ceratophyllum demersum, Cyperus sp., Echinocloa crus-galli,, Eliocharis palustris, Gallum sp., Hydrilla verticillata, Hydrocharis dubia, Lemna sp., Lycopus europaeus, Marsillea quadrifolia, Menyanthese trifolata, Myriophyllum spicatum, Myriophyllum verticillatum, Najas sp., Nasturtium officinale, Nelumbo nucifera, Nymphaea sp., Nymphoides peltatum, P. lucens, P. pucillus, Phragmites australis, Polygonum amphibium, Potamogeton natans, Potamogeton crispus, Ranunculus seeleratus, Rorripa islandica,Sagittaria sagitfo-*

Table 86 List of Protozoan taxa recorded at the investigated sites of the Anchar lake

S. No.	Protozoa taxa	Site-I	Site-II	Site-III
1	*ARCELLA DISCOIDES* EHRENBERG	+	+	+
2	*A. megastoma* Penard	+	+	+
3	*A. mitrata* Leidy	+	–	–
4	*A. vulgaris* Ehrenberg	+	+	+
5	*Bullinaria indica* Penard	+	+	–
6	*Centropyxis ecornis* Ehrenberg	+	+	–
7	*C. stellata* Wailes	+	+	+
8	*C. aerophila* Penard	+	+	–
9	*C. constricta* Ehrenberg	+	–	+
10	*Difflugia acuminata* Penard	+	+	+
11	*D. oblonga* Ehrenberg	–	–	+
12	*D. tuberculata* Wallich	–	–	+
13	*D.aurela* Ehrenberg	+	–	+
14	*D. corona* Ehrenberg	+	+	–
15	*D. lebes* Penard	+	+	+
16	*D. lobostoma* Leidy	+	+	–
17	*D. rubescens* Penard	+	+	–
18	*Euglypha laevis* Ehrenberg	–	+	–
19	*E. ciliata* Ehrenberg	+	+	+
20	*E. tuberculata* Dujardin	–	+	+
21	*Lesquereusia spiralis* Ehrenberg	–	–	+
22	*L. modesta* Ehrenberg	+	+	+
23	*Nebula dentistoma* Penard	+	–	+
24	*Paraeuglypha reticulata* Penard	+	+	–
25	*Paraeuglypha sp.*	–	+	+
26	*Paramecium sp.*	–	+	+
27	*Paraquadrula irregularis* Archer	+	+	+
28	*Pyxidicula scutella* Playfair	–	+	+
29	*Trinema sp.*	–	–	+
30	*Wailesella sp.*	–	+	+

including protozoans. Patnaik (1973) and Sajeev (1999) observed three peaks in the population density of protozoans in Chilka lake and in the Ropar headworks reservoir respectively, the one during the summer being the highest. Syal (1996) failed to find any definite seasonal trend in the distribution of protozoa in Satluj river and attributed this to constantly running water along which the protozoan are swept away (Table 87, Figs. 64 and 65).

The taxa which were recorded at site-I are *Arcella mitrata, A. discoides, A. vulgaris, A. megastoma, Bullinaria indica, Centropyxis stellata, C. ecornis, C.constricta, C. aerophila, Difflugia lebes, D. corona, D. aerula, D. accuminata, D. lobostoma, D. rubescens, Euglypha ciliate, Lesqueresia modesta, Nebula dentis-*

Table 87 Seasonal variations in the population density (Ind/L) of the protozoa at the investigated sites of the Anchar lake

Sites	Year	Autumn	Winter	Spring	Summer
Site-I	**2000–2001**	40	16	168	176
	2001–2002	96	48	104	112
	Average	68	32	136	144
Site-II	**2000–2001**	96	224	248	568
	2001–2002	88	144	232	456
	Average	92	184	240	512
Site-III	**2000–2001**	80	120	192	314
	2001–2002	24	80	104	120
	Average	52	100	148	217

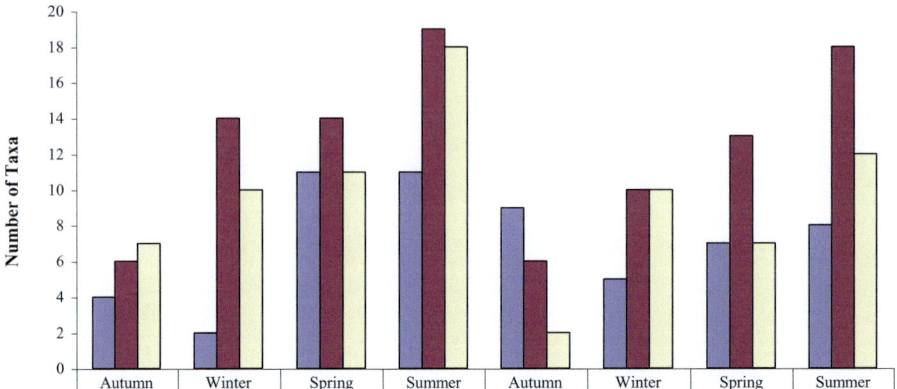

	Autumn	Winter	Spring	Summer	Autumn	Winter	Spring	Summer
Site-I	4	2	11	11	9	5	7	8
Site-II	6	14	14	19	6	10	13	18
Site-III	7	10	11	18	2	10	7	12

Fig. 64 Seasonal variations in species diversity of the Protozoa at the investigated sites of the Anchar lake

toma and Paraeuglypha reticulate and Paraquadrula irregularis (Table 88). This site lie where the river Sind feeds the Anchar lake and water here is fast flowing. The macrophytic vegetation is scarce. However, the littorals at this site are with willow plantation. This site is almost free from pollution source. Species diversity was observed to be lowest of all the sites. The population density was also observed to be lowest at this site. Sajeev (1999) in Ropar Headworks waters observed Protozoan species diversity and population density very low at the site which is bereft of aquatic plants and where water flow is swift. Similar suggestions were made by Ramakrishnaiah and Sarkar (1982) on Konar Reservoir of Bihar and by Bath (1996) in Harike Reservoir who opined that protozoans thrive well in relatively calm and un-disturbed water harboured with lot of aquatic plants.

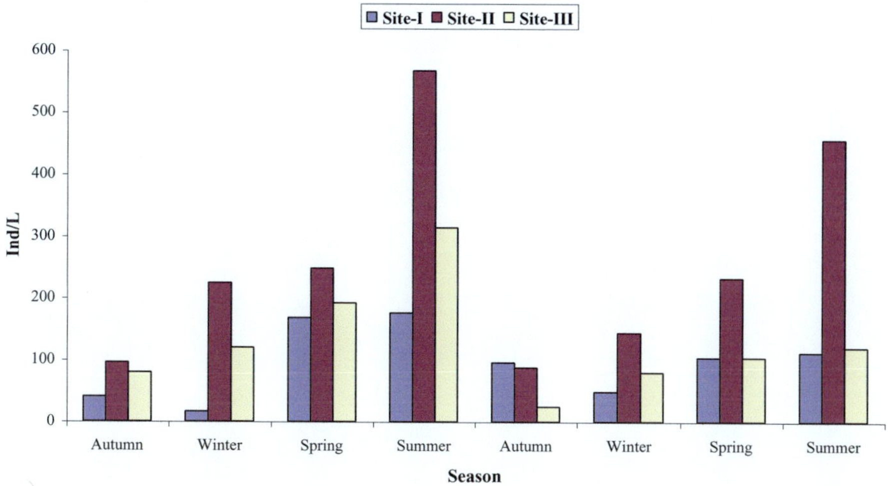

Fig. 65 Seasonal variations in the population density (Ind/L) of the Protozoa at the investigated sites of the Anchar lake

The protozoan taxa which were recorded at Site-II are *Arcella disciodes, A. vulgaris, A.megastoma, Bullinaria indica, Centropyxis stellata, C. aerophila, C. ecornis, Difflugia lebes, D. lobostoma, D. rubescens, D. acuminate, D. corona, Euglypha laevis, E.tuberculata, E. ciliate, Lesquereusia modesta, Paraeuglypha reticulata, P. araeuglypha sp.,Paramecium sp., Paraquadrula irregularis, Pyxidicula scutella* and *Wailesella* sp. respectively (Table 89). This site is located in the centre of the lake and is the deepest of all the sites. The site harbours rich growth of macrophytes. The site itself does not receive any sewage or agricultural run-off, however, the water coming from Site-III mixes at this site. The population density is highest of all the sites. Sajeev (1999) in Ropar headworks waters observed species diversity and population density very high at the site which is characterized by relatively less disturbed and calm waters and harbouring a rich growth of plants. Similar reports have also been given by Zafar (1959a), Philipose (1960), Venkateshwarlu (1969), Munawar (1970b), Ramakrishnaiah and Sarkar (1982), Bath (1996) and Kaur et al. (2003). Our findings are in agreement with all these authors.

The species which were recorded at Site-III are *Arcella discoides, A. vulgaris, A. megastoma, Centrophyxis constricta, C. stellata, Difflugia lebes, D. aurela, D. tuberculata, D. acuminata, D. oblonga, Euglypha ciliata, E. tuberculata, Lesquereusia modesta, L. spiralis, Nebula dentistoma, Paraeuglypha sp., Pyxidicula scutella, Paraquadrula irregularis, Paramecium sp., Trinema sp.* and *Wailesella sp.* (Table 90). This site is located at the littorals of the lake. Here water is shallow and highly turbid. This site recieves effluents from the adjoining agricultural fields, sewage from local catchment area, and above all effluents from the SKIMS hospital. The site is highly enriched with nutrients and harbours a rich macrophytic vegetation. The present study reveals high species diversity and population density of protozoans here. Sajeev (1999) observed high species diversity and population density

Table 88 Monthly variations in the population density (Ind/L) of the protozoa at the study site-I of the Anchar lake

Taxa	Sep	Oct	Nov	Dec	Jan	Feb	Mar	Apr	May	Jun	Jul	Aug	Sep	Oct	Nov	Dec	Jan	Feb	Mar	Apr	May	Jun	Jul	Aug
Arcella mitrata	–	–	–	–	–	–	+	+	+	++	++	+	–	–	–	–	–	–	–	–	–	+	+	++
A. discoides	+	+	–	–	–	–	–		+	–	–	–	++	++	–	–	–	–	+	+	–	–	–	–
A. megastoma	+	++	–	–	–	–	–	++	++	–	–	–	++	++	–	+	+	–	–	–	–	–	–	–
A. vulgairs	–	–	–	–	–	–	–	++	+	–	–	–	–	–	–	–	–	–	–	++	++	–	–	–
Bullinaria indica	–	–	–	–	–	–	–	++	–	–	+	+	–	–	–	–	–	–	+	+	+	–	++	++
C. aerophila	–	–	–	–	–	–	–	–	–	–	++	+	+	+	–	–	–	–	–	–	–	–	+	++
C. constricta	–	–	–	–	–	–	–	–	–	–	+	++	+	–	–	–	–	–	–	–	–	–	–	–
Centropyxis ecornis	–	–	–	+	+	–	–	–	–	–	++	++	–	–	–	+	–	–	–	–	–	–	++	–
Centropyxis stellata	–	–	–	–	–	–	–	–	++	–	–	–	–	–	–	–	–	–	–	+	–	–	–	–
D. accuminata	+	+	–	–	–	–	–	++	++	–	++	–	+	+	–	+	–	–	–	–	–	–	++	–
D. aurela	–	–	–	+	+	–	–	–		+	+++	++	+	–	–	+	–	–	–	–	–	–	++	+
D. corona	–	–	–	–	–	–	–	–		+	+	+	+	–	–	–	–	–	–	–	–	–	–	–
D. rubescens	–	–	–	–	–	–	–	–	++	–	–	–	–	–	–	–	–	–	–	+	–	–	–	–
Difflugia labostoma	–	–	–	–	–	–	–	–		–	+	+	–	–	–	–	–	–	–	–	–	–	+	+
Difflugia lebes.	–	–	–	–	–	–	–	++	+	–	–	.	–	–	–	–	–	–	–	+	–	–	–	–
Euglypha ciliata	+	+	–	–	–	–	–	–		–	–	–	+	–	–	–	–	–	–	–	–	–	–	–
Lesquereusia modesta	–	–	–	–	–	–	–	+		–	–	–	–	–	–	–	–	–	–	+	–	–	–	–
Nebula sp.	–	–	–	–	–	–	–	–		–	++	+	–	–	–	–	–	–	–	–	–	–	–	–
Paraeuglypha reticulata	–	–	–	–	–	–	–	–	+++	–	++	++	+	+	–	+	–	–	–	–	–	–	–	–
Paraquadrula irregularis	–	–	–	–	–	–	–	–		–	–	–	–	–	–	–	–	–	–	–	–	+	+	++

0–5 Ind/L = +
6–10 Ind/L = ++
11–15 Ind/L = +++
and above = ++++

Table 89 Monthly variations in the population density (Ind/L) of the protozoa at the study site-II of the Anchar lake

Taxa	Sep	Oct	Nov	Dec	Jan	Feb	Mar	Apr	May	Jun	Jul	Aug	Sep	Oct	Nov	Dec	Jan	Feb	Mar	Apr	May	Jun	Jul	Aug
Arcella discoides	+	++	–	+	+	–	–	++	++	+	++	–	+	+	–	–	–	–	–	–	–	–	–	–
Arcella megastoma	–	–	–	+	+	–	–	–	–	–	–	–	–	–	–	–	–	–	+	+	+	++	++	–
Arcella vulgaris	–	–	–	+	–	–	+	+	+	–	–	–	++	+		–	–	–	++	+	–	–	++	++
Bullinaria indica	–	–	–	–	–	–	–	–	–	+	+	–	–	–	–	–	–	–	–	–	–	–	–	–
Centropyxis aerophila	++	–	–	–	–	–	+	+	–	+	+	–	+	+	–	–	–	–	+	–	–	–	++	++
Centropyxis ecornis	++	+	–	+	–	–	++	+	+	++++	++	++	–	–	–	–	–	–	++	+	–	–	–	–
Centropyxis stellata	++	+	–	+	–	–	–	+	+	+++	++	++	+	++	–	+	+	–	++	–	++	++++	++++	+++
Difflugia acuminata	+	–	–	++	+	–	+	–	–	+++	++++	++++	+	+	–	+	+	–	++	–	–	++++	+++	+++
Difflugia corona	–	–	–	+	–	–	–	–	–	++	+++	–	–	–	–	+	–	–	–	–	–	+++	++	+
Difflugia lebes	–	–	–	+	–	–	+	++	+	+++	++	++	–	–	–	++	–	–	–	–	+	++++	++	++
Difflugia lobostoma	–	–	–	+	–	–	–	–	–	+	+	–	–	–	–	+	–	–	–	–	–	++	–	–
Difflugia rubescens	–	–	–	–	–	–	++	–	–	++	++++	++++	–	–	–	–	–	–	++	–	–	++++	++	–
Euglypha ciliata	–	–	–	–	–	–	+++	–	–	++	++	++	++	++	–	–	–	–	–	++	++	++	++	++
Euglypha laevis	–	–	–	–	–	–	–	–	–	++	++	++	–	–	–	–	–	–	–	–	–	++	++	–
Euglypha tuberculata	–	–	–	++	–	–	++	++	–	++++	++++	++	–	–	–	++	–	–	+	–	++	++	++	++

(continued)

Table 89 (continued)

Taxa	Sep	Oct	Nov	Dec	Jan	Feb	Mar	Apr	May	Jun	Jul	Aug	Sep	Oct	Nov	Dec	Jan	Feb	Mar	Apr	May	Jun	Jul	Aug
Lesquereusia modesta	–	–	–	++	++	–	++	–	++	++++	++	–	–	–	–	++	–	–	–	++	+	++++	+++	+
Paraeuglypha reticulata	–	–	–	++	++	–	+	+	+	++	++	+	–	–	–	++	+	–	–	++	–	++	++	++
Paraeuglypha sp.	–	–	–	–	–	–	–	–	–	++	++	++	–	–	–	–	–	–	–	–	–	++	+++	++
Paramecuim sp.	–	–	–	–	–	–	+	–	+	–	–	–	–	–	–	–	–	–	–	–	–	–	–	–
Paraquadrula irregularis	–	–	–	–	–	–	+	+	+	+++	++++	++++	–	–	–	–	–	–	–	++	++	++	+	–
Polyarthra vulgaris	–	–	–	++	++	–	–	–	–	++	+	–	–	–	–	++	–	–	–	–	–	–	+	++
Pyxidicula scutella	–	–	–	++	++	–	–	–	–	++	++	–	–	–	–	–	–	–	–	–	–	++	–	–
Wailesella sp.	++	–	–	–	–	–	–	–	–	–	–	–	–	–	–	++	–	–	–	–	–	–	–	–

0–5 Ind/L = +
6–10 Ind/L = ++
11–15 Ind/L = +++
and above = ++++

Table 90 Monthly variations in the population density (Ind/L) of the protozoa at the study site-III of the Anchar lake

Taxa	Sep	Oct	Nov	Dec	Jan	Feb	Mar	Apr	May	Jun	Jul	Aug	Sep	Oct	Nov	Dec	Jan	Feb	Mar	Apr	May	Jun	Jul	Aug
Arcella discoides	+	−	−	++	++	+	++	++	+	++	++	+				−	−	−	++	−	+		+	
Arcella megastoma	+	+	−	+	++	−	−	−	−	−	−	−	−	−	−	+	+	−	−	−	−	+	+	−
Arcella vulgaris	−	−	−	−	−	−	+	−	+	−	+	++	−	−	−	−	−	−	−	−	++	−	−	−
Centropyxis constricta	−	−	−	++	++	−	−	−	−	++	−	+	−	−	−	−	+	−	−	−	−	−	+	+
Centropyxis stellata	++	+	+	−	−	−	+	++	+++	−	+	−	−	−	−	+	−	−	−	++	−	−	+	−
D. tuberculata	−	−	−	+	−	−	−	−	−	++	++	+	+	++	−	−	−	−	−	−	−	−	−	−
Difflugia acuminata	−	−	−	−	−	−	−	+	−	−	++	−	−	−	−	−	−	−	−	−	−	+	+	++
Difflugia aurela	+	−	−	+	−	−	−	−	−	−	+++	++	−	−	−	+	+	−	−	−	−	++	−	−
Difflugia lebes	−	−	−	−	−	−	−	−	−	++	−	−	−	−	−	−	−	−	−	−	−	+	++	+
Difflugia oblonga	−	−	−	−	−	−	++	+	+	+	++++	++	−	−	−	−	+	−	−	+++	−	+	++	+
Euglypha ciliata	−	−	−	+	+	−	−	−	−	−	++	−	−	−	−	−	−	−	−	−	−	+	+	+
Euglypha tuberculata	+	−	−	+	+	+	−	−	−	−	++	−	−	−	−	−	−	−	−	−	−	+	+	+
Lesqueresia modesta	−	−	−	+	+	−	++	++	−	++++	+++	++	−	−	−	−	+	−	++	−	−	−	−	−
Lesqueresia spiralis	−	−	−	−	−	−	−	+	+	+++	+++	+	−	−	−	−	−	−	−	−	−	+	+++	+++

(continued)

Table 90 (continued)

Taxa	Sep	Oct	Nov	Dec	Jan	Feb	Mar	Apr	May	Jun	Jul	Aug	Sep	Oct	Nov	Dec	Jan	Feb	Mar	Apr	May	Jun	Jul	Aug
Nebula dentistoma	–	–	–	–	–	–	–	–	–	++	+	+	–	–	–	–	–	–	–	–	–	–	–	–
Paramecium sp.	–	–	–	–	–	–	++	++	+	–	–	–	–	–	–	–	–	–	+++	–	–	–	+	–
Paraquadrula irregularis	+	–	–	–	–	–	+	+	+	–	–	–	–	–	–	–	++	–	–	–	–	–	–	–
Pareuglypha sp.	+	–	–	–	–	–	–	–	–	–	++	++	–	–	+	+	+	–	–	–	–	–	–	–
Pyxidicula scutella	–	–	–	+	++	+	+	+	+	++	++	++	–	–	–	–	–	–	++	–	–	–	–	–
Trinema sp.	–	–	–	–	++	–	–	–	–	++	++	+	–	–	–	+	–	–	–	–	–	–	–	–
Wailesella sp.	–	–	–	–	–	–	+	+	–	++++	+	+	–	–	–	+	–	–	–	–	–	–	–	–

0–5 Ind/L = +
6–10 Ind/L = ++
11–15 Ind/L = +++
and above = ++++

of protozoa in water receiving organic waste discharges and harbouring rich growth of macrophytes in Ropar headworks reservoir. Hellawell (1986) opined that organic matter accelerates the growth of aquatic plants which in turn support zooplanktons. Kaur et al. (2003) observed high population density of protozoan at site characterised by rich growth of aquatic vegetation and nutrient rich waters. Similar observations were made by Zafar (1959a), Philipose (1960), Venkateshwarlu (1969), Munawar (1970a), and Bath (1996).

The taxa which were encountered at all the study sites of the lake included *Arcella discoides, A. vulgaris, A. megastoma, Centropyxis stellata, Diffulgia lebes, D. acuminata, Euglypha ciliata, Lesquereusia modesta* and *Paraquadrula irregularis*. Syal (1996) found *Arcella* sp. in both clean and polluted waters while Sukumaran et al. (1984) found it only in clean waters. Algarsamy et al. (1967) and Verma et al. (1978) discovered it in sewage contaminated waters. *Centropyxis* sp. was found in clean waters by Bilgrami and Dattamunshi (1979) while Syal (1996) found it in waters contaminated with sewage. David and Ray (1966) found it in both clean as well as polluted waters. *Difflugia* sp. was found to be present both in polluted and clean waters by David and Ray (1966), Verma and Dalela (1975) and Sexena and Mishra (1990). Algarsamy et al. (1967), Bilgrami and Dattamunshi (1979) and Kaur et al. (1996a, b) found it in waters contaminated with sewage. Syal (1996) found it in waters contaminated by industrial wastes. However, Sukumaran et al. (1984) found it only in clean waters. *Euglypha ciliata* was recorded at the sites which receive sewage and fertilizer factory waste by Syal (1996) in river Satluj. David and Ray (1966), however, reported its occurence in both clean as well as polluted waters.

Out of total of 20 protozoan taxa, 2 taxa were found exclusively at this site-I. These are *Arcella mitrata* and *Centropyxis ecornis*. Sukumaran et al. (1984) also found *Arcella* sp. in clean waters only. David and Ray (1966) and Bath (1996) found *Arcella* sp. in clean waters as well as in waters receiving effluence. Algarsamy et al. (1967) and Verma et al. (1978) discovered it in sewage waters. *Centropyxis sp.* was discovered both in clean waters and sewage receiving waters by David and Ray (1966), Bath (1996), and Syal (1996). Algarsamy et al. (1967) and Bilgrami and Dattamunshi (1979) on the other hand found it in waters contaminated with sewage. None of the protozoan taxon was found to be exclusively at site-II of the lake.

The taxa which were exclusively found at site-III during the study period are *Difflugia tuberculata, Difflugia oblonga, Lesquereusia spiralis* and *Trinema sp*. The presence of these protozoan taxa at this site suggests that these are tolerant towards pollution by sewage, agricultural run-off and hospital waste. Among the above mentioned protozoan taxa, *Difflugia* sp. was found to be ubiquitous in nature by David and Ray (1966) and Syal (1996), who found it in both clean and polluted waters. Algarsamy et al. (1967) and Bilgrami and Dattamunshi (1979) found it in waters contaminated by sewage. Verma and Dalela (1975) and Sexena and Mishra (1990) found it in waters contaminated by industrial wastes. On the other hand, Sukumaran et al. (1984) found it in clean waters. Presencce of *Trinema sp. and Lesquereusia spiralis*, indicates that these protozoans can thrive well in polluted waters. Bath (1996) also found *Trinema* sp. in the Satluj waters at the site which is polluted by industrial and domestic effluents. However, rest of two taxa were found to be absent in aquatic ecosystems of Punjab. (Syal 1996, Bath 1996; Sajeev 1999).

Previous studies on the Anchar lake do not include protozoans. (Balkhi and Yousuf 1991; Zutshi et al. 1980). The present study is the first one to report distribution pattern of Protozoa in the Anchar lake.

Rotifers

A total of 12 rotifer taxa were recorded in the Anchar lake during the entire study period. These include *Brachionus bidentada, B.calyciflorus, B.quadridentata, Cephalodella sp., Chromogaster ovalis, Epiphanies sp., Filinia longiseta, Gastropus sp., Keratella cochlearis, K.valga, Polyarthra vulgaris and Synchaeta sp*. Out of these 6 taxa were recorded at Site-I, 8 at Site-II, and 7 rotifer at Site-III (Table 91).

The rotifer fauna of Anchar lake shows only one peak in population density which was recorded during the summer at all the three sites. Sarwar and Parveen (1996) also recorded higher number of rotifers during summer in Khusalsar and Gilsar lakes of Kashmir. Zutshi et al. (1980) while studying nine lakes of Kashmir, recorded abundance of rotifers during both summer and winter in all the lakes excepting in Trigam lake where population peaks were observed during spring and autumn. Moore (1980), Nogrady (1983), Bath (1996), Kaur et al. (1996a, b) and Pulle and Khan (2003) also reported more rotifers at higher temperature. Gabriel et al. (1975) found that pattern of rotifer population density was different in different water bodies. He recorded three peaks during May, July and November in a mesotrophic lake, a single during summer in eutrophic lake and two distinct peak during spring and autumn in dystrophic lake. Biswas and Konar (2000) also recorded rotifer group dominating in winter at least polluted site and in summer at polluted site. Patnaik (1973) and Sajeev (1999) recorded three peaks in the Chilka lake and Ropar head works reservoir respectively. Vashist and Monga (1981) observed rotifer peak in

Table 91 List of Rotifer taxa recorded at the investigated sites of the Anchar lake

S. No.	Rotifer taxa	Site-I	Site-II	Site-III
1	*Brachionus bidentata* Pallus	–	–	+
2	*Brachionus calyciflorus* Pallus	–	+	+
3	*Brachionus quadridentata* Pallus	+	–	–
4	*Cephalodella sp.*	+	–	+
5	*Chromogaster ovalis* Lauterborn	+	–	–
6	*Epiphanes sp.*	–	+	+
7	*Filinia longiseta* Bory de St. Vincent	–	+	+
8	*Gastropus sp.*	+	+	–
9	*Keratella cochlearis* Bory de St. Vincent	+	+	+
10	*Keratella valga* Bory de St. Vincent	+	+	+
11	*Polyarthra vulgaris* Ehrenberg	–	+	–
12	*Synchaeta sp.*	–	+	–

January in Bhakra main line canal. Shukla (1995) observed rotifer peaks in December and May and fall in density in January and February in Govindgarh lake of Rewa (M.P). Sugunan (1995) and Khatri (1985) and Thomas and Azis (1996) recorded maximum rotifers during the monsoon season (Table 92, Figs. 66 and 67).

The taxa recorded at Site-I, include *Brachionus quadridentata, Cephalodella sp., Chromogaster ovalis, Gastropus sp., Keratella valga and K.cochlearis.* The water at this site is fast flowing and macrophytic vegetation is scarce. Any direct pollution source is absent here. The species diversity as well as population density of rotifers at this site was found to be the lowest of all the sites. Tonopi (1980) suggested that rotifers flourished well at places where plant growth was abundant as the latter provides food, shelter and retreat to them. Sajeev (1999) also recorded low species diversity and population density at sites of the Ropar wetland where current of water

Table 92 Seasonal variations in the population density (Ind/L) of the Rotifers at the investigated sites of the Anchar lake

Sites	Year	Autumn	Winter	Spring	Summer
Site-I	**2000–2001**	8	–	16	64
	2001–2002	20	–	–	48
	Average	14	0	8	56
Site-II	**2000–2001**	96	32	64	128
	2001–2002	80	16	24	104
	Average	88	24	44	116
Site-III	**2000–2001**	64	8	16	128
	2001–2002	16	8	8	72
	Average	40	8	12	100

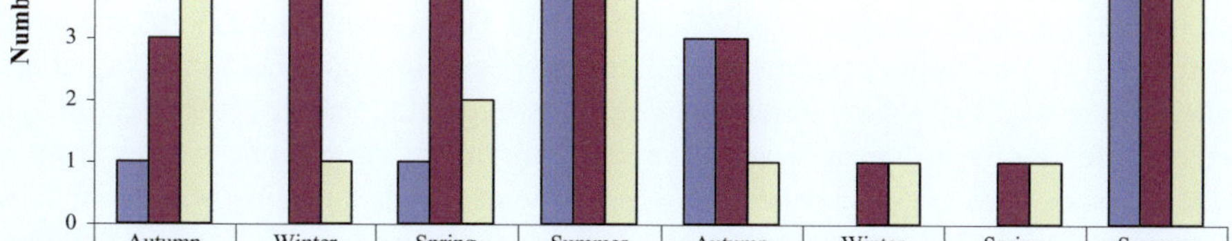

	Autumn	Winter	Spring	Summer	Autumn	Winter	Spring	Summer
Site-I	1		1	5	3			6
Site-II	3	4	4	5	3	1	1	7
Site-III	4	1	2	6	1	1	1	6

Fig. 66 Seasonal variations in the species diversity of the Rotifer at the investigated sites of the Anchar lake

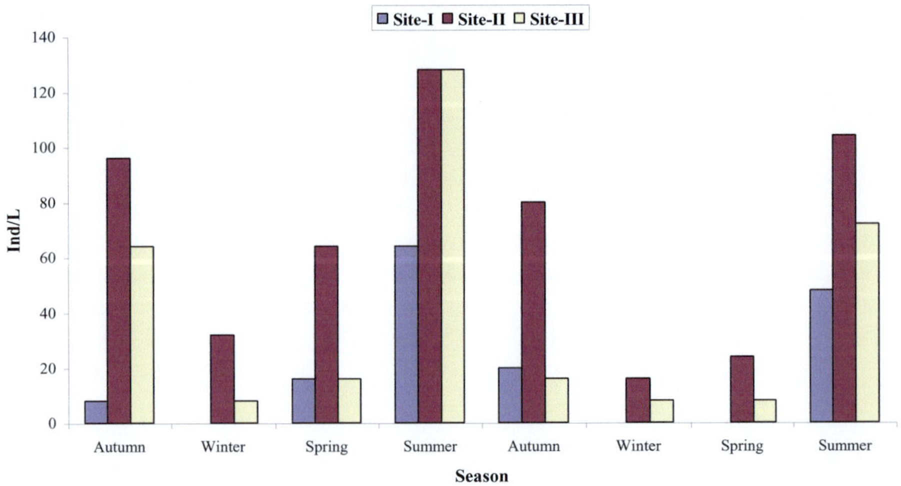

Fig. 67 Seasonal variations in the population density (Ind/L) of the Rotifers at the investigated sites of the Anchar lake

was fast and aquatic vegetation was either lacking or scarce. The low species diversity may also be due to comparatively less nutrients present at this site as evident from the physico-chemical studies. This has also been suggested by Jayangoudar (1980), Yousuf and Qadri (1985) and Bath and Kaur (1997) who opined that rotifers reproduce and grow well in shallow water areas having rich organic bottom and aquatic plants (Table 93).

At Site-II, the rotifer taxa recorded which include *Brachionus calyciflorus, Epiphanes sp., Filinia longiseta, Gastropus sp., Keratella cochlearis, K. valga, Polyarthra vulgaris and Synchaeta sp*. This site is located in the centre of the lake and is deepest of all the selected sites in the lake. Any direct source of pollution is absent, however, with the mixing of water coming from site-I and site-III in this region of the lake, the sewage, agricultural run-off and hospital wastes added at Site-III get diluted here. Physico-chemical studies have shown water to be moderately rich in nutrients at this site. This site is characterized by standing water harbouring rich growth of macrophytes. The species diversity and population density was found to be highest of all the sites. Balkhi and Yousuf (1991) and Syal (1996) recorded rotifers to be maximum in terms of number of species and population density at the sites where water was slow moving and densely vegetated Sajeev (1999) recorded a rich rotifer fauna in Ropar headworks reservoir receiving organic wastes from a guest house. The organic wastes enrich the water with nutrients which, in turn favour rapid multiplication. Furthermore, rich vegetation provide food and shelter to the growing rotifers. Similar observations have been recorded by Vashisht and Sra (1979), Pandit and Kaul (1981), Sexena and Mishra (1990), Kaur et al. (1996a, b) and Bath and Kaur (1997) (Table 94).

At Site-III, rotifer taxa recorded which include *Brachionus bidentata, B. calyciflorus, Cephalodella sp., Epiphanies sp., Filinia longiseta and Keratella valga and*

Table 93 Monthly variations in the population density (Ind/L) of the Rotifera at the study site-I of the Anchar lake

Taxa	Sep	Oct	Nov	Dec	Jan	Feb	Mar	Apr	May	Jun	Jul	Aug	Sep	Oct	Nov	Dec	Jan	Feb	Mar	Apr	May	Jun	Jul	Aug
Brachionus quadridentata	–	–	–	–	–	–	–	–	–	+	+	–	–	–	–	–	–	–	–	–	–	–	++	+
Cephalodella sp.	–	–	–	–	–	–	+	++	–	–	–	–	–	–	–	–	–	–	–	–	–	+	+	++
Chromogaster ovalis	–	–	–	–	–	–	–	–	–	–	++	+	–	–	–	–	–	–	–	–	–	+	–	–
Gastropus sp.	–	–	–	–	–	–	–	–	–	+	–	+	+	+	–	–	–	–	–	–	–	+	+	+
Keratella cochlearis	–	–	–	–	–	–	–	–	–	+	+	–	–	+	–	–	–	–	–	–	–	–	+	–
Keratella valga	+	+	–	–	–	–	–	–	–	+	++	–	–	+	+	–	–	–	–	–	–	–	+	+

0–5 Ind/L = +
6–10 Ind/L = ++
11–15 Ind/L = +++
and above = ++++

Table 94 Monthly variations in the population density (Ind/L) of the Rotifera at the study site-II of the Anchar lake

Taxa	Sep	Oct	Nov	Dec	Jan	Feb	Mar	Apr	May	Jun	Jul	Aug	Sep	Oct	Nov	Dec	Jan	Feb	Mar	Apr	May	Jun	Jul	Aug
Brachionus calyciflorus	–	–	–	+	–	+	+	–	++	++	++	+	–	–	–	+	+	–	++	+	+	+	++	+
Epiphanes sp.	–	–	–	–	–	–	+	+	++	–	–	–	–	–	–	–	–	–	–	–		+	–	+
Filinia longiseta	+	+	–	–	+	+	–	–	–	+	+	–	–	–	–	–	–	–	–	–	–	++	+	+
Gastropus sp.	–	–	–	–	–	–	–	–	–	+	++	+	+	+	++	–	–	–	–	–	–	–	–	–
Keratella cochlearis	+++	+++	++	–	+	–	–	+	–	++	++	++	+	++	+++	–	–	–	–	–	–	+	–	+
Keratella valga	++	+	–	–	–	–	–	–	–	–	–	–	++	+	+	–	–	–	–	–	–	–	+	+
Polyarthra vulgaris	–	–	–	–	+	+	+	–	+	–	–	–	–	–	–	–	–	–	–	–	–	–	+	–
Synchaeta sp.	–	–	–	–	–	–	–	–	–	+	++	+	–	–	–	–	–	–	–	–	–	–	+	+

0–5 Ind/L = +
6–10 Ind/L = ++
11–15 Ind/L = +++
and above = ++++

K. cochlearis. The population density as well as species diversity were found to be higher than the Site-I. This site is located at the littorals of the lake. Here water is shallow and highly turbid, and harbours a rich macrophytic vegetation. Further more, effluents from the adjoining agricultural fields, sewage from local catchments area and effluents coming out of the SKIMS hospital are added at this site. All these factors enrich the water with its nutritive contents which in turn favour rapid multiplication and growth of rotifers as these are known to grow well in hard and eutrophicated waters. (Vashisht and Sra 1979; Balkhi and Yousuf 1991; Pandit and Kaul 1981; Sexena and Mishra 1990; Kaur et al. 1996a, b; Syal 1996; Bath and Kaur 1997; Sajeev 1999).

Population density and species diversity of rotifers were found to be lower at Site-III as compared to Site-II. This may be because of higher level of nutrients at Site-III than at Site-II. Alam and Khan (1996) while studying dynamics of plankton communities of ponds in Aligarh, observed that population density of Zooplankters is adversely affected by washing products and detergents which contain high levels of phosphorus. Mukherjee (1925) discussed specifically the role of phosphorus and observed that influx of phosphorus in the water bodies dominates the effects of other nutrients and modifies the overall biological characteristics of the water body. In the present study also phosphrus was found to be higher at Site-III than at Site-II, and probably was the key nutrient in controlling the rotifer growth pattern in the anchar lake. *Keratella valga* was recorded from all the three sites of the studied lake indicating that this species can thrive well in polluted and un-polluted waters. Its highest population density was recorded at the Site-I. Arora, (1966b), Sladecek (1983) and Syal (1996) found this species from clean and polluted waters while Suduzuki et al. (1983) found it only in clean waters (Table 95).

The rotifer taxa exclusively found at the Site-I are represented by *Brachionus quadridentata*, and *Chromogaster ovalis*, Arora (1966b) also observed *B.quadridentata* in clean water. On the other hand, Sladecek (1983) recorded it in moderately polluted waters. Berzins (1949), George et al. (1966) Verma et al. (1978) reported it in heavily polluted waters and Syal (1996) recorded its presence in clean as well as polluted waters. The *Polyarthra vulgaris* and *Synchaeta* sp. were recorded only at Site-II indicating that these species can thrive well in moderately polluted waters. *Brachionus bidentata* was recorded only at Site-III indicating that this species can survive in organically rich water receiving effluence from sewage, agricultural fields and hospital wastes. Jarnefelt (1925) recorded the presence of this species in a variety of waters while Sladecek (1983) noted it only in moderately polluted waters. Syal (1996) observed the occurrence of this species in clean as well as polluted waters.

At study Site-II, the highest population density was recorded of the species *Brachionus calyciflorus* and *Filinia longiseta*. These species were found at Site-III also but their population density was lower. At Site-I, these were absent. These observations indicate that these species prefer polluted waters to unpolluted water. Furthermore, although they are able to tolerate high pollution load, but their growth is maximum in moderately polluted waters. The *Brachionus calyciflorus* was found to be present in mildly polluted waters by Pejler (1957), George et al. (1966),

Table 95 Monthly variations in the population density (Ind/L) of the Rotifera at the study site-III of the Anchar lake

Taxa	Sep	Oct	Nov	Dec	Jan	Feb	Mar	Apr	May	Jun	Jul	Aug	Sep	Oct	Nov	Dec	Jan	Feb	Mar	Apr	May	Jun	Jul	Aug
Brachionus bidentata	–	–	–	–	–	–	–	–	–	+	++	–	–	–	–	–	–	–	–	–	–	+	+	–
Brachionus calyciflorus	–	+	–	–	–	–	–	–	–	–	–	–	–	–	–	–	–	–	–	–	–	++	+	–
Cephalodella sp.	–	–	–	–	–	–	–	–	–	++	+++	+++	–	–	–	–	–	–	–	–	–	++	–	++
Epiphanes sp.	+	+	–	–	–	–	–	–	–	–	++	–	–	–	–	–	–	–	–	–	–	+	+	++
Filinia longiseta	+	+	–	–	+	+	–	+	–	+	++	++	–	–	–	–	–	–	+	+	–	+	+	+
Keratella cochlearis	–	–	–	–	–	–	–	+	–	–	+	–	–	–	–	–	+	–	–	–	–	–	–	–
Keratella valga	++	+	+	–	–	–	–	–	–	–	++	+	+	+	–	–	–	–	–	–	–	+	+	+

0–5 Ind/L=+
6–10 Ind/L=++
11–15 Ind/L=+++
and above=++++

Radwan (1976), Hakkari (1978), Sladecek (1983), Bilgrami and Datta Munshi (1985), Sinha (1987) and Sinha and Prasad (1988). Syal (1996) found it in waters receiving sewage outfalls and industrial effluence. Algarsamy et al. (1967), Arora (1966a, 1966b), De Smet and Evens (1972), Wilkens (1972), Dhanapathi (1977), Jyoti and Sehgal (1979) and Sandhu et al. (1984) found it in polluted waters and described it as an indicator of eutrophy. Berzins (1949), Pejler (1957), Arora (1966b), De Smet and Evens (1972), Radwan (1976), Hakkari (1978), Verma et al. (1978), Maemets (1983) and Syal (1996) also reported *Filinia longiseta* both in moderately and highly polluted waters, whereas its occurence only in moderately polluted waters was observed by George et al. (1966), Bilgrami and Datta Munshi (1985), Sinha (1987) and Sinha and Prasad (1988). On the other hand, Sladecek (1983) recorded its presence in both clean as well as moderately polluted waters and Shrivastava (1962) reported this from clean to highly polluted waters.

Zutshi et al. (1980) reported 9 rotifer taxa in the Anchar lake whereas in the present study, 12 taxa have been reported. The taxa like *Asplanchna sp., Lecane* sp., *Monostyla* sp. and *Trichocerca similes* found to be present in their studies have not been recovered during the present investigation.

In both the lakes, the rotifers showed a single peak in the population density during summer. The comparative study in the species composition of the two lakes show that *Brachionus calyciflorus* and *Filinia longiseta* can be considered as indicator of nutrient enrichment as these taxa were absent from the less polluted sites of the two lakes.

Crustacea

In all, 13 crustaceans belonging to the orders cladocera and copepoda were discovered. Out of these, 6 were found at site-I, 11 at site-II and 8 at site-III (Table 96).

The crustacean fauna recorded at the study sites include 8 cladocerans viz. *Moina sp., Pseudosida sp., Alonella sp., Daphnia pulex., Bosmina longirostris., Ceriodaphnia sp., Moina affins., Sida sp.* and 5 copepods viz. *Cyclops sp., Diaptomus sp., Mesocyclops sp., Macrocyclops sp., Canthocamptus sp.* thereby showing the dominance of cladocerans over copepods both in terms of species diversity and population density. Sharma and Michael (1987) opined that the zooplankton communities in India are invariably dominated by the species of cladocerans. Sarwar and Parveen (1996) also recorded dominance of cladocerans over copepod in two inter connected lakes of Srinagar. On the other hand Zutshi et al. (1980), Yousuf and Qadri (1985) and Yousuf (1996) for Kashmir lakes and Allan (1976), Singh et al. (1982), Ferguson (1982), Carmonze et al. (1983), Serruya and Pollingher (1983), Hecky (1984), Sharma and Patnaik (1985), Drenner et al. (1987), Payne (1987), Mengastou and Fernando (1991) for other freshwater bodies found that copepod were more abundant than cladocerans.

Crustacean population in general shows single peak in the population density during summer. The works of Dutta et al. (1954), George (1966), Battish (1992),

Table 96 List of Crustacean taxa recorded at the investigated sites of the Anchar lake

S. No.	Crustacean taxa	Site-I	Site-II	Site-III
1	**Pseudosida sp.**	+	+	+
2	*Alonella sp.*	+	+	+
3	*Daphnia pulex* Leydig	–	+	+
4	*Bosmina longirostris* Muller	–	+	–
5	*Ceriodaphnia sp.*	–	+	–
6	*Moina affins*	+	+	–
7	*Sida sp.*	–	+	–
8	*Cyclops sp.*	+	+	+
9	*Mesocyclops sp.*	–	+	+
10	*Macrocyclops sp.*	+	+	+
11	*Canthocamptus sp.*	–	+	–
12	*Moina sp.*	–	–	+
13	*Diaptomus sp.*	+	–	+

Table 97 Seasonal variations in the population density (Ind/L) of the Crustacea at the investigated sites of the Anchar lake

Sites	Year	Autumn	Winter	Spring	Summer
Site-I	**2000–2001**	4	4	3	18
	2001–2002	3	–	9	15
	Average	3.5	2	6	16.5
Site-II	**2000–2001**	9	2	42	88
	2001–2002	10	–	18	54
	Average	9.5	1	30	71
Site-III	**2000–2001**	6	10	21	55
	2001–2002	3	–	10	25
	Average	4.5	5	15.5	40

Patlas (1972), Nassar (1975), Agrawal and Gautam (1985), Saha and Pandit (1988), Bath (1996); Sarwar and Parveen (1996); Biswas and Konar (2000) and Sharma et al. (2000) have also revealed the maximum density of crustacean during summer. While Vasisht and Sharma (1975), O'Brien (1979), Kohli et al. (1982), Orcutt and Pace (1984) and Pulle and Khan (2003) recorded the maximum density of cladocera during winter. Syal (1996) failed to ascertain any definite seasonal trend. Bath (1996) recorded two population density peaks of cladocerans in the Harike lake. Sajeev (1999) recorded the total copepod population to rise thrice during 1 year. Patnaik (1973) and Ramakrishnaiah and Sarkar (1982) observed the presence of copepods throughout the year. Sukumaran et al. (1984) noted that copepods were present throughout the year, but were minimum in May and maximum in November in tanks of Maland region of Kataka (Table 97, Figs. 68 and 69).

At site-I, crustacean fauna consist of *Pseudosida sp.*, *Alonella sp. Moina affins,* (cladocera) and *Cyclops sp., Diaptomus sp., Macrocyclops sp.* (copepoda). Here water

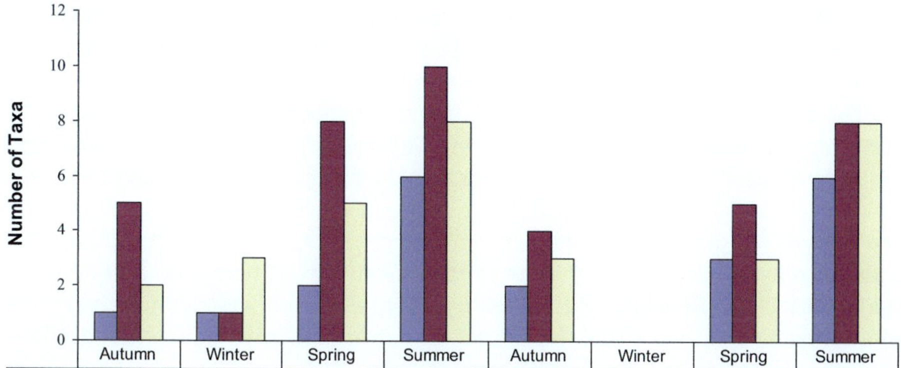

	Autumn	Winter	Spring	Summer	Autumn	Winter	Spring	Summer
Site-I	1	1	2	6	2		3	6
Site-II	5	1	8	10	4		5	8
Site-III	2	3	5	8	3		3	8

Season

Fig. 68 Seasonal variations in species diversity of the Crustacea at the investigated sites of the Anchar lake

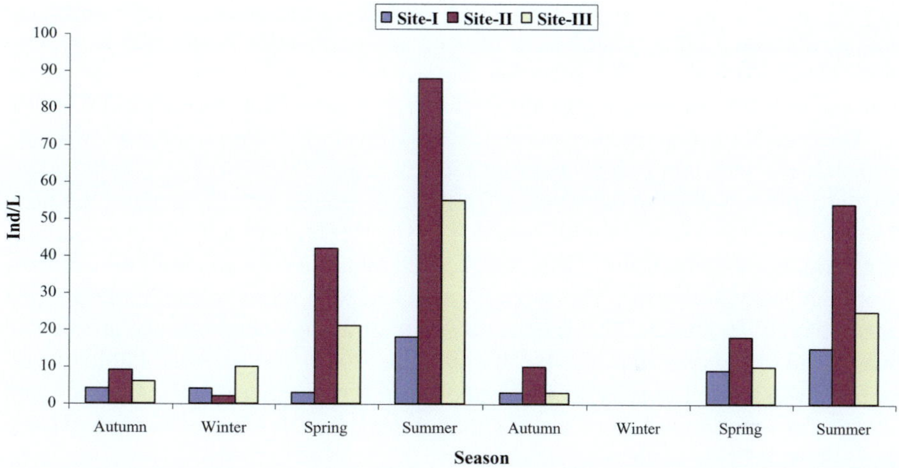

Fig. 69 Seasonal variations in population density (Ind/L) of the Crustacea at the investigated sites of the Anchar lake

was fast flowing and without any direct source of pollution. Macrophytic vegetation is scarce. The species diversity and population density at this site was found to be the lowest of all the sites. Presence of only a few aquatic plants accompanied by flowing water might be the reason for low species diversity and population density of crustacean at this site. Plants provide food, shelter and substratum to the crustaceans. Secondly increased silt accumulation might be another factor for the lowering of population diversity. Kaul et al. (1978) also observed lower population density and species diversity and suggested that silting is detrimental to the crustacean population in Anchar lake. Similar observations have been made by Tonopi (1980), Kaur et al. (1996a & b), Bath and Kaur (1997) and Bath (1996) (Table 98).

The crustacean population recorded at site-II include *Pseudosida sp., Alonella sp., Daphnia pulex., Bosmina longirostris., Ceriodaphnia sp., Moina affins, Sida sp.*(cladocera), *Cyclops sp., Mesocyclops sp., Macrocyclops sp.* and *Canthocamptus sp.*(copepoda). This site is located in the centre of the lake, has standing water and harbours rich growth of macrophytes. This site does not receive any sewage or agricultural run-off. However, the water coming from site-III brings in the pollutants added at the site-III which get diluted by its mixing with water coming from the site-I. Water chemistry at this site reveals the nutrients to be in moderate concentration, being higher than site-I and lower than site-III. These conditions seem to be favourable for crustacean growth as both population density and species diversity was found to be highest at this site. Dexter (1959), Holden and Green (1960), Macan (1963), Brooks and Dodson (1965), Krishnamoorthi and Visvesvara (1966), Jolly and Chapman (1966), Pahwa and Mehrotra (1966), Ray et al. (1966), Gideiri (1969), Hynes (1970), Kochsiek et al. (1971), Hall et al. (1976), Threlkeld (1979), O'Brien (1979), Tonopi (1980), Scullion et al. (1982), Singh (1982), Spencer et al. (1983), Orcutt and Pace (1984) Petts and Greenwood (1985), Saha and Pandit (1988), Bath (1996), Kaur et al. (1996a, b), Bath and Kaur (1997) and Sajeev (1999) suggested that crustacean usually inhabit slow moving and undisturbed water rich in vegetation. Unni and Naik (1997) recorded that crustacean population increased under nutrient rich conditions in the headwaters of the tropical River Narmada (Table 99).

The crustacean population recorded at site-III include *Moina sp., Pseudosida sp., Alonella sp., Daphnia pulex.*(cladocera), *Cyclops sp., Diaptomus sp., Mesocyclops sp.* and *Macrocyclops sp.*(copepoda). This site is located at the littorals of the lake. Here water is shallow and highly turbid. The site receives effluents from the adjoining agricultural fields, sewage from local catchments area and above all the effluents from the SKIMS hospital. The site gets enriched with nutrients and harbours rich macrophytic vegetation. The species diversity and population density at this site was found to be lower than site-II. Sharma et al. (2000) also recorded fall in crustaceans diversity and density in highly polluted waters. Das (1989) also recorded crustacean with limited species at highly polluted site. No species was found to be present exclusively at this site (Table 100).

The species which were exclusively recorded at site-II are *Canthocamptus sp.*(cladocera), *Bosmina longirostris, Sida sp. and Ceriodaphnia* sp.(copepoda) indicating their tolerance towards moderate pollution. Syal (1996) also found *Canthocamptus sp.* in moderately polluted waters while Das (1989) found it in highly polluted water. *Bosmina longirostris* was found to be present in clean as well as highly polluted waters by Das (1989), Syal (1996) and Bath (1996). However, Verma and Dalela (1975), Nanazato and Yasuno (1985) and Sharma and Michael (1987) found it in waters contaminated with sewage and industrial wastes. Jarnefelt (1925), Pejler (1957) and Sarwar and Parveen (1996) have recorded this species in eutropic waters. On the other hand, Jindal and Ghezta (1991) found it only in clean waters. *Ceriodaphnia* sp. was recorded in all types of water by Syal (1996) and Bath

Table 98 Monthly variations in the population density (Ind/L) of the Crustacea at the study site-I of the Anchar lake

Taxa	Sep	Oct	Nov	Dec	Jan	Feb	Mar	Apr	May	Jun	Jul	Aug	Sep	Oct	Nov	Dec	Jan	Feb	Mar	Apr	May	Jun	Jul	Aug
Alonella sp.	–	–	–	–	–	–	–	–	–	+	–	–	–	–	–	–	–	–	–	+	+	–	+	–
Cyclops sp.	–	–	+	+	–	–	–	–	–	+	+	–	–	–	–	–	–	–	+	+	–	–	+	+
Diaptomus sp.	–	–	–	–	–	–	–	+	–	–	–	+	–	–	–	–	–	–	–	–	–	–	+	–
Macrocyclops	–	–	–	–	–	–	–	–	–	–	+	+	–	–	–	–	–	–	–	+	–	+	–	–
Moina affins	–	–	–	–	–	–	–	–	–	+	–	–	+	–	–	–	–	–	–	–	–	+	–	–
Pseudosida sp.	–	–	–	–	–	–	–	+	–	+	+	–	+	–	–	–	–	–	–	–	–	–	+	+

0–5 Ind/L = +
6–10 Ind/L = ++
11–15 Ind/L = +++
and above = ++++

Table 99 Monthly variations in the population density (Ind/L) of the Crustacea at the study site-II of the Anchar lake

Taxa	Sep	Oct	Nov	Dec	Jan	Feb	Mar	Apr	May	Jun	Jul	Aug	Sep	Oct	Nov	Dec	Jan	Feb	Mar	Apr	May	Jun	Jul	Aug
Alonella sp.	+	−	−	−	−	−	−	−	−	+	+	++	+	−	−	−	−	−	−	−	−	+	+	−
Bosmina longirostris	−	−	−	−	−	−	−	+	−	−	+	−	−	−	−	−	−	−	+	+	−	−	−	+
Canthocamptus sp.	−	−	−	−	−	−	−	+	−	−	+	−	−	−	−	−	−	−	+	+	−	−	−	−
Ceriodaphnia	−	−	−	−	−	−	+	−	−	−	−	−	−	−	−	−	−	−	−	−	−	−	−	−
Cyclops sp.	−	−	+	−	−	−	+	−	+	−	++	++	−	+	−	−	−	−	−	−	−	+	+	+
Daphnia pulex	−	−	−	−	−	−	++	+	−	+	−	++	−	−	−	−	−	−	+	−	+	−	+	−
Macrocyclops	−	−	−	−	−	−	−	+	−	+	−	++	−	−	−	−	−	−	−	−	+	+	+	−
Mesocyclops sp.	−	−	−	−	−	−	+	−	+	−	+	+	−	−	−	−	−	−	−	−	−	++	+	−
Moina affins	+	−	−	−	−	+	−	−	−	+	+	−	+	−	−	−	−	−	−	−	−	−	+	+
Pseudosida sp.	+	−	−	−	−	−	−	+	−	++	+	+	−	−	−	−	−	−	−	−	+	+	+	−
Sida sp.	+	−	−	−	−	−	−	−	−	+	−	−	+	−	−	−	−	−	−	−	−	−	−	−

0–5 Ind/L = +
6–10 Ind/L = ++
11–15 Ind/L = +++
and above = ++++

Table 100 Monthly variations in the population density (Ind/L) of the Crustacea at the study site-III of the Anchar lake

Taxa	Sep	Oct	Nov	Dec	Jan	Feb	Mar	Apr	May	Jun	Jul	Aug	Sep	Oct	Nov	Dec	Jan	Feb	Mar	Apr	May	Jun	Jul	Aug
Moina sp.	–	–	–	–	–	+	–	–	+	+	–	–	–	+	–	–	–	–	–	–	–	+	+	–
Pseudosida sp.	+	–	–	–	–	–	–	–	–	+	+	+	–	+	–	–	–	–	–	–	+	+	–	–
Alonella sp.	–	–	–	–	–	–	–	–	–	+	+	+	–	–	–	–	–	–	–	–	–	+	+	–
Daphnia pulex	–	–	–	–	–	–	+	+	–	+	–	+	–	–	–	–	–	–	–	–	+	–	+	–
Cyclops sp.	–	–	–	+	–	–	+	+	+	+	++	+	–	–	–	–	–	–	+	+	–	–	–	+
Diaptomus sp.	–	–	–	–	–	–	–	+	–	–	+	–	–	–	–	–	–	–	–	–	–	+	–	–
Mesocyclops	–	–	–	–	–	+	–	+	–	+	+	–	–	–	–	–	–	–	–	–	–	–	+	+
Macrocyclops	–	–	+	–	–	–	–	–	–	+	++	–	+	–	–	–	–	–	–	–	–	–	+	+

0–5 Ind/L=+
6–10 Ind/L=++
11–15 Ind/L=+++
and above=++++

(1996) recorded this crustacean in all types of waters. Orcutt and Pace (1984) and Das (1989) found it to be present in highly polluted waters.

Moina sp. was found to be exclusively present at site-III indicating that can withstand high level of pollution. Unni and Naik (1997) also recorded this species under nutrient rich condition and labelled it as an indicator of eutrophication. Sajeev (1999) recorded this in the Ropar headworks at the site where water is shallow, slow moving and rich in macrophytes.

In both the lakes, crustacean comprised of 13 taxa belonging to cladocera and copepoda. Both the lakes showed single peak in population density observed during summer.

In the Dal lake species composition was similar at all the sites with difference of one taxon only at site-I where *Moinadaphnia sp.* was absent. In the Anchar lake however, species diversity was much variable at different sites with maximum at site-II (13) and lower at site-I (6) and site-III (8). Population density too was highest at site-II. Only four taxa were found to be common at all the sites of the Anchar lake. Four taxa were found specific to site –II and one to site-III. It seems that very high pollution load at site-III is deterrent for the growth of crustaceans. Zutshi et al. (1980) reported 5 taxa of crustacea in the Anchar lake whereas in the present study, 11 taxa have been reported revealing that many new taxa have made entry into the Anchar lake because of enrichment of water due to various anthropogenic activities.

Zoobenthos

Qualitative and quantitative analysis of zoobenthic invertebrate fauna revealed three groups viz. Annelida (5.5 %), Crustacea (16.6 %) and Insecta (77.0 %). Phylum Annelida was represented by a single taxon, *Placobdella* sp. (Hirudinea) while class Crustacea included *Gammarus* sp. (Amphipoda) and class Insecta was represented by 4 taxa namely *Stenonema* sp. (Ephemeroptera), *Enallagma* sp. (Odonata), *Chironomus* sp. (Diptera) and *Micropsectra* sp. (Diptera) thereby showing the dominance of insect fauna. Out of these, 3 were recorded at site-I, 4 at site-II and only 1 at site-III.

The zoobentic taxa in general showed a single peak in terms of species diversity and population density which was observed during spring. Singh and Sahai (1978), Dhillon et al. (1993a), Syal (1996) and Yousuf et al. (2002). Munawar (1970a, b), Das (1979), Rai and Duttamunshi (1979), Chowdhary (1984), Sharma (1986) and Kaushik et al. (1991) failed to record any seasonal trend in the population density of zoobenthic organisms (Figs. 70 and 71).

The zoobenthic taxa recorded at site-I include one crustacean represented by *Gammarus* sp. (25 %) and two insects, *Enallagma* sp. (37.5) and *Stenonema* sp.(nymphs only, 37.5 %) No annelid was recovered from this site. It was observed that the population density and species diversity was lower than site-II. This site is located where the river Sind feeds the Anchar lake and water here is fast flowing. The lake bed at this site is sandy and The macrophytic vegetation is scarce. The silt

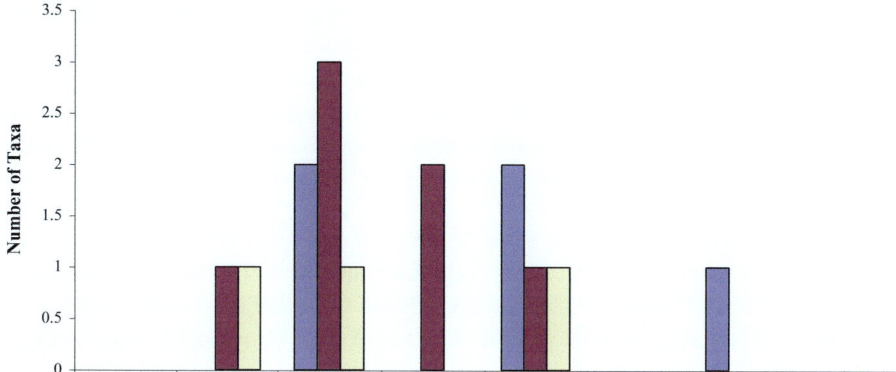

	Autumn	Winter	Spring	Summer	Autumn	Winter	Spring	Summer
Site-I			2		2		1	
Site-II		1	3	2	1			
Site-III		1	1		1			

Season

Fig. 70 Seasonal variations in the species diversity of Zoobenthos at the investigated sites of the Anchar lake

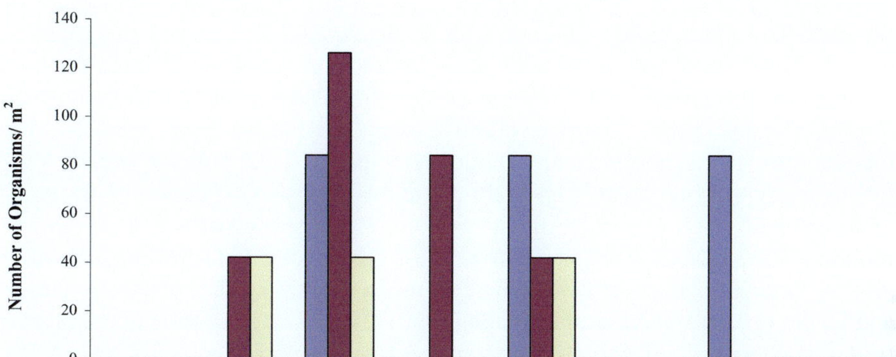

	Autumn	Winter	Spring	Summer	Autumn	Winter	Spring	Summer
Site-I			84		84		84	
Site-II		42	126	84	42			
Site-III		42	42		42			

Season

Fig. 71 Seasonal variations in the population density (number of organisms/m^2) of the Zoobenthos at the investigated Sites of the Anchar lake

accumulated here seems to be detrimental to the aquatic organisms as also reported by Kaul et al. (1978) and Pennak (1978). The nutrients available at this site are lowest of all the sites. Kumar (1996b) also observed the paucity of nutrients to be responsible for low density and diversity of zoobenthic fauna.

The zoobenthic fauna recorded at site-II included one annelid represented by *Placobdella* sp. (12.5 %), one crustacean *Gammarus* sp. (12.5 %) and larvae of two dipteran insects both belonging to the family Chironomidae, *Chironomus* sp.

(62.5 %) and *Micropsectra* sp. (12.5 %) The zoobenthic taxa were found to be maximum at this site in terms of species diversity and population density. This site is located in the centre of the lake and the water is stagnant here which seems favourable for greater zoobenthic diversity and density. Needham and Llyod (1916) and Kumar (1996b) also observed that such environmental conditions are favourable for greater production of zoobenthic fauna. This site harbours rich growth of macrophytes. Krecker (1939), Andrews and Hasler (1943), Tonopi (1980), Sharma (1988), Kaushik et al. (1991), Bath and Kaur (1998) and Sajeev (1999) have also opined that macrophytes provide protection to the benthos thrive well under these conditions. The site itself does not receive any sewage or agricultural run-off, however, the water coming from site-III and site-I mixes at this site and thus has a dilution effect on the water quality of the lake here. The increased zoobenthic diversity at this site as compared to site-III shows the decline in organic pollution load at this site. Mason (1981) has also observed an increase in the number of organisms at less polluted sites.

The only zoobenthic taxon found at site-III included one dipteran, *Chironomus* sp.(100 %) This site is located at the littorals of the lake. Here, water is shallow and highly turbid. This site receives effluents from the adjoining agricultural fields, sewage from local catchment area and above all effluents from the SKIMS hospital. The dissolved oxygen at this site ranged from 0.4 to 4.8 mg/l, which was minimum of all the sites. The site is highly enriched with nutrients and harbours rich macrophytic growth. These conditions seem to be deterrent for other zoobenthic fauna but conducible for survivability of *chironomus* sp. Shrivastava (1962), Das and Bisht (1979), Pandit (1980), Sharma (1986), and Das (1989) have reported that Chironomus sp. can survive under septic, deoxygenated and polluted waters. Very low diversity of macrobenthic invertebrates at this site is an indication of impact of anthropogenic stress caused by direct domestc discharges and run-off from the surrounding areas including hospital waste. These have resulted in substrate instability and has brought about a drastic change in the composition of zoobenthos. Dance and Hynes (1980) and Ajao and Fagado (1990) also observed decline in macrobenthos in waters receiving complex mixtures of domestic waste.

In the present study, no taxon was recorded to be distributed at all the studied sites. The zoobenthic taxa *Enallagma* sp.(37.5 %), and *Stenonema sp.*(37.5) were restricted at site-I only. According to Pennak (1978) *Enallagma* sp. is rarely found in polluted waters, while *Stenonema sp.* is found where there is plenty of oxygen. These suggestions explain the absence of these two taxa from site-II and III. The only taxon restricted to site-I (25 %) and site-II (12.5 %) is *Gammarus* sp. The presence of this taxon at these two sites indicates its preference for water which is less polluted to moderately polluted. Its percent contribution was higher at site-I, decreased at site-II and nil at site-III. Mellanby (1963) opined that this taxon prefers running and much oxygen containing waters. However, he has not mentioned the oxygen range. In the present study, the dissolved oxygen ranged from 5.8 to 9.5 at site-I, 2.8 to 8.9 at site-II and 0.4 to 4.8 at site-III. From this study it seems that the oxygen range at site-III seems to be the limiting factor for its survival.

Placobdella sp. (Hirudinea) and *Micropsectra* sp. (Diptera) were restricted to site-II only. Pennak (1978) has reported that *Placobdella* sp. can withstand some degree of pollution and found it at the places which are protected by macrophytes. The presence of this taxon at site-II indicates water to be moderately polluted. Schneider (1962), Serruya (1978), Allanson (1979) has established the presence of *Microspectra* sp. from oligotrophic and mesotrophic waters and also in waters polluted by sewage.

The taxon restricted only to site-II and III include *Chironomus* sp. Its present contribution was 62.5 % at site-II and 100 % at site-III. Gaufin and Tarzwell (1956), Curry (1962), Husainy (1965), Verma and Shukla (1969), King (1981), Sinha and Prasad (1988), Rao et al. (1991), Dhillon et al. (1993a, b) and Kumar (1996b) reported *Chironomus* sp. in waters polluted by sewage. Bath (1996) observed its presence in polluted water. David and Ray (1966), Das (1989) and Syal (1996) however, found this both in clean and polluted waters. Mandal and Moitra (1975), Kumar (1997) and Cowell and Vodopich (1981) opined that rich organic matter increase the production of chironomids.

Bioindicators In the Anchar lake, there has been an increase in calcium, magnesium, chloride, total alkalinity, total hardness, sodium, potassium, ammonical nitrogen, nitrate nitrogen and total phosphorus at site-III which is due to addition of agricultural run-off from the agricultural fields, sewage discharges from the catchment human habitation and above all the effluents from the SKIMS hospital. Due to high ionic concentration and presence of rich growth of macrophytes, water at the site-III is highly nutritive, more turbid, less transparent and shows higher TDS values. The wastes which are discharged at site-III have resulted in considerable decrease the dissolved oxygen. Site-II where water from site-I and Site-III mixes up, is moderately nutrient rich zone. The rich organic water at site-III has also resulted into rich protozoan fauna in terms of species diversity and population density. Among the zoobenthic organisms, the presence of only diptera at site-III also affirms that site is nutrient rich. The overall conditions depict that sewage, agricultural run-off and hospital waste have deteriorated the lake environment.

Based on distribution of zooplanktons and zoobenthos, certain bioindicators have been identified which are listed below:

Group	Clean water taxa (Site-I)	Polluted water taxa (Site-II and III)
Protozoa	–	*Difflugia tuberculata,*
		D. oblonga,
		Lesqueresia spiralis, Trinema sp.
Rotifera	***Chromogaster ovalis***	*B.calyciflorus,*
		Filinia longiseta
Crustacea	–	–
Zoobenthos	*Enallagma* sp. *and Stenonema* sp.	*Chlronomus* sp.

Fish Fauna of Dal and Anchar Lake

The ichthyofauna of Kashmir has attracted the attention of J.J. Heckel (1839) who for the first time described sixteen species of fishes from this region. Later on many new fish species were introduced into the aquatic systems of Kashmir making a total to 37 (Yousuf 1996).

During the present studies, a comparison of fish records in the Dal and the Ancher lake given by Yousuf (1996) and Department of Fisheries, J & K (2004) was made. The record collected reveal presence of total of 15 fish species in the Anchar lake, and 14 fish species in the Dal lake. The data reveals the disappearance of *Oreinus plagiostomus* (khont) from the Dal lake during the recent past (Tables 101 and 102).

Table 101 List of Zoobenthic Taxa recorded at the investigated sites of the Anchar Lake

TAXA	SITE-I	SITE-II	SITE-III
Enallagma sp.	+	–	–
Gammarus sp.	+	+	–
Stenonema sp.	+	–	–
Chironomus sp.	+	+	+
Micropsectra sp.	+	+	–
Placobdella sp.	+	+	–

Table 102 List of fish species from the Dal and the Anchar Lake recorded from Yousuf et al. (1996) to the present studies (2004)

S. no.	Fish fuana	Dal Lake		Anchar Lake	
		1996	2004	1996	2004
1.	*Cyprinus carpio specularis (mirror carp)*	+	+	+	+
2.	*Cyprinus carpio communis (Scale carp)*	+	+	+	+
3.	*Carassius carassius (Gund gaad)*	+	+	+	+
4.	*Schizothorax esocinus (Chhurru)*	+	+	+	+
5.	*S. niger (Sattur)*	+	+	+	+
6.	*S. curvifrons (Sattur)*	+	+	+	+
7.	*S. micropogon (Kashir gaad)*	+	+	+	+
8.	*S. planifrons (kashir gaad)*	–	–	+	+
9.	*Oreinus plagiostomus (khont)*	+	–	+	+
10.	*Crossocheilus diplochilus (Tettur)*	+	+	+	+
11.	*Puntius conchonius (Safed bacha)*	+	+	+	+
12.	*Botia birdii (Rama gurun)*	+	+	+	+
13.	*Gambusia affinis (Mahi gaad)*	+	+	+	+
14.	*Labeo dero (Roput)*	+	+	+	+
15.	*Nemachilus sp.*	+	+	+	+

Lake Management

On the basis of the physico-chemical parameters and biological features of the Dal and the Anchar lake, it has been recognised beyond doubt that both the lakes have deteriorated by addition of sewage, agricultural wastes and hospital wastes. These wastes add rich amounts of nutrients resulting in intense macrophytic growth which, in turn, add more decomposable material to the water accelerating the nutritive status of water. These conditions are evident by changes in the fauna and flora of the lakes with appearance of many pollution tolerant taxa of phytoplanktons, rotifers, crustaceans and annelids. The macrophytes choke the lake water restricting its free flow aeration. Agricultural practices in the catchment area and within the lake by constructing floating gardens not only add organic matter to water but also eat up the open water area of the lake. These problems created by anthropogenic activities are of serious nature and demand immediate attention for the restoration of natural conditions in these lakes.

The present investigation presents a very good example of water bodies that are undergoing rapid trophic evolution as a result of heavy nutrient loading due to ever increasing human interference. The problems of Dal and Anchar lakes due to urban development have been well recognized and several proposals have been suggested for the restoration of the lakes.

Dal Lake For the proper environmental management of the lake it is imperative to undertake an integrated systematic study of the lake environment, involving scientific, engineering, NGO's and other persons drawn from various disciplines. An interdisciplinary approach is needed to restore the damaged ecosystems from further deterioration. Two general approaches-passive and active management can be utilized. Passive management, with minimum human intervention, would help to maintain the lake in a semi-wilderness state. Plant and animal population would be allowed to fend for themselves. Active management would involve manipulating conditions in the lake to maximize the value of the systems for specific uses or goals, e.g. renewable resources, tourism, an ecosystem, or a combination of these. In view of the problems which demand immediate attention, various conservatory measures as management tools are suggested as follows:

Denudation of catchment area due to deforestation and heavy cattle grazing results in deposition of silt thereby raising the lake bed. Since the lake is situated in the low lying areas (1584 m altitude) and floodplains of river Jhelum and Sind, soil material that flowing water picks up is dropped when the water reaches a depression. A total volume of silt brought into the Dal lake has been estimated at 80,000 tonnes per year out of which 70 % is brought by the feeding channel, Telbal Nallah. In addition, another 40,000–50,000 tonnes of dead weed and allochthonous material are added every year (ENEX 1978).

It is recommended that the reforestation programme be accelerated and that positive steps be taken to control grazing on the slopes which are prone to soil erosion.

Agriculture is a major source of non-point pollution entering the lake ecosystem. It should, therefore, be considered a major management tool to attempt to reduce the non-point pollution from this source. There should be no agricultural activities in the vicinity of the lakes.

Floating gardens are the main source of agricultural waste that enter the Dal lake ecosystem. These should be removed from the lake so as to avoid enrichment of water with nutrients and pesticides. Furthermore, the floating gardens encroach upon the open area of the lake and restricts the free-flow of water. These should be removed to allow good aeration of water which is essential for sustenance of biodiversity. Further encroachments in the lake area should be totally stopped for which guidelines should be framed so that the construction of new houseboats, hotels and restaurants within and along the banks of the lake be regularised and construction within the lake should be totally banned.

The second dominant type of pollutant which threatens the lake ecosystem is organic matter added by domestic wastes emanating from the house boats, hotels and restaurants. The most effective measure to check this is by laying out well planned and effective sewerage pipelines. The second alternative is to divert waste water to sewage treatment plants before releasing into the lake system.

Aquatic plants have reached nuisance proportion in the lake. Some floating aquatic weeds, viz *Salvinia natans* and *Lemna* spp. have proliferated profusely in the lake. These macrophytes upon decomposition, add organic debris to the bottom of the lake system which besides enriching the lake water with organic matter, raise the lake bed. It is, therefore, important that dredging of the lake be done regularly to maintain the natural depth of the lake.

The problems of pollution in the investigated lake cannot be controlled only through technological measures without cognizance of the socio-economic conditions, including the basic needs of local population and potential of tourism in the area. A long term solution to the problem can be achieved only through an integrated feasible approach which regularizes the establishment meant for tourism and helps to prevent unplanned growth of settlements. The floating gardens are the earning beds for the economically poor section of the population. Removal of the gardens should be strictly followed and the alternative land should be provided to them for their sustenance.

Anchar Lake The lake is a typical sub-urban shallow water body receiving extensive quantities of domestic wastes and agricultural run-off all along its periphery. Over the years, large chunks of the lake have been encroached upon and are being utilized for the cultivation of willow plants. The Sind Nallah, the main source of water for the lake, brings in large quantities of silt which get lodged in the lake resulting in the shallowing of the lake. Above all the untreated effluent discharged from the SKIMS hospital has deteriorated the water quality of the lake. The data on its physico-chemical characters and biotic communities clearly reveal its eutrophic nature. The most important measures to be taken for the management of this water body include:

The process of extraction of sand and gravel in the upper reaches should be regularized so that the dislodged sand does not enter the lake. To control siltation and sedimentation, it is necessary to construct a silting basin prior to the entry of Sind Nalah into the Anchar lake to settle down the bulk of the sediment and insoluble nutrients.

The ever-increasing domestic wastes, agricultural run-off containing plant nutrients and hospital effluents need to be stopped from entering the lake. Further encroachments in the lake area should be totally stopped for which guidelines should be framed so that the construction of new houses within and along the banks of the lake be regularized, and construction within the lake should be totally banned.

The importance of shore zone (transition zone) has been emphasized throughout the world. (Jorgensen and Loffler 1990). The maintenance of a suitable protection zone between littorals and limnetic zone will meet all the specific goals in management.

A meaningful interpretation of the causes and mechanisms of perturbations in these lakes is only possible if periodic surveys of different ecological aspects are carried out. This emphasizes the need to monitor the lake environment on a regular basis.

Summary

The high altitude valley of Kashmir existing in the greater Himalaya abounds a vast array of freshwater bodies such as lakes, ponds, wetlands, springs, streams and rivers. These varied freshwater ecosystems are of great aesthetic, cultural, socio-economic and ecological value besides playing an important role in the conservation of genetic resources of both plants and animals. Among these, two lakes, the Dal and the Anchar lake situated at an altitude of 1584 m at *a.s.l.* within the geographical coordinates of *34.0° 7′ N* and *74.0° 52′ E and 34° 20′ and 34° 26′ N* latitude and 74° 82′ and 74° 85′ longitude respectively, have been subjected to over exploitation by man for economic reasons. Water areas have increasingly been used for establishment of hotels, houseboats and residential houses while floating gardens have been made to enhance agricultural production. Besides encroachment of the lake area, loads of sewage and agricultural wastes are added into the lake water thereby disturbing the lake environment. The present investigation was undertaken to study the impact of these human activities on the trophic status and biodiversity of these two takes.

In the Dal lake, the studies have been conducted at four sites keeping in consideration various human activities. Site-I lies in the open water area of the lake which has been taken as the reference site as it lacks any direct source of pollution. Site-II is located in the floating garden area receiving agricultural wastes while site-III near the houseboats and site-IV near the hotels, both receiving ample amount of sewage. In the Anchar lake, the studies have been conducted at three sites established at different ecological regions of the lake. Site-I lies where the Sind nallah enters the Anchar lake. This site lacks any direct source of pollution. Site-II is located in the centre of the lake and site-III is located where water receives agricultural run-off, sewage and hospital waste from the adjoining areas.

Detailed limnological studies in the Dal and the Anchar lake were conducted from September, 2000 to August, 2002. The physico-chemical analysis include the estimation of water temperature, depth, transparency, water turbidity, pH, dissolved oxygen, total alkalinity, total hardness, calcium, magnesium, iron, sodium, potassium, chloride, ammonical nitrogen, nitrate nitrogen, total phosphate phosphorus,

M. Jeelani, *Lake Ecology in Kashmir, India*,
DOI 10.1007/978-3-319-40880-4_5

silicates, total dissolved solids and conductivity. For the Dal lake, both surface and bottom water analysis was carried out. For biological studies, seasonal distribution, species composition and population dynamics of phytoplanktons, zooplanktons, macrophytes and zoobenthos have been undertaken. Besides fish fauna has been enlisted.

The Dal Lake

- The temperature of the lake water ranged from 4 °C to 27 °C during the entire study period showing not much variation at different sites within the lake. Water turbidity ranged from 7 to 21 NTU. The sites receiving agricultural run-off and sewage show an increase in the turbidity values. Water transparency ranged from 0.40 to 1.62 m with lower values at sites receiving agricultural run-off and sewage. Depth of the lake ranged from 0.8 to 2.80 m. The addition of the wastes and decomposition of macrophytes have raised the lake bed.
- The pH of water ranged from 7.0 to 9.3 both for surface and bottom waters. Due to buffering capacity of water, not much site specific changes were recorded. The conductivity values of the lake water ranged from 100μScm^{-1} to 530μScm^{-1} for surface waters and 100μScm^{-1} to 560μScm^{-1} for bottom waters with little difference across the water column. The conductivity values were higher where the lake water receive agricultural run-off and sewage from houseboats and hotels.
- The dissolved oxygen concentration of the lake ranged from 0.8 mg/L to 12.8 mg/L with lower values at sites receiving sewage from the houseboats and hotels, and higher values at floating garden area of the lake.
- The concentration of calcium ranged between 15.2 mg/L and 53.3 mg/L for surface water and 15.6 mg/L to 56.1 mg/L for bottom water. The magnesium concentration ranged from 2.2 mg/L to 30.5 mg/L for surface water and 3.4 mg/L to 31.4 mg/L for bottom water The total alkalinity fluctuated from 38 mg/L to 460 mg/L in surface water and 78 mg/L to 480 mg/L in bottom water. The total hardness ranged between 73 mg/L and 464 mg/L for surface water and 92 mg/L and 462 mg/L for bottom water. The values of all these chemical constituents of water were found to be higher at the sites receiving agricultural run-off from floating gardens and sewage from houseboats and hotels. On the basis of calcium, magnesium, total alkalinity and total hardness values, water of the Dal lake is categorised as moderately hard to very hard type.
- The chloride values ranged from 14 mg/L to 85 mg/L for surface waters and from 10 mg/L to 92 mg/L for bottom waters. The highest concentration of chloride was recorded at site receiving sewage from the hotels. The sodium concentration ranged from 2 to 18 mg/L for both surface and bottom waters. The potassium value ranged between 1 mg/L and 10 mg/L for surface water and 1 mg/L to 11 mg/L for bottom waters. The total dissolved solids ranged from 74 mg/L to 397 mg/L for surface water and 74 mg/L to 420 mg/L for bottom waters.

Sodium, potassium and total dissolved solids show higher values at the sites receiving agricultural run-off and sewage from the houseboats and hotels.

- The silicate values ranged from 0.1 mg/L to 6.0 mg/L for surface waters and 0.5 mg/L to 8.0 mg/L for bottom waters with higher values in bottom waters. However, no indication of increase in the concentration of silicate content with agricultural and sewage discharges in this study is revealed. The iron concentration ranged from 120 μg/L to 380 μg/L for surface water and 120 μg/L to 370 μg/L for bottom water without significant difference between surface and bottom waters. An increase has been recorded at sites which receive agricultural run-off and sewage.
- The ammonical nitrogen ranged from 104 μg/L to 395 μg/L for surface and 110 μg/L to 490 μg/L for bottom water showing higher values in the bottom waters. The concentration of nitrate nitrogen ranged from 113 μg/L to 910 μg/L for surface water and 215 μg/L to 850 μg/L for bottom waters with higher values at sites receiving agricultural waste and sewage. The total phosphate phosphorus values ranged from 120 μg/L to 850 μg/L for surface water and 180 μg/L to 851 μg/L for bottom water with higher value in bottom waters. Floating garden area showed maximum total phosphate phosphorous. Sewage receiving areas also show higher values. On the basis of nitrate nitrogen and total phosphate phosphorus concentration in water, the lake is observed to have acquired eutrophic status.

The physico-chemical analysis of water reveals overall enrichment of water of the Dal lake with calcium, magnesium, chloride, sodium, ammonical nitrogen, nitrate nitrogen and total phosphate phosphorus due to addition of agricultural run-off from the floating garden and sewage effluents from house boats and hotels. Due to high ionic concentration of these salts, water becomes less oxygenated, less transparent, more turbid and shows higher values of total alkalinity, total hardness, TDS and conductivity. Concentration of nitrate nitrogen and total phosphate phosphorus put the Dal lake in the category of highly productive and eutrophic water body.

- The study on phytoplanktons revealed 75 taxa of Bacillariophyceae, 34 of Chlorophyceae, 13 of Cynophyceae, 2 each of Dinophyceae and Euglenophyceae and 1 of Chrysophyceae. The Bacillariophyceae outnumbered the phytoplanktons both in terms of species diversity and population density. In the present study, the maximum population density of Bacillariophyceae was observed at site-IV thereby indicating that the lake at this site is nutrient rich. The species diversity was recorded maximum during autumn showing their affinity for cold water.
- The Chlorophyceae constituted the second largest group in terms of population density and species diversity with maximum population density during summer. The overall distribution pattern does not vary much at different study sites. The population density was recorded highest at site-II followed by site-I. The density at site-III and site-IV was recorded to be low.
- The maximum species diversity and population density of Cyanophyceae was recorded during summer. This group, too, does not show much variations in

species diversity at different sites. However, the population density was recorded higher at site-II and site-IV. The contribution of Euglenophyceae, Dinophyceae and Chrysophyceae towards total algal population was low. No seasonal trend was recorded for these groups.

- A total of 51 taxa of zooplankton belonging to protozoa, rotifera and crustacea have been recovered. Rotifera dominate the zooplankton followed by crustacea and protozoa. In all, 11 protozoan taxa were recorded with maximum species diversity and population density during summer. The studies based on protozoa indicate higher protozoan growth at site-II, III and IV which receive ample amounts of agricultural and sewage effluence.
- The rotifer fauna of the Dal lake comprised 27 taxa with single peak in population density during summer at all the studied sites. The population density and species diversity were lowest at site-I which was free from human interference, higher at site-III and IV and highest at site-II showing nutrient enrichment by agricultural run off and sewage to be conducive for rotifers to flourish.
- In all, 13 crustacean taxa were recorded with maximum species diversity and population density during summer. The species composition at different sites do not vary much but population density of crustacea was found to be higher at site-II, III and IV as compared to site-I due to nutrient enrichment by agricultural run off and sewage.
- The zoobenthic fauna was represented by the phylum annelida only. Both species diversity and population density were higher in summer. Site- specific distribiton of certain zoobenthic genera indicate nutrient enrichment of water at site-II, III and IV.
- Presence of certain bioindicator of eutrophy belonging to phytoplankton, zooplankton and zoobenthos at site-II, III and IV indicates that sewage and agricultural run off enrich the water with nutrients.

Anchar

- The temperature of the lake water ranged from 3 °C to 28 °C over a period of 2 years study. The lake is shallow due to deposition of silt coming from the Sind Nallah. The enrichment of nutrients have resulted in dense growth of macrophytes at site-II and III which upon death and decomposition have further raised the lake bed at these sites. Water transparency ranged from 0.1 to 1.8 m with lower values at the site receiving agricultural run-off and sewage discharges in addition to hospital waste. Water turbidity ranged from 2 to 32 NTU.
- The pH of water ranged from 6.9 to 8.6 indicating water to be on alkaline side. The dissolved oxygen concentration of the lake ranged from 0.5 to 9.5 mg/L with well oxygenated water entering the lake from the Nallah but getting highly depleted of its oxygen contents at site-III owing to addition of agricultural run-off, sewage and hospital wastes.

- The calcium ranged from 9.0 to 43.0 mg/L; magnesium, from 4.3 to 28.5 mg/L; total alkalinity, from 59 to 315 mg/L and total hardness, from 73 to 377 mg/L. The values of these constituents of water were found to be increased at sites receiving agricultural run-off, sewage and hospital wastes from adjoining areas.
- The sodium concentration ranged from 3.0 to 14 mg/L and potassium, from 2 to 9 mg/L. Both these cations have been found to be slightly higher at sites-II and III. The Iron concentration ranged from 100 to 600 μg/L with not much site-specific variations. The silicate ranged from 0.8 to 8.1 mg/L with lower values at site-III due to its utilization by diatoms.
- The concentration of chloride ranged from 9 to 85 mg/L with very high values at site receiving agricultural run-off and hospital waste. The ammonical nitrogen varied from 285 to 705 μg/L; nitrate nitrogen from 245 to 611 μg/L and total phosphate phosphorus from 200 to 505 μg/L. All these chemical constituents were found to be increased at site-III.
- Total dissolved solids ranged from 153 to 435 mg/L while conductivity values ranged from 205 to 580 μs/cm with higher values at site-III depicting high concentration of cations and anions at this site of the Anchar lake.
- The physico-chemical study has further revealed that various nutrients added at the site-III by agricultural run-off, sewage and hospital wastes get diluted at the site-II where water from site-I and III is mixed. All chemical constituents, therefore, show their minimum concentration at site-I, maximum at site-III and moderate at site-II.
- The zooplankton in the Anchar lake comprised of 55 taxa belonging to protozoa, rotifera and crustacea with protozoa to be the most dominating group among the zooplankton followed by rotifera and crustacea.
- A total of 30 protozoan taxa were recovered with species diversity and population density being highest in summer. Both species diversity and population density were recorded to be higher at site-II and site-III.
- In all, 12 rotifers taxa were recorded in this lake. The density and diversity of rotifera were recorded to be higher at site-II and III as compared to site-I with highest at site-II. A similar trend was observed in case of crustacean fauna which included 8 cladocerans and 5 copepods.
- It is evident from the of distribution pattern of zooplanktons that protozoan were maximum at site-III while rotifer and crustacean were maximum at
- site-II where the nutrient enrichment is moderate in concentration due to mixing of water coming from site-I and III. The site-III seems conducive for the growth of protozoa which outnumbered all other zooplankton here but were deterrent for the growth of some species of rotifera and crustacea.
- The zoobenthic invertebrate fauna in the lake comprised of three groups, viz. annelida, crustacea and ensecta with a single peak in the spring. Species composition was found to be variable at different sites with maximum species diversity and population density at site-II. Site-I was represented by crustacea, odonata and ephemeroptra; site-II by annelida, crustacea and diptera and site-III, by diptera only. The study further revealed dominance of clean water taxa at site-I, tolerant at site-II and bioindicator of eutrophy at site-III. The distribution pattern

gives a clear picture of the changes brought about in the trophic status of the lake by addition of agricultural waste, sewage and hospital waste.

- The macrophytes were represented by four life forms in both the lakes which include free floating, rooted floating, submerged and emergents. Importance value Index was found to be highest for emergents and lowest for free floating type of macrophytes. In the Anchar lake, too, Importance Value Index was found to be highest for emergent and lowest for free floating types. Importance value Index for emergent was higher in the Anchar lake than in the Dal lake while that of submerged was higher in the Dal lake than in the Anchar lake. These observations indicate the Dal lake water to be more conducive for the growth of submerged type of vegetation due to more light penetration.
- Correlation studies of both, the Dal and the Anchar lake revealed significant negative relationship of temperature with dissolved oxygen and total phosphate phosphorus while significant positive relationship of dissolved oxygen with nitrate nitrogen; of conductivity with calcium, magnesium, chloride, ammonical nitrogen, total dissolved solids, nitrate nitrogen, phosphate, silicate, sodium and potassium; of sodium with potassion; of calcium with magnesium, total hardness and total alkalinity and of magnesium with total hardness. The nitrate nitrogen showed a significant positive relationship with Cyanophyceae and Chrysophyceae while silicates showed a significant negative relationship with Bacillariophyceae. The correlations among different biological groups were recorded to be positive.

The present investigation presents a very good example of water bodies that are undergoing rapid trophic evolution as a result of excessive nutrient loading by ever increasing human activities causing threats to the ecology of the Dal and the Anchar lake. Proper conservation and management measures are required to save these water bodies from further degradation and to develop them into ecologically conserved ecosystems for which certain suggestions have been made.

For the Dal lake, it is recommended that floating gardens should be removed and further encroachments should be banned. Disposal of domestic wastes from adjoining houses, hotels and houseboat should be regularised. Dredging of the lake be done regularly to maintain the natural depth of the lake.

In the Anchar lake, the main source of water contamination is sewage, agricultural run-off and hospital waste. The process of extraction of sand in the upper reaches of the Sind Nallah should be regularised so that silt and sediments do not enter the lake. Necessary silting basin prior to the entry of Sind Nallah into the Anchar lake should be constructed. The need to monitor both the lake's environment on a regular basis in also emphasized.

Fishes and Maps

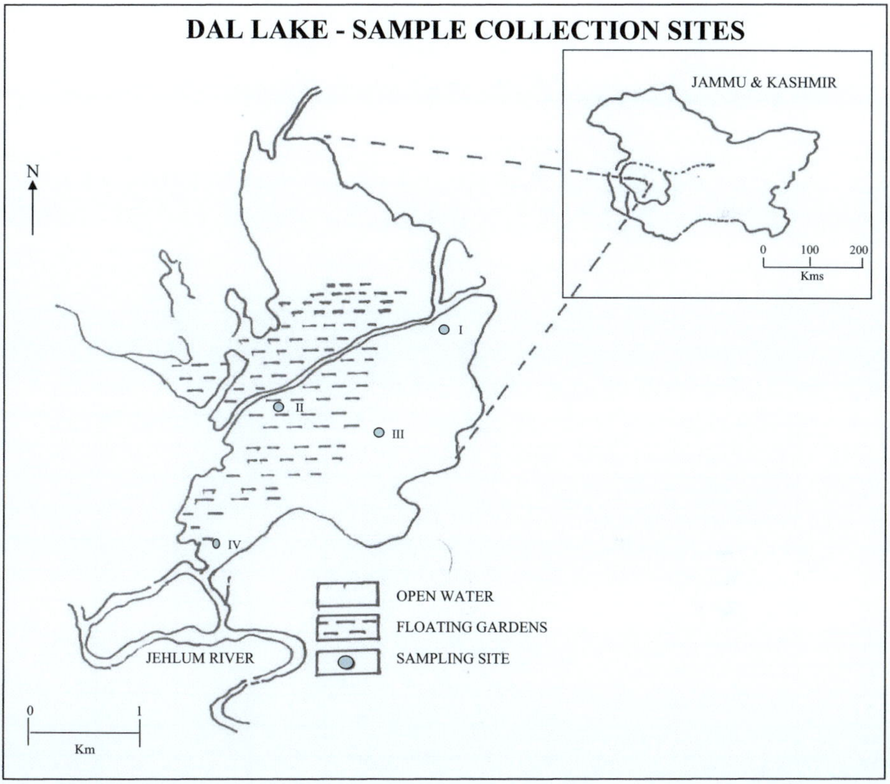

M. Jeelani, *Lake Ecology in Kashmir, India*,
DOI 10.1007/978-3-319-40880-4

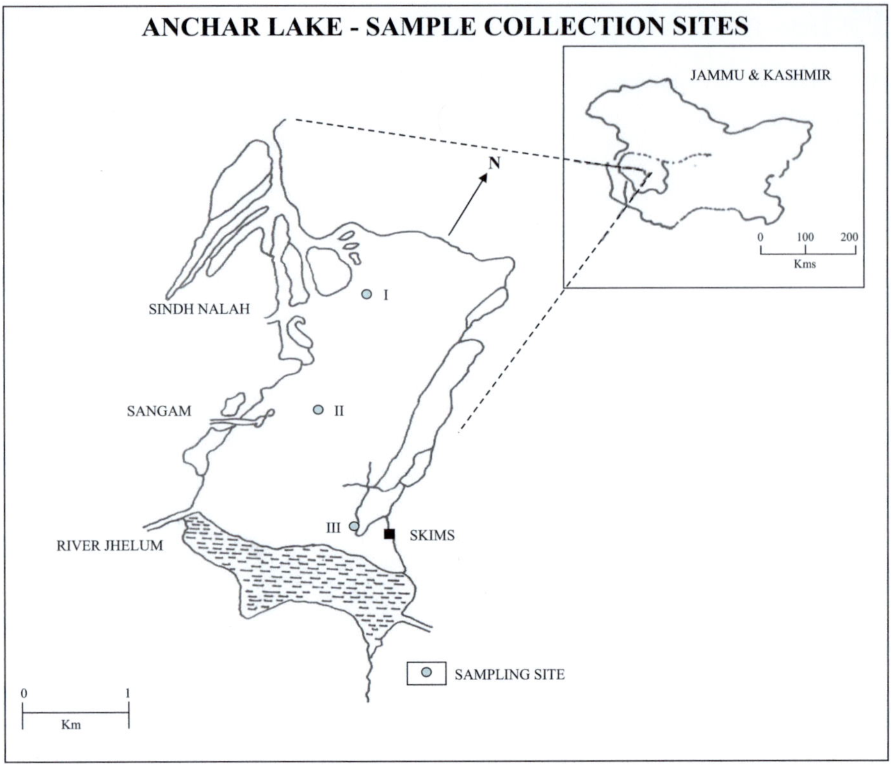

S. No.	Fish Fuana	Dal Lake		Anchar Lake	
		1996	2004	1996	2004
1.	*Cyprinus carpio specularis (mirror carp)*	+	+	+	+
2.	*Cyprinus carpio communis (Scale carp)*	+	+	+	+
3.	*Carassius carassius (Gund gaad)*	+	+	+	+
4.	*Schizothorax esocinus (Chhurru)*	+	+	+	+
5.	*S. niger (Sattur)*	+	+	+	+
6.	*S. curvifrons (Sattur)*	+	+	+	+
7.	*S. micropogon (Kashir gaad)*	+	+	+	+
8.	*S. Planifrons (kashir gaad)*	-	-	+	+
9.	*Oreinus plagiostomus (khont)*	+	-	+	+
10.	*Crossocheil us diplochilus (Tettur)*	+	+	+	+
11.	*Puntius conchonius (Safed bacha)*	+	+	+	+
12.	*Botia birdii (Rama gurun)*	+	+	+	+
13.	*Gambusia affinis (Mahi gaad)*	+	+	+	+
14.	*Labeo dero (Roput)*	+	+	+	+
15.	*Nemachilus sp.*	+	+	+	+

Correlation and Tables

Table 103 Showing correlation of water temperature with all other physico-chemical and biological parameters

	Dal lake								Anchar lake					
	I		II		III		IV		I		II		III	
	r	t	r	t	r	t	r	t	r	t	r	t	r	T
1. Secchi transparency	0.04	0.21	−0.005	0.35	−0.09	0.039	−0.60	0.048	−0.21	0.40	−0.06	0.65	0.05	0.62
2. Depth	−0.39	2.0	−0.08	0.56	−0.21	0.72	−0.29	0.82	−0.19	0.72	−0.01	0.05	0.38	0.13
3. Turbidity	0.42	2.18	0.19	0.48	0.04	0.71	0.39	0.85	0.10	0.20	−0.12	0.01	0.53	0.13
4. Dissolved Oxygen	−0.47	2.65	−0.37	2.32	−0.47	2.47	−0.50	2.59	−0.60	2.62	−0.49	2.12	−0.52	2.13
5. pH	0.49	0.81	0.06	0.94	0.23	1.21	0.29	1.11	−0.16	1.43	0.03	1.52	−0.09	1.23
6. Conductivity	−0.06	0.30	0.01	0.69	−0.60	3.84	−0.25	1.03	−0.45	2.96	−0.28	0.79	−0.47	2.84
7. Calcium	−0.28	1.37	−0.76	3.42	−0.31	0.97	−0.67	0.82	0.12	0.37	−0.13	0.45	0.08	0.92
8. Magnesium	0.27	1.34	−0.6	0.98	−0.32	0.42	−0.55	0.39	−0.08	1.40	−0.34	0.79	−0.13	0.81
9. Total alkalinity	0.09	0.46	−0.1	0.89	−0.33	0.98	−0.38	1.42	0.49	1.32	−0.02	1.09	−0.31	1.58
10. Chloride	0.29	0.97	0.05	0.87	−0.41	0.40	−0.28	0.20	0.08	0.48	−0.19	0.39	−0.37	0.84
11. Iron	0.39	1.99	0.15	1.20	−0.63	0.98	−0.12	1.04	−0.23	0.83	0.38	0.48	−0.20	0.94
12. Ammonical nitrogen	0.15	0.71	−0.13	0.94	−0.36	0.65	0.40	1.84	−0.27	0.89	−0.42	1.20	−0.67	0.98
13. Nitrate nitrogen	0.48	2.56	−0.3	1.42	0.33	1.45	0.07	0.98	−0.41	0.58	−0.35	1.87	−0.66	0.87
14. Total $PO_4 - P$	−0.56	3.19	−0.28	2.40	−0.13	2.35	−0.15	2.20	−0.34	2.51	−0.02	2.68	−0.007	2.59
15. Silicates	0.35	1.78	0.41	1.35	0.51	1.20	0.16	0.98	−0.001	1.43	−0.16	1.24	−0.02	0.57

(continued)

Table 103 (continued)

	Dal lake								Anchar lake					
	I		II		III		IV		I		II		III	
	r	t	r	t	r	t	r	t	r	t	r	t	r	T
16. Sodium	0.56	1.17	0.08	0.35	0.29	0.49	0.09	0.53	0.02	0.84	–0.09	0.74	–0.30	0.98
17. Potassium	0.29	1.44	0.01	0.48	0.27	0.23	–0.06	0.78	–0.19	0.99	–0.15	1.04	–0.19	1.06
18. Total hardness	0.14	0.90	0.21	0.85	0.31	1.01	0.28	0.95	–0.02	1.24	–0.39	1.05	–0.07	1.00
19. TDS	0.19	0.58	0.27	1.11	0.41	2.11	0.23	1.19	–0.45	1.43	–0.31	1.25	–0.47	0.98
20. Bacillariophyceae	0.35	0.86	0.21	1.20	0.49	0.98	0.53	1.03	–	–	–	–	–	–
21. Chlorophyceae	0.34	1.74	0.30	0.54	0.35	0.95	0.40	0.98	–	–	–	–	–	–
22. Cyanophyceae	0.41	2.17	0.41	1.40	0.29	1.35	0.31	1.44	–	–	–	–	–	–
23. Dinophyceae	0.24	1.19	0.13	1.31	0.15	0.85	0.19	1.02	–	–	–	–	–	–
24. Chrysophyceae	Absent		0.46	1.01	0.53	0.81	0.40	0.49	–	–	–	–	–	–
25. Euglenophyceae	0.10	0.47	0.29	0.29	0.41	1.20	0.39	1.03	–	–	–	–	–	–
26. Protozoa	0.09	0.42	0.45	1.43	0.30	1.23	-0.41	0.98	0.58	1.45	0.49	1.24	0.08	1.40
27. Rotifera	0.62	3.73	0.45	2.01	0.31	1.45	0.41	1.63	0.35	1.89	0.41	1.92	0.53	1.62
28. Cladocera	0.46	2.49	0.35	1.40	0.57	2.35	0.60	3.51	0.59	2.62	0.51	2.79	0.49	2.49
29. Copepoda	0.22	1.07	0.47	2.09	0.52	2.95	0.49	2.85	0.58	3.03	0.31	1.43	0.22	1.45

Table 104 Showing correlation of secchi transparency with all other physico-chemical and biological parameters

	Dal lake								Anchar lake					
	I		II		III		IV		I		II		III	
	r	t	r	t	r	t	r	t	r	t	r	t	r	t
2. Depth	0.28	1.38	0.75	2.56	0.09	0.48	0.73	1.94	0.99	3.45	0.30	0.45	0.39	1.68
3. Turbidity	−0.11	0.53	−0.36	1.48	−0.39	1.47	−0.44	2.10	−0.34	1.86	−0.2	0.94	−0.04	0.35
4. Dissolved Oxygen	−0.69	3.53	−0.42	1.84	0.34	1.24	0.15	1.11	−0.001	0.34	0.10	0.86	0.006	0.41
5. pH	0.75	4.37	−0.03	0.43	0.09	0.57	0.003	0.35	−0.22	1.00	0.07	0.41	0.34	1.58
6. Conductivity	0.38	1.96	0.47	1.94	0.14	0.94	0.59	2.10	0.28	1.84	−0.42	1.89	−0.02	0.40
7. Calcium	0.24	1.17	−0.06	0.43	−0.16	1.80	0.64	2.30	−0.19	0.94	−0.21	1.43	−0.36	1.87
8. Magnesium	−0.31	1.56	0.24	1.04	−0.13	1.43	0.60	2.06	−0.34	1.84	0.008	0.04	−0.28	0.87
9. Total alkalinity	0.31	0.63	0.37	1.50	−0.22	0.98	0.53	1.87	−0.17	0.98	0.04	0.45	−0.18	0.48
10. Chloride	0.14	0.66	0.29	0.86	0.14	0.43	0.24	0.58	−0.11	0.89	−0.12	0.56	−0.05	0.11
11. Iron	−0.21	1.0	0.16	0.53	−0.18	0.42	0.13	0.52	0.21	0.79	0.05	0.11	−0.01	0.24
12. Ammonical nitrogen	0.06	0.31	−0.31	1.46	−0.18	0.40	−0.36	1.69	0.32	1.01	−0.30	1.14	0.02	0.84
13. Nitrate nitrogen	−0.18	0.90	0.44	2.01	0.1	0.45	−0.01	0.16	0.75	2.56	0.19	1.43	0.17	1.13
14. Total $PO_4 – P$	−0.25	1.22	0.72	2.69	−0.11	1.11	−0.13	0.98	0.28	1.20	−0.06	0.48	0.20	0.52
15. Silicates	−0.06	0.28	−0.30	1.04	−0.2	0.90	−0.11	0.81	0.14	0.61	0.32	1.84	−0.41	1.99
16. Sodium	−0.36	1.85	0.47	2.11	0.007	0.43	−0.21	1.41	−0.38	1.68	−0.47	2.10	−0.45	2.15

(continued)

Table 104 (continued)

	Dal lake								Anchar lake					
	I		II		III		IV		I		II		III	
	r	t	r	t	r	t	r	t	r	t	r	t	r	t
17. Potassium	0.04	0.22	0.31	1.43	–0.18	0.90	–0.07	0.35	–0.32	1.43	–0.38	1.58	–0.06	0.43
18. Total hardness	0.21	1.01	0.16	0.98	0.21	0.95	0.14	043	–0.32	1.43	–0.09	0.15	–0.34	1.11
19. TDS	0.11	0.43	0.05	0.31	0.10	0.35	0.15	0.58	0.28	1.19	–0.42	2.10	–0.02	1.11
20. Bacillariophyceae	–0.08	0.39	–0.01	0.40	0.21	1.10	0.04	0.43	–	–	–	–	–	–
21. Chlorophyceae	-0.07	0.34	0.01	0.25	0.19	0.80	0.02	0.35	–	–	–	–	–	–
22. Cyanophyceae	–0.19	0.93	–0.05	0.15	0.11	0.69	0.30	0.72	–	–	–	–	–	–
23. Dinophyceae	0.39	2.04	–0.10	0.15	0.15	0.42	0.27	0.98	–	–	–	–	–	–
24. Chrysophyceae	–	–	–0.99	2.56	0.28	1.20	0.009	0.45	–	–	–	–	–	–
25. Euglenophyceae	0.05	0.27	–0.15	0.45	0.20	0.69	0.51	2.10	–	–	–	–	–	–
26. Protozoa	–0.19	0.95	0.14	0.42	0.15	0.68	0.25	0.29	0.09	0.35	0.11	0.86	0.09	0.43
27. Rotifera	–0.35	1.79	0.21	0.39	0.29	0.81	0.42	1.32	0.24	0.99	0.40	1.82	0.29	0.86
28. Cladocera	–0.39	2.01	0.13	0.54	0.35	1.04	0.21	2.56	0.36	1.50	0.25	0.85	0.09	0.45
29. Copepoda	–0.17	0.81	0.18	0.56	0.40	1.91	0.32	1.53	0.18	0.69	0.16	0.94	0.22	1.04

Table 105 Showing correlation of depth with all other physico-chemical and biological parameters

	Dal lake								Anchar lake					
	I		II		III		IV		I		II		III	
	r	t	r	t	r	t	r	t	r	t	r	t	r	t
3. Turbidity	–0.18	0.89	0.36	1.43	–0.03	0.45	–0.57	2.53	–0.35	1.11	–0.32	1.69	0.07	0.15
4. Dissolved Oxygen	–0.29	1.46	–0.31	1.24	–0.14	0.84	0.15	0.89	–0.03	0.35	0.39	1.74	0.37	0.98
5. pH	–0.06	0.28	–0.21	0.98	–0.21	0.96	0.20	0.91	–0.23	0.69	0.39	1.82	0.29	0.62
6. Conductivity	0.57	3.26	0.38	1.60	0.52	1.87	0.50	1.95	0.26	0.87	–0.21	0.67	–0.14	0.59
7. Calcium	0.54	3.05	0.05	0.42	0.04	0.45	0.23	0.57	–0.21	0.58	0.04	0.35	0.12	0.64
8. Magnesium	0.04	0.21	–0.01	0.35	0.16	0.41	0.40	1.62	–0.38	1.87	0.23	0.69	–0.35	1.64
9. Total alkalinity	0.15	0.74	0.52	1.69	0.43	2.42	0.34	1.50	–0.18	0.82	–0.23	0.74	0.06	0.19
10. Chloride	–0.10	0.50	0.54	1.92	0.33	1.80	–0.22	0.95	–0.04	0.43	–0.26	0.82	–0.48	2.10
11. Iron	–0.44	2.31	0.04	0.40	0.09	0.41	0.02	0.33	0.22	1.01	–0.01	0.35	–0.11	1.00
12. Ammonical nitrogen	0.16	0.76	–0.10	0.11	–0.46	1.96	–0.1	0.25	0.32	1.06	0.15	0.49	–0.57	1.87
13. Nitrate nitrogen	0.17	0.83	0.33	1.51	–0.15	0.91	–0.3	0.45	0.77	2.43	–0.26	0.84	–0.31	1.43
14. Total PO_4 – P	–0.44	2,32	0.63	2.17	–0.13	0.87	–0.2	0.61	0.27	0.98	–0.24	0.79	–0.001	0.10
15. Silicates	0.41	2.16	–0.04	1.69	–0.01	0.33	0.03	0.25	0.14	0.45	0.34	1.04	0.126	0.13

(continued)

Table 105 (continued)

	Dal lake								Anchar lake					
	I		II		III		IV		I		II		III	
	r	t	r	t	r	t	r	t	r	t	r	t	r	t
16. Sodium	–0.02	0.09	0.28	1.16	–0.14	0.54	–0.07	0.13	–0.39	1.56	–0.12	0.42	–0.164	0.25
17. Potassium	0.19	0.91	0.03	0.43	–0.05	0.10	0.13	0.29	–0.13	1.43	–0.22	0.68	–0.36	0.98
18. Total hardness	0.12	0.48	0.07	0.25	0.17	0.83	0.21	0.94	–0.35	1.68	0.09	0.35	–0.22	0.70
19. TDS	0.05	0.90	0.11	0.43	0.18	0.76	0.23	0.76	0.26	1.02	–0.15	0.49	–0.13	0.65
20. Bacillariophyceae	0.17	0.81	0.23	0.98	0.27	1.04	0.12	0.43						
21. Chlorophyceae	0.11	0.53	0.13	0.43	0.13	0.69	0.25	0.75						
22. Cyanophyceae	0.05	0.25	0.12	0.23	0.19	0.52	0.22	0.95						
23. Dinophyceae	0.03	0.16	0.18	0.40	0.39	1.62	0.20	1.01						
24. Chrysophyceae	–	–	0.66	2.13	0.35	1.52	0.19	0.82						
25. Euglenophyceae	0.18	0.89	0.10	0.43	0.10	0.48	0.12	0.49						
26. Protozoa	0.01	0.07	0.01	0.10	0.19	0.44	0.21	0.58	0.09	0.21	0.13	0.29	0.11	0.32
27. Rotifera	0.28	1.40	0.20	0.45	0.15	0.35	0.25	0.98	0.19	0.59	0.17	0.89	0.12	0.53
28. Cladocera	0.21	1.04	0.17	0.69	0.16	0.50	0.32	1.32	0.52	1.67	0.32	1.04	0.10	0.14
29. Copepoda	0.37	1.86	0.13	0.50	0.21	0.69	0.01	0.45	0.10	0.45	0.15	0.35	0.17	0.25

Table 106 Showing correlation of turbidity with all other physico-chemical and biological parameters

	Dal lake								Anchar lake					
	I		II		III		IV		I		II		III	
	r	t	r	t	r	t	r	t	r	t	r	t	r	t
4. Dissolved Oxygen	0.16	0.76	−0.21	0.91	−0.16	0.41	−0.35	0.85	−0.32	1.07	−0.11	0.19	−0.09	0.35
5. pH	0.30	1.49	−0.1	0.50	0.04	0.35	−0.10	0.45	0.21	0.89	−0.59	1.94	−0.32	1.01
6. Conductivity	0.18	0.90	0.41	0.91	0.02	0.40	−0.20	0.51	0.31	1.20	0.35	1.58	−0.64	2.16
7. Calcium	−0.26	−1.26	−0.37	1.43	0.25	0.98	−0.05	0.14	−0.38	0.67	−0.11	0.74	0.39	0.96
8. Magnesium	0.02	0.11	0.25	0.75	0.20	0.85	−0.30	0.95	−0.04	0.37	0.23	0.54	0.21	0.84
9. Total alkalinity	−0.25	−1.21	0.45	1.23	0.26	0.81	−0.07	0.25	0.35	1.11	−0.30	1.35	−0.23	1.45
10. Chloride	0.01	0.06	0.07	0.45	0.03	0.23	0.30	0.91	0.39	0.35	0.69	2.65	−0.16	1.43
11. Iron	−0.01	−0.06	0.54	1.82	0.09	0.33	0.12	0.21	0.33	1.87	−0.25	0.80	−0.17	0.50
12. Ammonical nitrogen	−0.25	−1.22	−0.61	2.10	0.08	0.35	0.05	0.15	−0.60	2.50	0.31	1.41	−0.25	0.72
13. Nitrate nitrogen	−0.11	−0.55	0.03	0.55	0.15	0.31	0.6	2.10	−0.62	2.43	0.37	1.62	−0.16	0.89
14. Total $PO_4 – P$	0.33	1.64	0.47	2.52	0.34	1.45	0.10	0.98	0.70	3.50	0.25	1.07	0.20	0.98
15. Silicates	0.30	1.51	0.05	0.45	0.37	1.30	−0.05	0.69	−0.31	1.89	−0.56	2.10	−0.09	0.45

(continued)

Table 106 (continued)

	Dal lake								Anchar lake					
	I		II		III		IV		I		II		III	
	r	t	r	t	r	t	r	t	r	t	r	t	r	t
16. Sodium	0.28	1.37	0.38	1.64	−0.13	0.54	0.11	0.45	0.53	2.15	0.10	0.35	−0.03	0.14
17. Potassium	0.36	1.84	0.28	1.20	0.04	0.95	−0.2	1.02	0.30	1.05	0.58	2.13	0.11	1.04
18. Total hardness	0.11	0.48	0.14	0.79	0.20	0.53	0.25	0.92	−0.15	0.78	0.23	0.83	0.30	1.05
19. TDS	0.19	0.85	0.15	0.62	0.09	0.14	0.25	0.79	0.30	1.01	0.36	1.25	−0.64	1.03
20. Bacillariophyceae	0.24	1.20	0.31	0.98	0.17	0.40	0.11	0.19						
21. Chlorophyceae	0.11	0.52	0.25	0.78	0.20	0.51	0.21	0.80						
22. Cyanophyceae	0.41	2.15	0.04	0.14	0.19	0.59	0.29	0.78						
23. Dinophyceae	0.17	0.03	0.66	0.49	0.30	1.11	0.45	1.27						
24. Chrysophyceae			0.78	3.14	0.35	1.46	0.30	1.29						
25. Euglenophyceae	0.26	1.29	−0.57	2.14	0.40	2.11	0.11	0.59						
26. Protozoa	0.18	0.86	0.11	0.87	0.23	1.04	0.25	0.91	0.01	0.43	0.11	0.53	0.27	0.33
27. Rotifera	0.19	0.92	0.18	0.49	0.21	0.37	0.22	0.41	0.15	0.40	0.10	0.27	0.21	0.29
28. Cladocera	0.04	0.22	0.17	0.63	0.16	0.53	0.21	0.21	0.21	0.32	0.16	0.28	0.10	0.40
29. Copepoda	0.33	1.68	0.31	1.27	0.01	0.21	0.04	0.30	0.04	0.41	0.06	0.25	0.14	0.87

Table 107 Showing correlation of dissolved oxygen with all other physico-chemical and biological parameters

	Dal lake								30 pt					
	I		II		III		IV		I		II		III	
	r	t	r	t	r	t	r	t	r	t	r	t	r	t
5. pH	0.26	1.27	0.07	0.43	0.13	0.80	0.29	0.98	0.13	0.56	0.26	0.78	0.17	0.52
6. Conductivity	−0.1	0.5	−0.07	0.9	−0.22	0.54	−0.3	0.76	−0.06	0.09	−0.37	0.98	−0.33	0.99
7. Calcium	−0.20	0.90	−0.38	1.35	−0.27	1.05	−0.08	0.47	0.51	1.57	0.16	1.01	0.39	1.35
8. Magnesium	−0.2	1.3	−0.22	1.3	−0.30	1.14	−0.04	0.53	0.55	1.69	0.72	1.85	−0.07	0.53
9. Total alkalinity	0.02	0.12	0.14	0.53	−0.38	0.98	−0.01	0.13	−0.27	0,56	−0.25	0.79	0.11	0.58
10. Chloride	0.18	0.86	0.04	0.54	0.04	0.34	−0.27	0.69	0.03	0.43	0.13	0.35	−0.20	0.25
11. Iron	0.04	0.22	0.19	0.57	0.44	0.99	0.44	1.43	0.22	0.98	0.01	0.23	0.23	0.56
12. Ammonical nitrogen	0.29	1.45	0.15	0.98	0.43	1.95	0.21	0.87	0.52	1.89	0.57	2.01	−0.24	0.74
13. Nitrate nitrogen	0.46	2.46	0.23	2.13	0.36	2.19	0.4	2.13	0.27	2.01	0.22	1.75	0.002	0.50
14. Total $PO_4 - P$	0.02	0.09	0.03	0.74	0.21	0.82	−0.01	0.51	0.41	1.43	0.09	0.43	0.11	0.65
15. Silicates	−0.4	−2.2	0.27	0.78	−0.34	1.82	−0.52	2.15	0.42	2.17	0.32	1.87	−0.14	0.72
16. Sodium	−0.4	−2.1	−0.04	0.27	0.17	0.80	−0.36	1.42	0.05	0.23	0.19	0.48	−0.27	0.58
17. Potassium	−0.01	−0.06	−0.19	0.69	−0.10	0.43	−0.26	0.68	0.33	1.49	0.22	0.79	−0.14	2.10
18. Total hardness	0.21	1.10	0.33	1.43	0.35	1.51	0.30	1.01	0.56	1.79	0.63	2.09	0.08	0.56
19. TDS	0.30	1.41	0.14	0.98	0.21	1.00	0.25	1.09	0.06	0.79	0.39	1.43	0.34	1.23
20. Bacillariophyceae	0.3	1.6	0.09	0.58	0.20	0.97	0.26	1.47						
21. Chlorophyceae	0.58	3.3	0.17	0.98	0.58	2.13	0.47	2.51						
22. Cyanophyceae	0.3	2.0	0.29	1.43	0.64	2.10	0.56	2.31						
23. Dinophyceae	0.01	0.07	0.02	0.48	0.21	0.95	0.31	1.80						
24. Chrysophyceae			0.55	1.72	0.58	1.64	0.62	1.82						
25. Euglenophyceae	0.08	0.38	0.09	0.65	0.68	2.51	0.65	2.38						
26. Protozoa	0.37	1.91	0.40	1.87	0.25	1.10	0.20	0.48	0.19	0.39	0.01	0,10	0.15	0.30
27. Rotifera	0.56	3.21	0.15	0.35	0.17	0.69	0.18	0.69	0.03	0.15	0.28	0.81	0.90	3.49
28. Cladocera	0.5	3.4	0.25	0.72	0.31	0.87	0.19	0.38	0.11	0.19	0.13	0.28	0.14	0.30
29. Copepoda	0.07	0.36	0.14	0.32	0.21	0.47	0.23	0.49	0.05	0.13	0.10	0.23	0.11	0.33

Table 108 Showing correlation of pH with all other physico-chemical and biological parameters

	Dal lake								Anchar lake					
	I		II		III		IV		I		II		III	
	r	t	r	t	r	t	r	t	r	t	r	t	r	t
6. Conductivity	−0.30	1.49	−0.007	0.35	−0.08	0.25	−0.18	0.58	0.22	0.72	0.02	0.15	0.09	0.17
7. Calcium	−0.19	0.95	−0.02	0.42	0.08	0.51	−0.41	0.94	−0.09	0.63	0.33	0.54	−0.09	0.13
8. Magnesium	−0.31	1.53	−0.10	0.50	−0.09	0.98	−0.11	0.95	0.08	0.40	0.15	0.25	−0.54	1.25
9. Total alkalinity	0.13	0.61	0.31	1.51	0.10	0.50	0.006	0.14	0.15	0.35	0.06	0.45	0.15	0.25
10. Chloride	0.13	0.62	−0.01	0.25	−0.20	0.36	−0.27	0.49	−0.02	0.15	−0.34	0.98	−0.11	0.15
11. Iron	−0.08	0.42	−0.13	0.65	0.004	0.25	−0.55	1.25	0.26	0.95	0.46	1.04	−0.19	0.58
12. Ammonical nitrogen	0.06	0.31	0.39	0.95	−0.17	0.15	0.29	0.59	−0.08	0.25	0.03	0.35	0.03	0.34
13. Nitrate nitrogen	0.27	1.34	−0.005	0.21	0.18	0.35	−0.33	0.78	−0.26	0.58	−0.34	0.69	−0.02	0.13
14. Total PO_4 – P	0.26	1.30	−0.23	0.80	0.009	0.10	−0.50	2.10	−0.27	0.14	−0.29	0.57	0.17	0.34
15. Silicates	0.01	0.06	0.02	0.41	0.09	0.10	−0.43	1.80	−0.03	0.13	0.30	1.01	0.35	1.27
16. Sodium	0.37	1.92	−0.03	0.16	0.13	0.25	−0.42	2.41	0.10	0.81	0.16	1.01	−0.06	0.97
17. Potassium	0.40	2.0	0.06	0.43	0.29	1.04	−0.28	1.15	0.24	1.69	−0.18	0.89	−0.34	1.69
18. Total hardness	0.13	0.84	0.25	0.99	0.17	0.78	0.19	0.72	0.03	0.15	0.16	0.43	−0.45	2.43
19. TDS	0.11	0.98	0.17	1.01	0.21	1.45	0.23	1.56	0.22	0.98	0.11	0.35	0.09	0.19
20. Bacillariophyceae	0.14	0.68	0.39	2.15	0.65	2.56	0.61	2.41						
21. Chlorophyceae	0.31	1.54	0.31	1.64	0.44	2.45	0.53	2.15						
22. Cyanophyceae	0.16	0.78	0.64	2.56	0.62	3.57	0.60	3.42						
23. Dinophyceae	0.13	0.65	0.08	0.91	0.15	1.01	0.21	1.43						
24. Chrysophyceae			0.66	2.58	0.45	1.62	0.51	1.92						
25. Euglenophyceae	0.01	0.07	0.16	1.01	0.14	0.49	0.10	0.25						
26. Protozoa	0.23	1.11	0.11	0.19	0.16	0.23	0.17	0.35	0.01	0.11	0.05	0.10	0.10	0.37
27. Rotifera	0.01	0.08	0.15	0.47	0.03	0.15	0.24	0.69	0.23	0.64	0.22	0.48	0.06	0.10
28. Cladocera	0.27	1.33	0.03	0.14	0.18	0.73	0.14	0.60	0.12	0.53	0.19	0.62	0.03	0.14
29. Copepoda	0.20	0.95	0.07	0.15	0.36	1.98	0.24	1.04	0.18	0.81	0.06	0.19	0.24	0.84

Table 109 Showing correlation of conductivity with all other physico-chemical and biological parameters

	Dal lake								Anchar lake					
	I		II		III		IV		I		II		III	
	r	t	r	t	r	t	r	t	r	t	r	t	r	t
7. Calcium	0.53	2.95	0.24	2.09	0.29	2.10	0.62	2.15	0.51	2.57	0.07	1.35	0.17	2.15
8. Magnesium	0.39	2.02	0.21	2.30	0.36	2.19	0.68	2.59	0.16	1.56	0.40	2.30	0.15	1.29
9. Total alkalinity	0.22	1.08	0.39	1.52	0.53	1.62	0.43	1.84	0.12	0.95	0.24	0.84	0.33	0.75
10. Chloride	–0.03	0.16	0.36	2.56	0.50	2.95	0.35	2.13	0.14	2.50	0.60	2.85	0.44	2.10
11. Iron	0.35	1.76	0.28	1.25	0.33	1.62	0.39	1.59	0.56	1.87	0.17	0.98	0.54	1.24
12. Ammonical nitrogen	0.05	0.245	0.17	0.98	0.16	1.87	0.36	2.52	0.11	1.52	0.70	2.18	0.33	2.15
13. Nitrate nitrogen	0.28	1.41	0.33	1.49	0.24	1.20	0.24	1.62	0.21	1.52	0.50	2.58	0.39	2.62
14. Total $PO_4 - P$	0.06	0.29	0.48	2.53	0.01	1.43	0.19	1.29	0.03	0.58	0.24	0.98	0.25	1.23
15. Silicates	0.44	2.34	0.23	1.98	0.37	2.19	0.23	1.92	0.07	0.58	0.17	0.98	0.01	0.57
16. Sodium	0.29	–1.46	0.05	0.17	0.36	1.89	0.12	0.43	0.02	0.12	0.10	0.29	0.10	0.31
17. Potassium	0.03	0.16	0.14	0.32	0.24	0.39	0.12	0.32	0.06	0.01	0.42	1.82	0.15	0.40
18. Total hardness	0.31	1.05	0.45	2.09	0.53	2.53	0.50	2.10	–0.28	1.16	0.40	2.13	–0.18	0.98
19. TDS	0.97	3.41	0.99	4.35	0.98	4.13	0.99	4.50	0.99	4.36	0.98	3.46	0.99	4.31
20. Bacillariophyceae	–0.11	0.52	–0.18	0.91	0.14	0.87	0.18	0.89						
21. Chlorophyceae	–0.28	1.41	–0.19	0.92	0.30	1.04	0.16	0.56						
22. Cyanophyceae	0.13	0.63	0.17	0.87	0.01	0.56	0.11	0.51						
23. Dinophyceae	0.10	0.49	–0.29	0.90	0.09	0.40	0.17	0.67						
24. Chrysophyceae	Absent		–0.77	2.53	0.24	1.04	0.16	0.70						
25. Euglenophyceae	0.01	0.08	–0.03	0.10	0.17	0.12	0.11	0.41						
26. Protozoa	–0.03	0.18	0.23	0.70	0.48	2.10	0.21	0.87	0.19	0.73	0.13	0.58	0.18	0.62
27. Rotifera	–0.19	0.92	0.14	0.60	0.23	0.65	0.12	0.42	0.10	0.35	0.005	0.10	0.15	0.41
28. Cladocera	–0.38	1.96	0.23	0.82	0.20	0.72	0.27	1.01	0.33	1.42	0.23	1.01	0.16	0.82
29. Copepoda	–0.34	1.72	0.05	0.12	0.16	0.39	0.23	0.68	0.11	0.52	0.24	0.87	0.07	0.23

Table 110 Showing correlation of calcium with all other physico-chemical and biological parameters

	Dal lake								Anchar lake					
	I		II		III		IV		I		II		III	
V/S	r	t	r	t	r	t	r	t	r	t	r	t	r	t
8. Magnesium	0.45	2.41	0.39	2.51	0.42	2.10	0.75	2.10	0.74	2.49	0.50	2.80	0.54	2.75
9. Total alkalinity	0.54	3.07	0.26	2.90	0.40	2.59	0.62	2.80	0.33	2.59	0.59	2.87	0.35	2.95
10. Chloride	0.09	0.44	0.13	0.43	−0.05	0.49	0.68	0.95	0.08	0.80	−0.08	0.50	−0.32	0.30
11. Iron	0.008	0.03	−0.36	0.04	−0.01	0.50	0.41	0.69	−0.20	0.75	0.22	0.63	0.26	0.82
12. Ammonical nitrogen	0.20	0.96	0.19	0.84	0.11	0.64	−0.52	0.55	0.20	0.83	0.04	0.53	−0.33	0.63
13. Nitrate nitrogen	−0.08	0.41	0.22	0.50	−0.51	0.43	0.35	0.69	−0.02	0.89	−0.28	0.80	−0.10	0.93
14. Total $PO_4 - P$	0.02	0.11	−0.3	0.01	0.10	0.40	0.02	0.20	0.15	0.29	0.20	0.58	−0.13	0.53
15. Silicates	−0.41	0.67	−0.1	0.01	−0.19	0.53	−0.05	0.49	0.36	0.81	−0.21	0.61	0.28	0.49
16. Sodium	−0.12	0.57	−0.3	0.81	−0.31	0.86	−0.16	0.46	0.12	0.70	0.20	.90	0.08	0.81
17. Potassium	−0.28	1.40	−0.2	0.66	0.05	0.83	−0.04	0.74	0.21	0.77	−0.19	0.81	−0.33	0.77
18. Total hardness	0.51	2.13	0.43	2.10	0.42	2.09	0.50	2.10	0.87	2.43	0.68	2.19	0.76	2.49
19. TDS	0.23	1.13	0.18	1.52	0.25	1.69	0.31	1.83	−0.50	2.15	0.12	1.01	−0.16	1.16
20. Bacillariophyceae	0.11	0.56	0.25	0.87	0.17	0.69	0.13	0.50						
21. Chlorophyceae	0.08	0.41	0.33	1.68	0.15	0.19	0.19	0.38						
22. Cyanophyceae	0.15	0.75	0.34	1.86	0.15	0.45	0.05	0.13						
23. Dinophyceae	0.15	0.74	0.27	1.01	0.02	0.23	0.21	0.45						
24. Chrysophyceae			0.58	2.15	0.12	0.43	0.13	0.58						
25. Euglenophyceae	0.09	0.46	0.18	0.40	0.05	0.13	0.11	0.23						
26. Protozoa	−0.07	0.37	0.10	0.23	0.005	0.11	0.20	0.45	0.13	0.87	0.12	0.69	0.26	0.56
27. Rotifera	−0.14	0.70	0.20	0.29	0.11	0.32	0.01	0.20	0.20	0.58	0.12	0.40	0.12	0.39
28. Cladocera	−0.17	0.85	0.12	0.85	0.28	0.98	0.09	0.30	0.33	1.58	0.26	1.01	0.13	0.91
29. Copepoda	−0.16	0.77	0.10	0.79	0.13	0.87	0.01	0.20	0.30	1.62	0.25	1.01	0.14	0.89

Table 111 Showing correlation of magnesium with all other physico-chemical and biological parameters

	Dal lake								Anchar lake					
	I		II		III		IV		I		II		III	
	r	t	r	t	r	t	r	t	r	t	r	t	r	t
9. Total alkalinity	0.23	1.14	0.27	1.15	0.65	1.82	0.67	1.34	0.38	1.02	−0.13	0.95	0.06	0.80
10. Chloride	−0.02	0.12	−0.1	0.51	0.47	0.92	0.48	1.02	0.03	0.80	0.28	0.57	−0.09	0.59
11. Iron	0.30	1.50	0.27	1.20	0.05	1.11	0.40	1.15	0.02	1.41	−0.008	0.49	0.07	0.53
12. Ammonical nitrogen	−0.20	1.0	−0.34	1.01	−0.16	0.98	−0.32	0.33	0.13	0.45	0.62	0.97	0.06	0.43
13. Nitrate nitrogen	−0.33	1.67	0.23	0.98	−0.58	1.64	0.008	0.23	−0.16	0.40	0.23	0.79	−0.01	0.65
14. Total $PO_4 - P$	0.19	0.93	−0.04	0.51	−0.36	0.99	−0.41	1.04	−0.05	1.51	0.02	0.85	−0.15	0.95
15. Silicates	0.10	0.47	−0.16	0.59	−0.32	0.63	−0.04	0.48	0.30	0.59	0.04	0.89	−0.10	0.74
16. Sodium	−0.05	0.27	0.10	0.94	−0.67	0.53	−0.32	0.91	0.27	0.87	0.17	0.80	0.23	0.95
17. Potassium	−0.47	2.56	0.23	0.85	−0.51	0.51	0.01	0.83	0.44	0.99	0.21	1.50	0.31	0.99
18. Total hardness	0.53	2.43	0.62	2.51	0.48	2.13	0.51	2.10	0.97	3.01	0.92	2.95	0.95	2.85
19. TDS	0.21	1.11	0.18	1.23	0.15	1.20	0.18	1.31	0.16	1.30	0.46	2.50	0.15	1.10
20. Bacillariophyceae	0.06	0.28	0.18	0.23	0.09	0.10	0.17	0.14						
21. Chlorophyceae	0.008	0.64	0.28	0.48	0.13	0.32	0.12	0.24						
22. Cyanophyceae	0.17	0.85	0.29	0.68	0.15	0.40	0.21	0.80						
23. Dinophyceae	0.19	0.94	0.05	0.11	0.21	0.39	0.05	0.15						
24. Chrysophyceae			0.93	2.62	0.56	2.13	0.61	2.32						
25. Euglenophyceae	0.01	0.04	0.02	0.11	0.11	0.17	0.21	0.34						
26. Protozoa	−0.26	−1.31	0.09	0.10	0.20	0.33	0.23	0.49	0.30	0.53	0.16	0.45	0.13	0.67
27. Rotifera	0.11	0.55	0.16	0.45	0.21	0.78	0.10	0.55	0.27	0.45	0.18	0.55	0.23	0.61
28. Cladocera	−0.05	−0.26	0.08	0.11	0.10	0.33	0.01	0.12	0.20	0.45	0.10	0.35	0.21	0.49
29. Copepoda	−0.26	−1.28	0.31	1.61	0.29	1.43	0.23	1.01	0.04	0.11	0.02	0.15	0.008	0.03

Table 112 Showing correlation of total alkalinity with all other physico-chemical and biological parameters

	Dal lake								Anchar lake					
	I		II		III		IV		I		II		III	
	r	t	r	t	r	t	r	t	r	t	r	t	r	t
10. Chloride	0.40	2.06	0.74	2.50	0.61	2.11	0.64	2.13	0.16	2.01	0.14	1.98	0.22	2.03
11. Iron	0.16	0.79	0.12	0.45	0.01	0.89	0.38	0.34	0.10	0.56	0.26	0.43	0.47	0.94
12. Ammonical nitrogen	0.13	0.64	0.32	0.98	0.15	0.58	0.27	0.63	0.41	0.64	0.42	0.82	0.06	0.13
13. Nitrate nitrogen	−0.31	1.55	0.33	0.94	−0.31	0.54	0.13	0.92	−0.31	1.04	−0.16	0.49	−0.24	0.32
14. Total $PO_4 - P$	0.18	0.87	0.43	0.58	−0.13	0.68	−0.14	0.63	−0.49	0.72	0.22	0.58	0.01	0.89
15. Silicates	−0.2	1.0	−0.07	0.56	−0.19	0.23	−0.28	0.59	−0.04	0.36	−0.25	0.90	−0.40	0.99
16. Sodium	−0.2	1.0	0.003	0.50	−0.35	0.58	−0.34	0.72	0.26	0.60	0.10	0.52	0.30	0.68
17. Potassium	−0.24	1.18	−0.23	0.98	0.001	0.40	−0.03	0.23	0.07	0.25	−0.40	1.25	−0.17	0.72
18. Total hardness	0.11	0.53	0.13	0.83	0.19	0.95	0.25	1.04	−0.39	2.41	0.07	0.11	0.17	0.37
19. TDS	0.09	0.23	0.10	0.62	0.07	0.21	0.15	0.35	−0.12	0.45	−0.27	0.98	0.33	1.65
20. Bacillariophyceae	0.04	0.22	0.29	1.04	0.72	3.15	0.52	2.15						
21. Chlorophyceae	0.29	1.45	0.33	1.68	0.92	4.17	0.42	2.10						
22. Cyanophyceae	0.19	0.92	0.07	0.45	0.74	3.10	0.61	3.01						
23. Dinophyceae	0.05	0.27	0.25	0.68	0.15	0.45	0.22	0.44						
24. Chrysophyceae			0.98	3.45	0.23	0.68	0.25	0.37						
25. Euglenophyceae	0.41	2.15	−0.19	1.11	0.88	2.68	0.11	0.22						
26. Protozoa	0.19	0.95	0.23	1.04	0.30	1.42	0.14	0.58	0.19	0.86	0.24	0.85	0.27	1.01
27. Rotifera	0.23	1.1	0.28	1.11	0.27	1.61	1.26	1.50	0.10	0.79	0.20	0.87	0.18	0.69
28. Cladocera	−0.15	0.72	0.17	1.10	0.26	1.59	0.20	0.99	0.05	0.45	0.01	0.33	0.20	0.45
29. Copepoda	−0.18	0.89	0.19	1.11	0.11	0.43	0.05	0.51	0.15	0.33	0.11	0.25	0.16	0.55

Table 113 Showing correlation of chloride with all other physico-chemical and biological parameters

	Dal lake								Anchar lake					
	I		II		III		IV		I		II		III	
	r	t	r	t	r	t	r	t	r	t	r	t	r	t
11. Iron	0.01	0.05	–0.18	1.40	0.12	0.98	0.41	0.96	0.23	0.62	0.02	0.58	0.50	1.23
12. Ammonical nitrogen	0.29	1.42	0.21	1.32	–0.17	1.04	–0.49	1.56	0.05	0.99	0.37	1.04	0.45	1.34
13. Nitrate nitrogen	0.53	2.99	0.26	1.56	0.25	0.98	0.57	1.87	0.03	0.35	0.67	2.13	0.42	2.56
14. Total PO_4 – P	0.28	1.37	0.36	1.87	–0.16	0.90	0.02	0.35	–0.08	0.21	0.43	1.98	–0.11	0.72
15. Silicates	0.005	0.02	–0.20	0.51	–0.37	1.87	–0.19	0.94	–0.06	0.35	–0.58	2.56	–0.07	0.52
16. Sodium	0.28	2.42	0.18	0.68	0.50	2.50	0.25	1.35	0.20	0.51	0.15	0.49	0.06	0.30
17. Potassium	0.24	1.20	0.54	2.54	0.44	2.15	0.25	1.05	0.28	1.20	0.60	2.91	0.21	1.47
18. Total hardness	0.10	0.43	0.15	0.65	0.13	0.55	0.17	0.79	0.05	0.10	0.31	1.90	–0.18	0.79
19. TDS	0.12	0.53	0.09	0.15	0.10	0.43	0.09	0.18	0.14	0.43	0.58	2.13	0.44	2.08
20. Bacillariophyceae	–0.32	1.61	–0.03	0.16	0.09	0.14	0.08	0.14						
21. Chlorophyceae	–0.15	0.71	–0.08	0.20	0.14	0.58	0.13	0.43						
22. Cyanophyceae	0.16	0.78	0.32	0.97	0.15	0.32	0.21	0.14						
23. Dinophyceae	–0.11	0.55	–0.03	0.13	0.21	0.54	0.11	0.35						
24. Chrysophyceae			–0.73	2.13	0.13	0.41	0.14	0.53						
25. Euglenophyceae	–0.25	1.21	0.08	0.59	0.24	0.64	0.15	0.98						
26. Protozoa	0.06	0.28	0.09	0.90	0.16	1.01	0.13	0.58	0.23	0.65	0.21	0.95	0.15	0.35
27. Rotifera	0.11	0.55	0.01	0.21	0.10	0.31	0.13	0.45	0.05	0.10	0.09	0.13	0.15	0.32
28. Cladocera	–0.01	0.06	0.14	0.96	0.09	0.54	0.10	0.31	0.21	0.51	0.18	0.45	0.05	0.13
29. Copepoda	0.10	0.49	0.05	0.12	0.23	0.98	0.27	1.16	0.15	0.74	0.09	0.23	0.10	0.29

Table 114 Showing correlation of iron with all other physico-chemical and biological parameters

	Dal lake								Anchar lake					
	I		II		III		IV		I		II		III	
	r	t	r	t	r	t	r	t	r	t	r	t	r	t
12. Ammonical nitrogen	0.03	0.17	–0.47	2.05	0.19	1.11	–0.46	1.97	0.07	0.48	–0.33	1.20	0.03	0.40
13. Nitrate nitrogen	–0.2	–1.10	0.23	0.97	0.05	0.33	0.47	1.89	0.15	0.45	–0.22	0.78	0.32	1.68
14. Total PO_4 – P	0.04	0.19	0.24	0.88	0.03	0.11	0.04	0.13	–0.21	0.58	–0.05	0.33	–0.05	0.37
15. Silicates	–0.54	3.0	–0.13	0.55	–0.34	1.10	0.42	2.03	0.52	2.14	–0.08	0.41	–0.13	0.78
16. Sodium	0.05	0.25	0.54	2.14	–0.13	0.41	0.29	0.78	0.23	0.68	0.35	1.01	–0.29	0.79
17. Potassium	–0.23	1.11	0.32	1.68	–0.18	0.89	0.12	0.49	0.27	0.99	0.05	0.43	–0.33	1.11
18. Total hardness	0.05	0.51	0.10	0.61	0.12	0.79	0.09	0.35	–0.05	0.14	0.05	0.17	0.14	0.55
19. TDS	0.29	1.00	0.31	1.13	0.29	0.98	0.35	1.30	0.55	2.16	–0.13	1.11	0.54	2.56
20. Bacillariophyceae	0.02	0.13	0.39	1.64	0.09	0.48	0.21	0.78						
21. Chlorophyceae	0.47	2.55	0.25	0.43	0.57	2.13	0.64	2.39						
22. Cyanophyceae	0.006	0.22	0.08	0.53	0.23	0.65	0.20	0.43						
23. Dinophyceae	0.26	1.31	0.46	1.87	0.13	0.43	0.14	0.55						
24. Chrysophyceae			0.92	4.53	0.40	2.54	0.25	1.68						
25. Euglenophyceae	0.41	2.1	0.44	2.70	0.51	2.81	0.64	3.01						
26. Protozoa	–0.05	0.26	0.2	0.49	0.04	1.2	0.03	0.59	0.28	0.59	0.13	0.53	0.13	0.45
27. Rotifera	0.30	1.49	0.92	2.99	0.32	1.31	0.04	1.03	0.05	0.98	0.13	0.14	0.14	0.55
28. Cladocera	0.20	0.97	0.87	3.54	0.19	1.19	0.09	0.05	0.01	0.96	0.23	0.29	0.15	0.69
29. Copepoda	0.18	0.87	0.43	2.13	0.29	0.90	0.20	0.90	0.29	1.04	0.12	0.31	0.19	0.59

Table 115 Showing correlation of ammonical nitrogen with all other physico-chemical and biological parameters

	Dal lake								Anchar lake					
	I		II		III		IV		I		II		III	
	r	t	r	t	r	t	r	t	r	t	r	t	r	t
13. Nitrate nitrogen	0.22	1.07	0.10	1.04	–0.23	1.23	–0.3	1.45	0.67	2.54	0.43	2.15	0.80	3.16
14. Total $PO_4 – P$	–0.02	0.10	–0.29	1.47	0.41	2.13	–0.01	0.43	0.77	3.45	–0.24	1.13	0.19	0.80
15. Silicates	–0.23	1.11	–0.23	1.23	0.04	0.25	–0.13	0.43	0.35	0.89	0.09	0.25	–0.17	0.57
16. Sodium	–0.16	0.77	–0.42	2.10	0.43	1.97	–0.3	0.87	–0.49	2.10	0.23	1.23	0.31	1.75
17. Potassium	–0.09	0.44	–0.48	2.19	0.33	1.83	–0.1	0.80	–0.02	0.41	0.48	2.10	0.50	2.17
18. Total hardness	0.12	0.55	0.01	0.12	0.09	0.19	0.11	0.43	0.16	0.70	0.49	2.15	–0.06	0.72
19. TDS	0.11	0.44	0.21	0.98	0.19	0.86	0.13	0.42	0.12	0.58	0.75	3.15	0.33	1.47
20. Bacillariophyceae	0.20	0.96	0.19	0.79	0.21	0.91	0.17	0.75						
21. Chlorophyceae	0.25	0.11	0.16	0.82	0.09	0.40	0.13	0.25						
22. Cyanophyceae	0.14	0.70	0.32	1.39	0.12	0.70	0.21	1.21						
23. Dinophyceae	0.14	0.66	0.39	1.43	0.05	0.40	0.15	0.45						
24. Chrysophyceae			0.97	4.13	0.53	2.13	0.56	2.40						
25. Euglenophyceae	0.01	0.05	0.48	2.17	0.21	1.43	0.30	1.87						
26. Protozoa	0.09	0.42	0.43	2.15	0.15	0.78	0.51	2.13	0.23	1.11	0.21	0.92	0.31	1.35
27. Rotifera	0.12	0.58	0.12	0.73	0.21	0.65	0.49	2.11	0.15	0.53	0.32	1.52	0.15	0.35
28. Cladocera	0.33	1.6	0.17	0.53	0.15	0.58	0.39	1.14	0.21	0.79	0.27	1.14	0.12	0.97
29. Copepoda	0.18	0.86	0.22	1.43	0.25	0.99	0.21	0.87	0.15	0.43	0.20	0.43	0.05	0.13

Table 116 Showing correlation of nitrate nitrogen with all other physico-chemical and biological parameters

	Dal lake								Anchar lake					
	I		II		III		IV		I		II		III	
	r	t	r	t	r	t	r	t	r	t	r	t	r	t
14. Total $PO_4 - P$	–0.15	0.74	0.32	1.43	0.16	0.82	0.04	0.13	0.64	2.13	0.11	0.43	0.17	0.58
15. Silicates	0.27	1.36	–0.68	2.14	0.26	1.13	–0.009	0.10	0.23	0.82	–0.01	0.45	–0.27	0.98
16. Sodium	0.52	2.9	0.56	2.99	0.60	3.15	–0.03	0.49	0.58	2.62	0.09	0.51	0.15	1.16
17. Potassium	0.47	2.55	0.15	1.14	0.49	2.53	0.36	2.10	0.24	1.59	0.29	1.68	0.28	1.82
18. Total hardness	0.19	0.68	0.21	0.90	0.23	0.98	0.29	1.11	–0.13	0.45	0.16	0.69	–0.04	0.11
19. TDS	0.18	0.65	0.21	0.95	0.29	1.01	0.35	1.25	0.21	0.33	0.45	2.11	0.39	2.14
20. Bacillariophyceae	0.3	1.47	–0.13	0.78	0.16	0.84	0.32	1.87						
21. Chlorophyceae	0.2	0.97	–0.10	0.14	0.24	0.54	0.47	2.13						
22. Cyanophyceae	0.52	2.13	0.48	2.17	0.47	2.10	0.39	1.98						
23. Dinophyceae	–0.20	0.95	0.10	0.40	0.16	0.41	0.39	1.80						
24. Chrysophyceae			0.63	2.13	0.52	2.51	0.61	2.72						
25. Euglenophyceae	–0.06	0.32	0.04	0.45	0.62	2.59	0.74	3.13						
26. Protozoa	0.009	0.04	0.05	0.40	0.09	0.14	0.21	0.45	0.18	0.40	0.15	0.39	0.21	0.14
27. Rotifera	0.17	0.82	0.10	0.35	0.17	0.39	0.23	0.49	0.21	0.47	0.23	0.61	0.01	0.21
28. Cladocera	0.43	2.25	0.13	0.72	0.10	0.51	0.09	0.13	0.17	0.42	0.05	0.13	0.09	0.12
29. Copepoda	0.02	0.11	0.33	0.99	0.15	0.43	0.21	0.58	0.27	0.68	0.15	0.33	0.19	0.21

Table 117 Showing correlation of total phosphate phosphorus with all other physico-chemical and biological parameters

	Dal lake								Anchar lake					
	I		II		III		IV		I		II		III	
	r	t	r	t	r	t	r	t	r	t	r	t	r	t
15. Silicates	−0.04	0.22	−0.17	0.58	−0.44	1.40	−0.26	0.68	−0.12	0.49	−0.55	1.04	−0.06	0.17
16. Sodium	0.04	0.19	0.44	1.04	0.23	0.94	0.40	0.98	0.64	0.50	0.17	0.39	0.11	0.51
17. Potassium	0.11	0.54	0.13	0.41	0.03	0.93	0.25	1.04	0.28	0.91	0.01	0.93	−0.09	0.51
18. Total hardness	0.11	0.94	0.19	1.01	0.21	1.15	0.35	1.29	0.008	0.14	0.16	0.43	−0.16	0.62
19. TDS	0.09	0.14	0.11	0.53	0.15	0.69	0.21	0.98	0.036	0.12	0.18	0.32	0.25	0.58
20. Bacillariophyceae	0.28	1.38	0.04	0.94	0.12	1.45	0.4	1.92						
21. Chlorophyceae	0.08	0.38	0.08	0.94	0.21	0.56	0.35	0.83						
22. Cyanophyceae	0.56	3.22	0.14	2.13	0.63	2.09	0.54	2.43						
23. Dinophyceae	0.06	0.32	0.18	0.41	0.9	0.53	0.41	0.98						
24. Chrysophyceae			0.72	0.39	0.31	0.38	0.25	0.97						
25. Euglenophyceae	0.21	1.02	0.16	1.04	0.31	0.98	0.25	0.65						
26. Protozoa	0.12	0.59	0.13	0.68	0.19	0.97	0.05	0.35	0.14	0.45	0.21	0.53	0.09	0.25
27. Rotifera	0.41	2.11	0.23	0.79	0.21	0.81	0.36	1.04	0.10	0.33	0.15	0.47	0.17	0.52
28. Cladocera	0.21	1.05	0.09	0.14	0.19	0.50	0.43	1.89	0.21	0.79	0.23	0.81	0.05	0.12
29. Copepoda	0.13	0.64	0.12	0.42	0.28	0.72	0.11	0.32	0.30	0.74	0.19	0.32	0.21	0.42

Table 118 Showing correlation of silicates with all other physico-chemical and biological parameters

	Dal lake								Anchar lake					
	I		II		III		IV		I		II		III	
	r	t	r	t	r	t	r	t	r	t	r	t	r	t
16. Sodium	0.23	1.10	–0.39	1.43	0.44	1.98	0.66	2.73	–0.10	0.43	–0.15	0.39	0.31	1.11
17. Potassium	0.29	1.46	–0.06	0.43	0.37	1.25	0.61	2.13	0.24	1.15	–0.30	1.31	–0.38	1.61
18. Total hardness	0.11	0.49	0.17	0.68	0.21	0.79	0.19	0.79	0.33	1.41	–0.13	0.45	0.02	0.13
19. TDS	0.08	0.11	0.10	0.42	0.12	0.53	0.01	0.12	–0.07	0.04	–0.15	0.69	0.01	0.74
20. Bacillariophyceae	–0.41	2.56	–0.40	2.13	–0.39	2.22	–0.43	2.46						
21. Chlorophyceae	–0.10	0.47	–0.02	0.15	0.11	0.45	0.01	0.11						
22. Cyanophyceae	0.38	1.96	–0.02	0.13	0.10	0.47	0.20	0.87						
23. Dinophyceae	–0.28	1.41	–0.32	0.98	0.05	0.14	0.10	0.35						
24. Chrysophyceae			–0.05	0.13	0.15	0.45	0.21	0.98						
25. Euglenophyceae	–0.23	1.15	–0.34	0.74	0.22	1.14	0.21	1.13						
26. Protozoa	–0.06	0.28	0.16	0.54	0.01	0.10	0.11	0.33	0.15	0.54	0.16	0.73	0.12	0.54
27. Rotifera	0.13	0.63	0.20	0.64	0.19	0.63	0.20	0.74	0.04	0.15	0.10	0.33	0.06	0.11
28. Cladocera	0.14	0.69	0.10	0.33	0.20	0.55	0.11	0.74	0.09	0.15	0.21	0.83	0.23	0.94
29. Copepoda	–0.10	0.47	0.09	0.16	0.21	0.43	0.12	0.33	0.13	0.51	0.24	0.68	0.19	0.58

Table 119 Showing correlation of sodium with all other physico-chemical and biological parameters

	Dal lake								Anchar lake					
	I		II		III		IV		I		II		III	
	r	t	r	t	r	t	r	t	r	t	r	t	r	t
17. Potassium	0.59	3.50	0.54	2.56	0.77	3.05	0.71	2.98	0.38	2.51	0.65	3.01	0.50	3.05
18. Total hardness	0.21	0.43	0.18	0.35	0.21	0.40	0.24	0.69	0.24	1.43	0.20	1.11	0.20	1.21
19. TDS	0.09	0.21	0.11	0.25	0.14	0.35	0.29	0.68	0.01	0.78	0.14	0.98	−0.10	0.43
20. Bacillariophyceae	0.21	1.03	0.02	0.41	0.13	0.35	0.09	0.11						
21. Chlorophyceae	0.33	1.64	0.12	0.53	0.16	0.94	0.14	0.86						
22. Cyanophyceae	0.26	1.29	0.04	0.11	0.08	0.13	0.09	0.14						
23. Dinophyceae	0.24	1.16	−0.09	0.22	0.13	0.44	0.02	0.13						
24. Chrysophyceae			−0.52	2.43	−0.43	2.11	−0.51	2.31						
25. Euglenophyceae	0.27	1.32	0.29	1.43	0.21	1.41	0.23	1.53						
26. Protozoa	−0.13	0.62	0.23	1.14	0.21	1.09	0.11	0.98	0.01	0.43	0.008	0.11	0.05	0.14
27. Rotifera	0.56	3.21	0.43	2.11	0.51	2.13	0.52	2.18	0.21	0.11	0.19	0.43	0.21	0.33
28. Cladocera	0.62	3.71	0.42	2.35	0.29	0.43	0.45	2.00	0.11	0.99	0.21	0.87	0.31	1.14
29. Copepoda	0.54	3.03	0.22	1.22	0.21	1.33	0.35	1.35	0.10	0.43	0.09	0.43	0.14	0.80

Table 120 Showing correlation of potassium with all other physico-chemical and biological parameters

	Dal lake								Anchar lake					
	I		II		III		IV		I		II		III	
	r	t	r	t	r	t	r	t	r	t	r	t	r	t
18. Total hardness	0.30	1.05	0.15	0.73	0.19	0.98	0.21	1.09	0.39	1.43	0.18	0.73	0.12	0.56
19. TDS	0.11	0.78	0.21	0.98	0.01	0.21	0.10	0.43	0.06	0.69	0.44	1.69	0.15	0.73
20. Bacillariophyceae	0.18	0.87	0.07	0.45	0.12	0.62	0.13	0.72						
21. Chlorophyceae	0.009	0.04	0.01	0.36	0.05	0.11	0.08	0.23						
22. Cyanophyceae	0.04	0.22	0.13	0.41	0.03	0.21	0.11	0.33						
23. Dinophyceae	0.38	1.95	0.06	0.33	0.21	0.50	0.12	0.64						
24. Chrysophyceae			0.38	1.11	0.11	0.34	0.19	0.54						
25. Euglenophyceae	0.21	1.02	0.24	0.68	0.10	0.13	0.09	0.19						
26. Protozoa	0.14	0.71	0.14	0.45	0.008	0.11	0.08	0.13	0.11	0.33	0.09	0.14	0.10	0.44
27. Rotifera	0.15	0.75	0.14	0.35	0.01	0.22	0.02	0.11	0.05	0.22	0.12	0.19	0.06	0.21
28. Cladocera	0.18	0.86	0.06	0.29	0.16	0.89	0.09	0.42	0.22	0.78	0.12	0.68	0.18	0.44
29. Copepoda	0.21	1.05	0.11	0.33	0.18	0.78	0.17	0.66	0.16	0.41	0.14	0.23	0.15	0.22

Table 121 Showing correlation of total hardness with all other physico-chemical and biological parameters

	Dal lake								Anchar lake					
	I		II		III		IV		I		II		III	
	r	t	r	t	r	t	r	t	r	t	r	t	r	t
19. TDS	0.66	3.01	0.21	0.72	0.35	0.98	0.21	0.68	−0.28	0.94	0.45	2.07	−0.17	0.53
20. Bacillariophyceae	0.01	0.35	0.05	0.12	0.02	0.16	0.12	0.53						
21. Chlorophyceae	0.11	0.43	0.13	0.32	0.14	0.32	0.21	0.62						
22. Cyanophyceae	0.05	0.11	0.10	0.21	0.09	0.12	0.10	0.48						
23. Dinophyceae	0.01	0.17	0.11	0.31	0.10	0.21	0.30	1.02						
24. Chrysophyceae	0.09	0.16	0.31	0.69	0.27	0.45	0.10	0.32						
25. Euglenophyceae	0.18	0.43	−0.21	0.59	0.23	0.69	0.10	0.35						
26. Protozoa	0.28	0.69	0.15	0.21	0.25	0.82	0.04	0.18	0.09	0.11	0.18	0.41	0.03	0.15
27. Rotifera	0.11	0.35	0.08	0.35	0.15	0.49	0.02	0.11	0.29	0.58	0.06	0.13	0.30	0.21
28. Cladocera	0.14	0.43	0.07	0.29	0.11	0.21	0.27	0.89	0.22	0.41	0.04	0.11	0.10	0.31
29. Copepoda	−0.21	0.52	0.19	0.47	0.15	0.42	0.11	0.21	0.13	0.31	0.15	0.43	0.28	0.25

Table 122 Showing correlation of total dissolved solids with biological parameters

	Dal lake								Anchar lake					
	I		II		III		IV		I		II		III	
	r	t	r	t	r	t	r	t	r	t	r	t	r	t
20. Bacillariophyceae	0.51	2.06	0.13	0.53	0.12	0.40	0.38	1.06						
21. Chlorophyceae	0.28	0.63	0.20	0.76	0.05	0.30	0.19	0.41						
22. Cyanophyceae	0.16	0.98	0.13	0.43	0.19	0.49	0.14	0.25						
23. Dinophyceae	0.10	0.78	0.15	0.60	0.03	0.50	0.11	0.20						
24. Chrysophyceae	–	–	0.09	0.14	0.10	0.62	0.02	0.10						
25. Euglenophyceae	0.10	0.27	0.11	0.44	0.17	0.52	0.06	0.01						
26. Protozoa	0.02	0.47	0.01	0.17	0.23	0.71	0.10	0.15	0.13	0.28	0.15	0.65	0.10	0.40
27. Rotifera	0.21	0.58	0.11	0.33	0.16	0.42	–0.10	0.20	0.11	0.41	0.17	0.73	0.16	0.75
28. Cladocera	0.35	0.67	0.13	0.41	0.15	0.41	0.01	0.13	0.09	0.11	0.12	0.55	0.07	0.19
29. Copepoda	0.19	0.44	0.18	0.53	0.12	0.39	0.15	0.21	0.14	0.33	0.20	0.80	0.30	0.89

Table 123 Showing correlation of bacillariophyceae with biological parameters

	Dal lake								Anchar lake					
	I		II		III		IV		I		II		III	
	r	t	r	t	r	t	r	t	r	t	r	t	r	t
21. Chlorophyceae	0.28	1.39	0.39	1.48	0.48	2.01	0.30	1.62						
22. Cyanophyceae	0.20	0.98	0.21	0.98	0.39	1.58	0.29	1.10						
23. Dinophyceae	0.29	1.46	0.31	1.62	0.41	1.80	0.20	0.91						
24. Chrysophyceae			0.12	0.58	0.15	0.45	0.19	0.35						
25. Euglenophyceae	0.15	0.73	0.18	0.55	0.15	0.47	0.20	0.55						
26. Protozoa	–0.27	1.31	0.19	0.48	0.25	0.10	0.21	0.56						
27. Rotifera	0.09	0.44	0.10	0.29	0.09	0.10	0.18	0.94						
28. Cladocera	0.09	0.46	0.21	0.78	0.05	0.30	0.16	0.40						
29. Copepoda	–0.16	0.77	0.24	1.01	0.16	0.81	0.09	0.13						
21. Chlorophyceae	0.68	3.15	0.19	1.01	0.11	0.43	0.21	0.98						

Table 124 Showing correlation of chlorophyceae with biological parameters

	Dal lake								Anchar lake					
	I		II		III		IV		I		II		III	
	r	t	r	t	r	t	r	t	r	t	r	t	r	t
22. Cyanophyceae	0.27	1.32	0.29	1.05	0.31	1.41	0.41	0.98						
23. Dinophyceae	0.32	1.59	0.30	0.97	0.12	0.35	0.18	0.55						
24. Chrysophyceae			0.17	0.48	0.09	0.13	0.10	0.18						
25. Euglenophyceae	0.4	2.09	0.45	2.16	0.51	2.30	0.43	2.13						
26. Protozoa	0.17	0.85	0.14	1.01	0.08	0.15	0.13	0.45						
27. Rotifera	0.53	2.96	0.64	3.51	0.61	3.01	0.58	2.96						
28. Cladocera	0.59	3.44	0.55	3.61	0.51	2.98	0.41	2.10						
29. Copepoda	0.30	1.52	0.33	1.68	0.21	1.01	0.31	0.92						
21. Chlorophyceae	0.19	0.98	0.21	1.01	0.21	1.05	0.11	0.79						

Table 125 Showing correlation of cyanophyceae with biological parameters

	Dal lake								Anchar lake					
	I		II		III		IV		I		II		III	
	r	t	r	t	r	t	r	t	r	t	r	t	r	t
23. Dinophyceae	0.09	0.44	0.10	0.53	0.13	0.62	0.22	0.80						
24. Chrysophyceae			0.08	0.10	0.12	0.49	0.05	0.13						
25. Euglenophyceae	0.34	1.70	0.21	0.78	0.009	0.13	0.14	0.56						
26. Protozoa	0.29	1.42	0.14	0.68	0.20	0.79	0.12	0.89						
27. Rotifera	0.56	3.18	0.50	2.01	0.45	1.98	0.40	2.08						
28. Cladocera	0.45	2.39	0.64	3.01	0.40	2.10	0.41	1.98						
29. Copepoda	0.09	0.46	0.10	0.86	0.10	0.79	0.08	0.11						
21. Chlorophyceae	0.13	0.90	0.22	1.22	0.19	0.98	0.11	0.56						

Table 126 Showing correlation of dinophyceae with biological parameters

	Dal lake								Anchar lake					
	I		II		III		IV		I		II		III	
	r	t	r	t	r	t	r	t	r	t	r	t	r	t
24. Chrysophyceae			0.12	1.01	0.22	1.52	0.19	1.33						
25. Euglenophyceae	0.55	3.10	0.11	0.41	0.12	0.58	0.09	0.13						
26. Protozoa	0.10	0.47	0.01	0.13	0.07	0.11	0.02	0.09						
27. Rotifera	0.15	0.71	0.09	0.10	0.10	0.25	0.09	0.12						
28. Cladocera	0.07	0.35	0.10	0.14	0.07	0.09	0.11	0.10						
29. Copepoda	0.24	1.17	0.11	0.33	0.20	0.45	0.13	0.48						
21. Chlorophyceae	0.08	0.72	0.15	0.98	0.07	0.11	0.18	0.79						

Table 127 Showing correlation of chrysophyceae with biological parameters

	Dal lake								Anchar lake					
	I		II		III		IV		I		II		III	
	r	t	r	t	r	t	r	t	r	t	r	t	r	t
25. Euglenophyceae			0.01	0.32	0.10	0.45	0.17	0.68						
26. Protozoa			0.20	0.62	0.17	0.81	0.21	0.97						
27. Rotifera			0.19	0.58	0.08	0.16	0.10	0.47						
28. Cladocera			0.15	0.87	0.04	0.11	0.08	0.10						
29. Copepoda			0.24	1.11	0.19	0.99	0.21	1.01						
21. Chlorophyceae			0.22	1.29	0.19	0.86	0.33	1.56						

Table 128 Showing correlation of euglenophyceae with biological parameters

	Dal lake								Anchar lake					
	I		II		III		IV		I		II		III	
	r	t	r	t	r	t	r	t	r	t	r	t	r	t
26. Protozoa	0.11	0.55	0.09	0.19	0.19	0.90	−0.18	0.87						
27. Rotifera	0.43	2.25	0.51	1.98	0.46	1.67	0.58	2.62						
28. Cladocera	0.15	0.73	0.11	0.19	0.13	0.45	0.09	0.11						
29. Copepoda	0.31	1.56	0.21	0.62	0.21	0.58	0.22	0.60						

Table 129 Showing correlation of protozoa with biological parameters

	Dal lake								Anchar lake					
	I		II		III		IV		I		II		III	
	r	t	r	t	r	t	r	t	r	t	r	t	r	t
1. Rotifera	0.55	2.50	0.65	3.13	0.55	2.75	0.61	2.61	0.48	2.10	0.39	2.40	0.49	2.81
2. Cladocera	0.49	2.56	0.39	2.01	0.41	2.13	0.45	2.13	0.38	2.06	0.29	1.86	0.35	2.10
3. Copepoda	0.33	2.03	0.55	2.45	0.65	2.49	0.39	2.10	0.45	2.13	0.38	2.49	0.45	2.50

Table 130 Showing correlation of rotifera with biological parameters

	Dal lake								Anchar lake					
	I		II		III		IV		I		II		III	
	r	t	r	t	r	t	r	t	r	t	r	t	r	t
1. Cladocera	0.73	5.10	0.65	3.15	0.60	3.45	0.61	2.96	0.48	2.50	0.39	2.15	0.41	2.13
2. Copepoda	0.60	3.56	0.71	4.15	0.69	3.89	0.65	3.10	0.62	2.86	0.60	2.75	0.72	4.15

Table 131 Showing correlation of cladocera with biological parameters

	Dal lake								Anchar lake					
	I		II		III		IV		I		II		III	
	r	t	r	t	r	t	r	t	r	t	r	t	r	t
1. Copepoda	0.49	2.67	0.51	2.13	0.80	4.15	0.41	2.10	0.53	3.14	0.49	2.51	0.51	2.15

References

A.P.H.A. (1989). *Standard methods for the examination of water and wastewater* (17th ed.). Washington, DC: A.P.H.A.

Abbasi, S. A., Arya, D. S., Hameed, A. S., & Abbasi, N. (1996). Water quality of a typical river of Kerela: Punnurpuzha. *Pollution Research, 15*(2), 163–166.

Abel, P. D. (1989). *Water pollution biology*. Chichester: Ellis Horwood Ltd.

Adak, M. D., & Purohit, K. M. (2001). Status of surface and ground water quality of Mandiakudar – Part III: Correlaton coefficient and regression equations. *Pollution Research, 20*(2), 227–232.

Adholia, U. N., Kulshrestha, S. K., & Bhatnagar, A. (1991). *Pollution status of river Khan, Kshipra and Chambal with reference to the abiotic characteristics*. In Proceedings of national symposium planning forenvironment sustainable development. New Delhi.

Adoni, A. D., & Joshi, G. (1987). Physico-chemical regime of three fresh water bodies in and around Sagar (M.P.). *Geobios New Reports, 6*(1), 43–46.

Agrawal, S. K., & Gautam, P. G. (1985). Potamozooplankton periodicity in river Chambal at Kota. *Acta Ecologica, 7*, 9–11.

Agrawal, D. K., Gaur, S. D., Tiwari, I. C., Narayanswami, N., & Marwah, S. M. (1976). Physico-chemical characteristics of Ganges water in Varansi. *Indian Journal of Environmental Health, 18*, 201–206.

Ahmad, M. S. (1996). Ecological survey of some algal flora of polluted habitats of Darbhanga. *Journal of Environment and Pollution, 3*(3 & 4), 147–151.

Ajao, E. A., & Fagade, S. O. (1990). A study of sediment and communities in Lagos Lagoon, Nigeria. *Oil and Chemical Pollution, 19*, 123–130.

Ajmal, M., Razuiddin, & Khan, A. V. (1985). Physico-chemical aspects of pollution in Kalinadi. IAWPC. *Technical Annual,* 12, 106–114.

Alam, A., & Khan, A. A. (1996). Dynamics of plankton communities in four freshwater lentic ecosystems in relation to varying dominant biota. *Pollution Research, 15*(3), 289–291.

Algarsamy, S. R., Abdulappa, M. K., & Bopardikar, M. V. (1967). High rate deep stabilization pond, Part-II. Studies on the faunal assemblage of the system. *Environmental Health, 9*, 306–316.

Alikunhi, K. H. (1956). Indian Farming. *6*, 7, 23.

Alikunhi, K. H. (1957). Fish culture in India. *Farm Bulletin*. New Delhi, 20, ICAR, 6.

Allan, J. F. (1976). Life history patterns in zooplankton. *American Naturalist, 110*, 165–180.

Allanson, B. R. (1979). *Lake Sibaya*. (Monographiae Biologicae, Vol 36, pp. 364). The Hague: W. Junk.

Andrews, J. D., & Hasler, A. D. (1943). Fluctuations in the animal populations of the littoral zone in lake Nendota. *Transactions of Wisconsin Academy of Sciences, Arts and Letters, 35*, 175–185.

M. Jeelani, *Lake Ecology in Kashmir, India*,
DOI 10.1007/978-3-319-40880-4

Angino, E. E., Magnuson, L. M., & Stewart, G. F. (1992). Effects of urbanisation on storm water run-off quality: A limited experiment, Naismith Ditch, Lawrence, Kansas. *Water Resources Research, 8*, 135–140.

Anjana, S., & Kanhere, R. R. (1998). Seasonal dynamics of phytoplankton population in relation to abiotic factors of a fresh water pond at Barwani (M.P.). *Pollution Research, 17*(2), 133–136.

Anonymous. (1991). *Animal resources of India: Protozoa to mammalia. State of the art*. Calcutta: Zoological Survey of India, 694 pages.

Ansari, K. K., & Prakash, S. (2000). Limnological studies on Tulsidas Tal of Tarai region of Balrampur in relation to fisheries. *Pollution Research, 19*(4), 651–655.

Ara, S., Khan, M. A., & Zargar, M. Y. (2003). Physico-chemical characteristics of Dal lake water of Kashmir valley, India. *Indian Journal of Environment and Ecoplanning, 7*(1), 47–50.

Aravinda, H. B., Manjappa, S., & Puttaiah, E. T. (1998). Correlation coefficients of some physico-chemical parameters of river Thunga Bhadra, Karantaka. *Pollution Research, 17*(4), 371–375.

Arber, A. (1920). *Water plants: A study of aquatic angiosperms*. London: Cambridge University Press.

Arora, H. C. (1966). Rotifers as indicators of trophic nature of environments. *Hydrobiologia, 27*(1–2), 146–159.

Arora, H. C. (1996). Responses of rotifera to variations in some ecological factors. *Proceedings of the Indian Academy of Sciences, 58*, 57–63.

Badola, S. P., & Singh, H. R. (1981). Hydrobiology of the river Alaknanda of the Garhwal Himalaya. *Indian Journal of Ecology, 8*(2), 269–276.

Bagde, U. S., & Verma, A. K. (1985a). Physico-chemical characteristics of water of J. N.U. lake at New Delhi. *Indian Journal of Ecology, 12*(1), 151–156.

Bagde, U. S., & Verma, A. K. (1985b). Limnological studies on J.N.U. lake, New Delhi, India. *Bulletin of Botanical Society Sagar, 32*, 16–23.

Bagnoulus, F. & Meher Homji, V. M. (1959). *Bioclimatic types of South-East Asia Travaux dela section scientific et. technique*. Institute Francais De Pondicherry, 227 pp.

Balkhi, M. H. & Yousuf, A. R. (1991). *Distribution pattern of rotifera in the freshwaters of Kashmir*. In Proceedings of National Conference on Agricultural Sciences in India. New Delhi.

Balkhi, M. H., Yousuf, A. R., & Qadri, M. Y. (1984). Rotifera of Anchar lake during summer and winter. *Geobios New Reports, 3*(2), 163–165.

Bandela, N. N., Vaidya, D. P., Lomte, V. S., & Shivanikar, S. V. (1999). The distribution pattern of phosphate and nitrogen forms and their interrelationship in bowl dam water. *Pollution Research, 18*(4), 411–414.

Bandopadhyay, S., & Gopal, B. (1991). Ecosystem studies and management problems of a coastal lagoon. The lake Chika. In B. Gopal & V. Asthana (Eds.), *Aquatic sciences in India* (pp. 117–172). New Delhi: Indian Association of Limnology and Oceanography.

Banerjee, A. C., & Roychowdhuri, N. C. (1966). Observations on some physico-chemical features of Chilka lake. *Indian Journal of Fisheries, 13*(1–2), 359–429.

Barnes, I. (1965). Geochemistry of Birch Creek Thyo Country, California, a travertine depositing creek in an arid climate Geochim et Cosmochin. *Acta, 29*(3), 85–112.

Baruah, B. K., & Das, M. (2001). Study on plankton as indicator and index of pollution in aquatic ecosystem receiving paper mill effluent. *Indian Journal of Environmental Sciences, 5*(1), 44–46.

Bath, K. S. (1996). *Limnological studies on the Harike wetland ecosystem*. Ph.D. thesis. Patiala: Punjabi University.

Bath, K. S., & Kaur, H. (1997). Aquatic insects as bioindicators at Harike reservoir in Punjab India. *Indian Journal of Environmental Science., 2*(1), 133–138.

Bath, K. S., & Kaur, H. (1998). Seasonal distribution and population dynamics of aquatic insects in Harike reservoir (Punjab). *Journal of Ecobiology, 10*(1), 43–46.

Battish, S. K. (1968). A study of the taxonomy and seasonal variations of Rotifers and Crustaceans. M. Sc. (Hons.) thesis approved by Punjabi University, Chandigarh.

Battish, S. K. (1992). *Freshwater zooplankton of India*. New Delhi: Oxford and IBH Publishing Co.

Bennet, G. W. (1971). *Management of lakes and ponds*. Reinhold: Van Nostrand.

Berzins, B. (1949). Zur Limnologie der Seen Sudostlettlands. *Schweizerische Zeitschrift für Hydrologie, 11*, 583–607.

Bhatt, U. G., Neelakanthan, B., Kusuma, N., & Naik, U. G. (1988). Environmental characteristics of the marine and estuarine habitats of Karnoar - an overview. *Journal of Indian Fisheries Association, 18*, 401.

Bhatt, L. R., Lacoul, P., Lekhak, H. D., & Jha, P. K. (1999). Physico-chemical characteristics and phytoplanktons of Taudaha lake, Kathmandu. *Pollution Research, 18*(4), 353–358.

Bhavani, T., Sarma, P. N., & Krishna, D. (2000). Correlations among physico-chemical properties of effluents of various chemical process industries. *Pollution Research, 19*(4), 601–607.

Bhowmick, B. N., & Singh, A. K. (1985). Phytoplankton population in relation to physico-chemical factors of river Ganga at Patna. *Indian Journal of Ecology, 13*(1), 138–145.

Bhuyan, B. R. (1970). Physico-chemical qualities of the water of some ancient tanks in Sibsaqar, Assam. *Environmenal Health, 12*, 129–134.

Bilgrami, K. S., & Datta Munshi, J. S. (1985). *Ecology of river Ganges. Impact of human activities and conservation of biota*. Final Tech. Ref. Man and Bios. Prog., Post graduate department of Botany, Bhagalpur Univ. Bhagalpur.

Bilgrami, K. S., & Dattamunshi, J. S. (1979). *Limnological survey and impact of human activities on the river Ganges*. Technical report. MAB Project-5, pp. 1–91.

Bilgrami, K. S., Sheo, K., Sahay, S. S., & Sanjib, K. (1994). Diel Variations in abiotic and biotic factors of the river Ganga at Bhagalpur, Bihar. *Environment and Ecology, 12*(2), 374–378.

Bisht, K. L. (1986). Impact of environmental parameters on biota of the river Pinder of Garhwal Himalaya. In: *Fish and their environment* (pp. 91–95). (Haridwar). India.

Biswas, K., & Calder, C. C. (1937). *Handbook of common water and marsh plants of India and Burma, 1936*. Manager of Publications, Delhi, 140 pages + 32 plates.

Biswas, B. K., & Konar, S. K. (2000). Impact of waste disposal on plankton abundance and diversity in the river Ganga at Hathidah (Bihar). *Pollution Research, 19*(4), 633–640.

Blinn, D. W., Fredericksen, A., & Korte, V. (1980). Colonization rates and community structure of diatoms on three different rocks substrata in a lotic system. *British Phycological Journal, 15*, 303–310.

Blum, J. L. (1956). The ecology of river algea. *Botanical Review, 22*, 291–334.

Blum, J. L. (1957). An ecological study of the algal of the Satieh river Michigan. *Hydrobiologia, 9*(4), 361–408.

Boznaik, E. G., & Kennedy, L. L. (1968). Periodicity and ecology of the plankton in an oligotrophic and eutrophic lake. *Canadian Journal of Botany, 46*(10), 1259–1271.

Brinkurst, R. O., & Jamison, B. G. M. (1971). *Aquatic oligochaeta of the world*. Edinburgh: Oliver and Boyd.

Brooks, J. L., & Dodson, L. (1965). Predation body size and composition of plankton. *Science, 150*, 28–35.

Bruce, A. (1958). Report on a biological and chemical investigation of the waters in the Aven and Hethcata rivers. *Reprinted by the pollution advisory council Morimo*. New Zealand: Department Wellington

Brusven, M. A., & Prather, K. (1974). Influence of stream sediments on macrobenthos. *Journal of the Entomological Society of British Columbia, 71*, 25–32.

Bulusu, K. R., Arora, H. C., & Aboo, K. M. (1967). Certain observation on self-purification of Khan river and its effect on Kshipra river. *Environmental Health, 9*, 275–295.

Cairns, J. J., Heath, A. G., & Parker, B. C. (1975). The effects of temperature upon the toxicity of chemicals to aquatic organisms. *Hydrobiologia, 47*(1), 135–171.

Carmonze, J. P., Durang, J. R., & Leveque, C. (Eds). (1983). *Lake chad: Ecology and productivity of shallow tropical ecosystem monographical biological* (vol 53) (pp. 575). The Hague: Dr. W. Junk. Publishers.

Carter, H. J. (1856b). Notes on the fresh-water Infusoria of the Island of Bombay. *Annals and Magazine of Natural History Series, 2*(18), 115–132.

Cairns, J., Jr., & Dickson, K. L. (1971). A simple method for biological assessment of the effect of waste discharges on the aquatic bottom dwelling organisms. *Journal of Water Pollution Control Federation. pp, 775*, 5–12.

Cassie, V. (1979). Algea in relation to water quality. In P. A. Mulcock (Ed.), *A review of some biological methods for the assessment of water quality with special reference to New Zealand* (Technical Report, Vol. 18, pp. 1–46). Wellington: National water and soil conservation organisation.

CBCWP. (1982). *Basin Sub-basin inventory of water pollution: The Yamuna sub-basin (April, 1977 - December, 1978)*. New Delhi: Central Board for the Prevention and Control of Water Pollution.

Chakraborty, D., & Konar, S. K. (2000). Studies on the hydrobiology of river Damodor from Raniganj region (W.B.) with reference to phytoplankton. *Indian Journal of Environment and Ecoplanning, 3*(3), 485–490.

Chakraboarty, R. D., Roy, P., & Singh, S. B. (1959). A quantitative study of the plankton and physico-chemical conditions of the river Yamuna at Allahabad in 1954–55. *Indian Journal of Fisheries, 6*(1), 186–203.

Chapman, D., & Kimstach, V. (1992). In D. Chapman (Ed.), *The selection of water quality variables in water quality assessment*. London/New York/Tokyo: Chapman and Hall.

Chauhan, R. (1988). Seasonal abundance of zooplankton in Rewalsar lake. (H.P.). *Geobios New Reports, 7*, 117–121.

Chowdhary, S. K. (1984). Studies on bio-ecology of aquatic insects of Sind and Lidder streenss of Kashmir. *Indian Journal of Ecology, 11*(1), 160–165.

Clarke, F. W. (1924). The data of geochemistry (5th Ed). *U.S. Geological Survey Bulletin, 770*, 841. Washington DC: U. S. Govt. printing office.

Cole, G. A. (1975). *Text book of limnology.* Saint louis: The C.V. Mosby company.

Conner, H. E. (1977). *The poisonous plants of New Zealand* (p. 247). Wellington: Government Printer.

Cook, C. D. K. (1996). *Aquatic and wetland plant of India*. Oxford: Oxford University Press. 385 pages.

Cordery, I. (1976). Some effects of urbanisation on streams. *Civil Engineering Transactions, 18*, 7–11.

Cowell, B. C., & Vodopich, D. S. (1981). Distribution and abundance of benthic macro invertebrates in a sub tropical Florida lake. *Hydrobiologia, 78*, 97–105.

Crayton, W. M., & Sommerfeld, M. R. (1978). Phytoplankton of lower Colorado river, Grand Canyon region. *Journal of the Arizona-Nevada Academy Science, 13*, 19–24.

Cummins, K. W. (1966) A review of future problems in benthic ecology (pp. 2–51). In Organism-substrate relationships in streams. Spec. Publ. No.4. Pymatuning Laboratory of ecology, University of Pittsburgh, 145 pp.

Curry, L. L. (1962). A survey of environmental requirements for the midge (Diptera-Tendipedidae). In C. M. Tarzweli, P. H. S. Usdhew, & A. Robert (Eds.), *Biological problems in water pollution. Transactions of 3rd seminar*. Cincinnati: Taft Sanitary Engineering Centre.

Curtis, J. T. (1959). *The vegetation of Wisconsin: An ordination of plant communities*. Madison: University of Wisconsin Press.

Daborn, G. R. (1976). Physical and Chemical factors of a vernal temporary pond in Western Canada. *Hydrobiologia, 51*(1), 33–38.

Dahiya, S., Kaur, A., Garg, V. K., & Jain, N. (2000). Qualification of fluoride in groundwater in rural area of Tosham subdivision, district Bhiwani-Haryana. *Pollution Research, 19*(3), 417–419.

Dance, K. W., & Hynes, H. B. N. (1980). Some effects of agricultural land use on stream insect communities. *Environmental Pollution Series A, 22*, 19–28.

Das, S. M. (1979). Effects of increased land use upon freshwater in the Central Himalayas. *Procedings of Water Technical, 11*(6), 195–208.
Das, S. M. (1989). *Handbook of limnology and water pollution*. New Delhi: South Asian Publishers Pvt. Ltd.
Das, S. M., & Bisht, R. S. (1979). Ecology of some hemiptera and coleoptra of Kumaun lakes. *Indian Journal of Ecology, 6*(1), 35–40.
Das, S. M., & Pandey, J. (1978). Some physico-chemical and biological indicators of pollution in lake Nainital, Kumaun (U.P.). *Indian Journal of Ecology, 5*(1), 7–16.
Das, S. M., & Subla, B. A. (1970). The Pamir Kashmir theory of origin, evolution and distribution of Kashmir fishes with their general ecology. *Kashmir Science, 7*, 1–15.
Das, S. M., Daftari, S., Singh, H., Akhtar, S., Chowdhuri, S., & Ahmad, N. (1969). Studies on organic production in high altitude lakes of Kashmir-The general ecology and zooplankton of Kashmir lakes. *Kashmir Science, 6*(1-2), 1–22.
Das, A. C., Baruah, B. K., Baruah, D., & Gupta, S. S. (2003). Study on wetlands of Guwahati city - 2. Water quality of rivers and drains. *Pollution Research, 22*(1), 117–119.
Datta Munshi, J. S., & Singh, D. K. (1991). Physico-chemical profile of river Ganga at Kahalgaon, Bihar, India. In V. Asthana & B. Gopal (Eds.), *Aquatic sciences in India* (pp. 83–92). New Delhi: Indian association for limnology and oceanography.
David, A. (1956). *Studies on the pollution of Bhadra river fisheries at Bhadravati Mysore with industrial effluent*. In Proceedings of National Institute of Sciences India, *22*, 132–139.
David, A., & Ray, P. (1966). Studies on the pollution of the river Daha (N-Bihar) by sugar and distillery wastes. *Environmental Health, 8*, 6–35.
David, L. K., James, L., Plafkin, J., Cairns, J. R., & Rex, L. L. (1981). Qualitative characterization of aquatic environments using diatom life form strategies. *Transactions of the American Microscopical Society, 100*, 2.
De Smet, W. H. O., & Evens, F. M. J. C. (1972). A hydrobiological study of the polluted river Lieve (Ghent, Belgium). *Hydrobiologia, 31*(1), 91–154.
Desikachary, T. V. (1959). *Cyanophyta*. New Delhi: Indian Council of Agricultural Research.
Dexter, R. W. (1959). Anostraca: 558–571. Quoted from Ward. H. B. and Whipple, G.C. (1959).
Dhanapathi, M. V. S. S. S. (1977). Studies on the distribution of *Brachionus calyciflorus* in India. *Archive Hydrobiology Beihefte Ergebnisse Limnology, 8*, 226–229.
Dhembare, A. J., & Pondhe, G. M. (1997). Correlations of ground water quality parameters of Sonal area (Maharashtra). *Pollution Research, 16*(3), 189–190.
Dhillon, S. S., Kaur, H., Bath, K. S., Mander, G., & Syal, J. (1993a). The impact of effluence of on the zooplankton population of Beas river. *Journal Of Ecotoxicology and Environmental Monitoring, 3*(2), 177–120.
Dhillon, S. S., Kaur, H., Bath, K. S., Mander, G., & Syal, J. (1993b). Zooplankton fauna of the river Beas. *Environment and Ecology, 11*(4), 916–919.
Dickman, M. (1975). A comparison between the effects of sporadic and continuous loading from sulphate paper mills on attached algae in the Ottawa river near Ottawa, Canada. In *Proceedings of the 10th Canada symposium. Water Pollution Research Canada* (pp. 65–72).
Drenner, R. W., Hambright, K. D., Vinyard, G. L., Gophen, M., & Pollingher, U. (1987). Experimental study of size selective phytoplankton grazing by a fitter feeding chchild and cichlids effect on plankton community structure. *Linology and Oceongraphy, 32*, 1138–1144.
Drew, F. (1875). *The Jammu and Kashmir territories. A geographical account with maps and plates*. London: Stanford.
Drischel, H. (1940). Chloride sulphate and Nitratgehalt der almospherischen. *Niedershclage in Bad Reiners and obserschreiber hau in Vergleich zei bisher bekannton Western an derer orte Balneologe., 7*, 321–334.
Dudani, B. K., Kumar, S., Jha, V., & Pandey, S. (1986). Seasonal variations in physico-chemical parameters of a Makhana fish pond at Darbhanga North Bihar. *Journal of Mandel, 3*(3), 171–174.

Dunn, B. (1962) Hydrology of the Iron dequott Greek basis Rochester, New York. Open file research paper No. 40, US Geological survey and New York State Department of Health.

Dutta, N. A., Malhotra, J. C., & Bose, B. B. (1954). Hydrology and seasonal fluctuations of the plankton in the Hooghly estuary. *Proceedings of Indo Pacific Fish Communities, 11*(5), 35–47.

Edmondson, W. T. (1959). Trophic relations of the zooplankton. *Transactions of the American Microscopical Society, 76*, 225–246.

Elliot, I. J. (1976). Seasonal changes in the abundance and distribution of planktonic rotifers in Grasmere (English Lake District). *Fresh Water Biology, 8*, 144–166.

Ellis, M. M., Westfall, B. A., & Ellis, M. B (1946). Determination of water quality. Fish and wild life service. *U. S. Department of Interior Resource Report, 9*, 122.

Enex. (1978). *Study of the pollution of Dal lake, Srinagar, Kashmir*. Enex Consortium, New Zealand.

Esmaeili, H. R., & Johal, M. S. (2005). *Study of physico-chemical parameters of water of Gobindsagar reservoir, India*. In: M. S. Johal (Ed.), Proceedings of the National Seminar "New trends in Fishery Development in India". Chandigarh: Punjab University, Feb. 16–18, 173–177.

Farrow, G. E., Syritski, J. P. M., & Tunnicliffe, V. (1983). Suspended particulate loading on the macrobenthos in a highly turbid Fjord, Knight Inlet, Bristish Columbia. *Canadian Journal of Fisheries and Aquatic Sciences, 40*, 273–288.

Ferguson, J. E. (1982). Studies on the zooplankton of lake Torkana. In A. J. Hopson (Ed.), *Lake Torkana - a report on the findings of lake Torkana project 1972–1975*. London: Quersecs. Devel. Admin.

Forel, F. A. (1901). Handbuch der Seenkunde: Allgemeine Limnologie. *Stuttgart, 3*, 114–249.

Freiser, H., & Fernando, Q. (1966). *Ionic equilibra in analytical chemistry*. New York: Wiley and Sons.

Gabriel, B., Kowalczyk, C., Lecewlez, W., Radwan, S., Wojeiechowska, W., & Wojciechowshi, I. (1975). Influence of abiotic environmental factors on plankton in lakes of different trophy. *Polskh Archiwum Hydrobiology, 22*(2), 123–139.

Ganapati, S. V. (1940). The ecology of a temple tank containing a permanent bloom of *Microcystis aeruginosa* (kutz) Henfr. *Journal of the Bombay Natural History Society, 42*(1), 65–77.

Ganapati, S. V. (1956). Hydrobiological investigations of the Hope reservoirs and of Thambaraparni river at Papanasam Tirunalveli Dt. Madras state India. *Geographical Journal, 31*, 1–2.

Garg, V. K., Sharma, I. S., & Bishnoi, M. S. (1998). Fluoride in underground waters of Uklana town, district Hisar, Haryana. *Pollution Research, 17*(2), 149–152.

Gaufin, A. R., & Tarzwell, C. M. (1956). Aquatic invertebrates as indictors of organic polution in Lytte Creak. *Sewage and Industrial Wastes, 28*, 906–924.

George, M. G. (1966). Comparative plankton ecology of five fish tanks in Delhi, India. *Hydrobiologia, 27*(1-2), 81–108.

George, M. G., Qasim, S. Z., & Siddiqi (1966). A limnological survey of river Kali with reference to fish mortality. *Environmental Health, 8*(4), 262–269.

Ghose, N. C., & Sharma, C. B. (1988). Effect of drain water on the physcio-chemical and bacteriological characters of river Ganga at Patna, Bihar. In R. K. Trivedy (Ed.), *Ecology and pollution of Indian rivers* (pp. 255–269). New Delhi: Asian Publishing House.

Gibbs, R. J. (1970). Mechanisms controlling world water chemistry. *Science, 170*, 1088–1089.

Gideiri, Y. B. A. (1969). The development and distribution of plankton in the northern part of the White Nile. *Hydrobiologia, 33*(3-4), 369–378.

Goel, P. K., & Chavan, V. R. (1991). *Limnological study of a polluted fishery tank at Kolhapur*. In Proceedings of national symposium planning for environment sustainable development. New Delhi.

Goel, P. K., Gopal, B., & Trivedy, R. K. (1980). Impact of sewage on freshwater ecosystem Part-II. Physicochemical characteristics of water and their seasonal changes. *International Journal of Ecology and Environmental Science, 6*, 97–116.

Golterman, H.L., & Clymo, R. S. (1969). *Methods for chemical analysis for freshwater. IBP. Hand book No. 8*. Oxford: Blackwell Scientific Publications.

Gonzalves, E. A., & Joshi, D. B. (1946). Freshwater algae near Bombay. 1. The seasonal succession of algae in tank of Bandra. *Journal of the Bombay Natural History Society, 46*(1), 154–176.

Gopal, B. (1990). Indian subcontinent and the aquate habitats. In *Ecology and management of aquate vegetation in the Indian subcontinent* (pp. 7–28). Netherlands: Kluwer Academic Publ.

Govindan, V. S., & Sundaresan, B. B. (1979). Seasonal succession in algae flora in polluted region of Adyar river. *Indian Journal of Enviornmental Health, 21*(2), 131–142.

Gujarathi, S. A., & Kanhere, R. R. (1998). Seasonal dynamics of phytoplankton population in relation to abiotic factors of a fresh-water pond at Barwani (M.P.). *Pollution Research, 17*(2), 133–136.

Gunale, V. R., & Balakrishnan, M. S. (1981). Biomonitoring of eutrophication in the Pavana, Mula and Mutha rivers flowing through Poona. *Indian Journal of Environmental Health, 23*(4), 316–321.

Gunther, R. E. R. (1936). A report on the Oceanographical investigations in the Peru coastal current. *Discovery Report, 13*, 107–276.

Gupta, A. K., & Saxena, G. C. (1997). Nitrate contamination in ground waters of Agra and its correlation with various water quality parameters including heavy metals. *Pollution Research, 16*(3), 155–157.

Haberman, J. (1983). A comparative study of zooplankton in two lakes of Estonia. Proceedings of the Estonian Academy of Sciences Biology. *Ecology, 46*(4), 225–245.

Hakkari, L. (1978). On the productivity and ecology of zooplankton and its role as food for fish in some lakes in Central Finland. *Biological Research Reports, University of Jyvaskyla , 4*, 3–87.

Hall, D. J., Threlkeld, S. T., Burns, C. N., & Crowby, P. H. (1976). The size efficiency hypothesis and the size structure of zooplankton communities. *Annual Review of Ecologic and Systematics., 7*, 177–208.

Hannan, H. H., & Young. W. J. (1974). The influence of a deep storage reservoir on the physicochemical limnology of central Texas river. *Hydrobiologia, 44*(2–3), 177–207.

Harold, C. H. M. (1934). 29th Annual *report Metropolitan water Board.* London.

Harshey, D. K., Patil, S. G., & Singh, D. F. (1982). Limnological studies on a tropical fresh water fish tank of Jabalpur, India. *Geobios New Reports., 1*(2), 98–102.

Harvey, H. W. (1940). Nitrogen and phosphorus required for the growth of phytoplankton. *Journal of Marine Biological Association UK*, 24, 115–123.

Hasan, M. Z., & Pandey, S. P. (1985). Heavy metals in urban water supplies of Nagpur. In *Current pollution researches in India*. Karad: Environmental Publication.

Heckel, J. J. (1839) *Fisches aus Cashmir*. London.

Hecky, R. E. (1984). African lakes and their tropic efficiencies: a temporal perspective. In D. G. Dewey & J. R. Stricler (Eds.), *Tropic interactions in aquatic ecosystems* (pp. 405–420). Boulder: Westview Press.

Hejney, S., & Husuk, S. (1978). Higher plant communities. In J. Kvet & D. Dkyjova (Eds.), *Plant littoral ecosystems*. New York: Springer.

Hellawell, J. M. (1986). *Biological indicators of freshwater pollution and environmental management* (p. 546). England: Elsevier Applied Science Publishers Ltd.

Hentschel, E. (1928). Die Grundzuge der planktonverteilung in Sudailantischen Ozean. *Internationale Revue der gesamten Hydrobiologia Hydrographie, 21*, 1–6.

Heron, J. (1961). Phosphorus adsorption by lake sediments. *Limnology and Oceanography, 6*, 338.

Heurck, H. V. (1896). *Diatomaceae*. London: William Wesley and Sons.

Holden, M. J., & Green, J. (1960). The hydrology and plankton of river Skoto. *Journal of Animal Ecology, 29*(1), 68–84.

Hooker, J. D. (1875). *Flora of British India* (Vol. 1–7). Kent: Reeve.

Husainy, S. U. (1965). Limnological studies of the departmental pond at Anamalainagar. *Environmental Health, 7*, 24–31.

Hussainy, S. U. (1967). Studies on limnology and primary production of a tropical lake. *Hydrobiologia, 30*, 335–352.

Hutchinson, G. E. (1941). Limnological studies in connectient IV. The mechanism of intermediary metabolism in stratified lakes. *Ecological Monographs, 2*, 21–60.

Hutchinson, G. E. (1957). A treatise on limnology, I. geography, physics and chemistry (vol. 2). New York: Wiley.

Hutchinson, G. E. (1969). Eutrophication past and present. In *Eutrophication causes, consequences and correctiveness* (pp. 17–26). Washington, DC: National Academy of Sciences.

Hynes, H. B. N. (1970). The ecology of running waters (pp. 555). Toronto: Toronto University Press.

I.S.I. (1986). *Indian standard institution. Tolerance limints for inland waters used for pubic water supplies*. New Delhi, Ist - 2296.

Islam, M. F., & Islam, S. (1996). Surface water quality of water bodies of chalan beel area in Nator district, Bangladesh. *Journal Bulletin of Environmental Science, 14*, 1–3.

Israili, A. W. (1991). Occurrence of heavy metals in Ganga river water and sediments of western U. P. *Pollution Research, 10*(2), 103–109.

Jain, C. K., & Sharma, M. K. (1997). Relationship among water quality parameters of ground water of Jammu District. *Pollution Research, 16*(4), 241–246.

Jameel, A. A. (1998). Physico-chemical studies in Uyyakondan, C. water of river Cauvery. *Pollution Research, 17*(2), 111–114.

Jameson, J., & Rana, B. C. (1996). Pollution status of the river complex Sabermati at Kheda region of Gujrat I. Physico-chemical characters. *Pollution Research, 15*(1), 53–55.

James, L., Vijayram, K., & Loganathan, P. (1990). Assessment of hospital wastes and their impact on water quality of Uyyakondan channel, Tiruchirappalli, Tamil Nadu. *Journal of Current Biosciences, 7*, 13–17.

Jamil, K. (1990). Physico-chemical characteristics of water bodies infested with water hyacinth. Suggestions you useful Bio-technologies. In V. P. Agrawal & P. Das (Eds.), *Recent trends in limnology* (pp. 281–286). Muzafarnagar: Society of bioscience.

Jarnefelt, H. (1952). Zur limnologic einiger Gewasser Finlands. *Annales Zoologici Societatis Zoologicae-Botanicae Fennicae Vanamo, 2*, 185–356.

Jayangoudar. (1980). Hydrobilogical studies on the Ajwa reservoir, the source of raw water supply to the Baroda water works. *Hydrobiologia, 72*, 113–123.

Jhingran, A. G. (1991a). *Fish in relation to water quality paper.* Presented at *Indo-Dutch Meet*. On *Biomon*, *Yard Dev*. New Delhi.

Jhingran, A. G. (1991b). Challenging frontiers in freshwater fisheries in India. In B. Gopal & V. Asthana (Eds.), *Aquatic sciences in India*. New Delhi: Indian Association for Limnology and Oceanography.

Jindal, R., & Ghezta, R. K. (1991). *Limnology of Sukhna lake, Chandigarh.* In Proceedings of national conference on aquatic sciences in India. New Delhi

Jindal, R., & Kumar, R. (1993). Limnology of a freshwater pond of Nalagarh. I. Physico-chemical complexes. In N. K. Shastree (Ed.), *Advances in limnology* (pp. 107–112). Delhi: Narendra Publishing House.

Jindal , R. & Singh, H. (2005). Ecology of benthos of some freshwater ecosystem of Punjab. In M.S. Johal (Ed.), 2005 Proceedings of national seminar 'New Trends in Fishery Development in India' (pp. 179–184). Chandigarh: Panjub University.

Johal, M. S., Armantrout, N. B. and Arunachalam, M. (2005). In M. S. Johal (Ed.), *Proceedings of the national seminar new trends in fishery development in India*. Chandigarh: Punjab University, Feb., 16–18, 35–41.

Jolly, V. H., & Chapman, V. A. (1966). A preliminary biological study of the effects of pollution on farmers Creek and Cox's river, New South Wales. *Hydrobiologia, 27*, 160–187.

Jorgensen, S. E., & Loffler, H. (1990). *Guidelines of lake management* (vol. 3). Lake shore management. Japan: ILECF and UNEP, 174 pp.

Juday, C., & Birge, E. A. (1933). The transparency, the colour and the specific conductance of the lake waters of north eastern Wisconsin. *Transactions of Wisconsin Academy of Sciences, Arts and Letters, 28*, 205–259.

Jyoti, M. K., & Sehgal, H. (1979). Ecology of rotifera of Surinsar. A subtropical, freshwater lake in Jammu (J & K), India. *Hydrobiologia, 65*(1), 23–32.

Kannan, K. (1991). *Fundamental of environmental pollution*. New Delhi: S. Chand and company Ltd.

Kant, S., & Kachroo, P. (1972). Phytoplankton population dynamics and distribution in two adjoining lakes in Srinagar. I. Macroflora in relation to phytoplankton. *Proceedings of the National Academy of Science, India, 37*, 163–188.

Kant, S., & Kachroo, P. (1974). Limnological studies of Kashmir lakes IV. Seasonal dynamics of phytoplankton in the Dal and Nagin. *Proceedings of Indian National Science Academy B, 40*, 77–99.

Kant, S., & Kachroo, P. (1975). Limnological studies in Kashmir lakes. III. Diurnal movements of phytoplankton. *Journal of the Indian Botanical Society, 54*(½), 9–12.

Kant, S., & Kachroo, P. (1977). Limnological studies of Kashmir lakes. I. Hydrological features, composition and periodicity of phytoplankton in the Dal and Nagin lakes. *Phykos, 16*(1-2), 77–97.

Kar, G. K., Mishra, P. C., Dash, M. C., & Das, R. C. (1987). Pollution studies in river 1b, III: Plankton population and primary productivity. *Indian Journal of Environmental Health, 29*(4), 322–329.

Katti, R. J., Venkatesha Moorthy, K. S., Kumar, B. M., Ronald, K. D. S., & Shanthanagouda, A. H. (2002). Macrobenthos in relation to sediment characteristics in the Arabian sea off Chitrapur, receiving treated industrial effluents- pre and post discharge study. *Pollution Research, 21*(2), 195–201.

Kaul, V. (1970). Production and ecology of some macrophytes of Kashmir lakes. *Hydrobiologia, 12*, 63–69.

Kaul, V. (1977). Limnological survey of Kashmir lakes with reference to trophic status and conservation. *Indian Journal Ecology and Environmental Sciences, 3*, 29–44.

Kaul, S. (1979). *Contribution to the analysis of some fresh water ecosystems with special reference to biogeochemical cycling*. Ph.D. thesis. Srinagar: Kashmir University.

Kaul, V., & Handoo, J. K. (1980). Water characteristics of some freshwater bodies of Kashmir. *Current Trends in Life Sciences, 9*, 221–246.

Kaul, S., & Kachroo, P. (1973). Limnological studies in Kashmir lakes-Dal and Nagin distribution of phytoplankton. *Proceedings of National Academy of Sciences, 839*, 632–645.

Kaul, V., & Zutshi, D. P. (1965). Some ecological considerations of floating islands in Srinagar lakes. *Proceedings of the National Academy of Science, India, 36*, 273–281.

Kaul, V., & Zutshi, D. P. (1979). Is Dal dying? *Science Reporter, 16*(11), 734–736.

Kaul, V., Fotedar, D. N., Pandit, A. K., & Trisal, C. L. (1978). A Comparative study and plankton populations of some typical freshwater bodies of Jammu and Kashmir state. In D. N. Sen & R. P. Bansal (Eds.), *Environmental physiology and ecology of plants*. Dehra Dun: B. Singh, M. Pal Singh.

Kaur, H., Dhillon, S. S., Bath, K. S., & Mander, G. (1996a). Abiotic and biotic components of a fresh water pond of Patiala (Punjab). *Pollution Research, 15*(3), 253–256.

Kaur, H., Dhillon, S. S., Bath, K. S., & Mander, G. (1996b). Analysis of the elements polluting river Ghaggar in the region of Punjab. *Journal of Environment and Pollution, 3*(2), 65–68.

Kaur, H., Dhillon, S. S., Bath, K. S., & Mander, G. (1997). A Statistical approach to interrelationship between protozoans and physico-chemical factors at Harike reservoir (Punjab-India). *Indian Journal of Environmental Sciences, 1*(1), 47–50.

Kaur, H., Dhillon, S. S., Bath, K. S., & Mander, G. (1999). Correlation of Rotifers with physico-chemical factors at Harike Reservoir (Punjab - India). In K. Vijay Kumar (Ed.), *Freshwater ecosystem of India*. Delhi: Daya Publishing House.

Kaur, H., Syal, J., & Dhillon, S. S. (2001). Water quality index of the river Satluj. *Pollution Research, 20*(2), 199–204.

Kaur, H., Syal, J., & Dhillon, S. S. (2002). Impact of fertilizer factory wastes on physico-chemical and biological features of Satluj River. In A. Kumar (Ed.), *Ecology of polluted waters* (vol II). New Delhi: A.P.H. Publishing Corporation.

Kaur, H., Syal, J., & Dhillon, S. S. (2003). Impact of fertilizer factory wastes on physicochemical and biological features of Satluj river. In A. Kumar (Ed.), *Aquatic ecosystems* (pp. 71–81). New Delhi: A.P.H. Publishing Corporation.

Kaushik, S., & Saksena, D. N. (1999). Physico-chemical limnology of certain water bodies of central India. In K. Vijay Kumar (Ed.), *Freshwater ecosystem of India*. Delhi: Daya Publishing House Delhi.

Kaushik, B., Sharma, S., & Saxena, D. N. (1991). Ecological studies of certain polluted lentic waters of Gwalior region with reference of aquatic insect communities. In N. K. Shastree (Ed.), *Current trends in limnology-I* (pp. 185–200).

Khabade, S. A., Mule, M. B., & Sathe, S. S. (2003). Studies on physico-chemical parameters of Lodhe water reservoir from Tasgaon Tahsil (Maharashtra). In A. Kumar (Ed.), *Aquatic ecosystems* (pp. 201–206). New Delhi: A.P.H. Publishing corporation.

Khalique, A., & Afser, M. R. (1995). Physico-chemical analysis of river Ganges at Farakka. *Biojournal, 7*(1–2), 101–105.

Khalof, A. K., & Mac Donald, L. J. (1975). Hydrobiologia, 47, 310–318. Quoted from Patra, A.K. and Nayak, M (1983).

Khan, M. A. (1979). Some observations on trophic evolution of monomictic flood-plain lake. *Indian Journal of Ecology, 6*, 27–32.

Khan, M. A. (1984). Studies on primary productivity and some hydrochemical aspects of a polluted water-body of the Himalayan Jhelum Valley. *Acta Hydrochimica et Hydrobiologica, 12*, 539–548.

Khan, A. A. (1993). Benthic fauna in relation to pollution: A case study at river Chambal near Kota in central India. *Environment International, 19*, 597–610.

Khan, M. A., & Bhatt, G. H. (2000). Biological invasion and red water phenomenon in lake Manasbal of Kashmir valley. *Pollution Research, 19*(1), 113–117.

Khan, M. A., & Zutshi, D. P. (1979). Relative contribution by nano and net plankton towards primary production of two Kashmir Himalayan lakes. *Journal of the Indian Botanical Society, 58*, 263–267.

Khan, M. A., & Zutshi, D. P. (1980 a & b). Primary productivity and trophic status of a Kashmir Himalayan lake. *Hydrobiologia, 68*, 3–8.

Khan, M. A., Panwar, R. S., Mathur, A., & Jetly, R. (1998). Studies on plankton communities and primary production in middle stretch of river Ganga in context of cultural eutrophication. *Journal of the Inland Fisheries Society of India, 30*(1), 47–68.

Khanna, D. R., & Bhutiani, R. (2003). Ecological status of Sitapur pond at Hardwar (Uttaranchal), India. *Indian Journal of Environment and Ecoplanning, 7*(1), 175–178.

Khare, A., Varma, R. L., Hazela, K. P., & Pandey, G. N. (1979). *An ecological study of river Ganga at Kanpur. I. Influence of pollutants on Macro-organisms*. In Proceedings of symposium on environmental biology (pp. 383–395). The Acadamy of Environmental Biology, India.

Khatri, T. C. (1985). Physico-chemical feature of Idukki reservoir Kerala. During pre-monsoon period. *Environment and Ecology, 3*, 134–137.

Khatri, S. K., & Dhankhar, R. (2003). Evaluation of sewage waste water as a source of irrigation and manure of Rohtak city of Harayana. *Pollution Research, 22*(4), 549–552.

King, J. M. (1981). The distribution of invertebrate communities a small South African river. *Hydrobiologia, 83*, 43–65.

Klimourcz, H. (1973). Microfauna of activated sludge. Part-III. The effects of physico-chemical factors on the occurrence of microfauna in the annual cycle. *Acta Hydrobiologia, 15*, 167–188.

Kochsiek, K. A., Wilhm, J. L., & Morrison, R. (1971). Species diversity of net zooplankton and physico-chemical conditions in Keystone reservoir, Oklahoma. *Ecology, 6*, 1119–1125.

Kohli, M. P. S., Jhakar, N. K., & Munnet, S. K. (1982). Seasonal changes in plankton population of some freshwater ponds at Patna, Bihar (India). *Journal of the Inland Fisheries Society of India, 14*(1), 69–76.

Kohli, M. P. S., Jhakar, N. K., & Munnet, S. K. (1982). Seasonal changes in plankton population of some freshwater ponds at Patna, Bihar (India). *Journal of the Inland Fisheries Society of India, 14*(1), 69–76.

Kolkwitz, R., & Marsson, M. (1908). Oekologie der pflanzlichen Sprolien. *Ber Bot Ges, 26*, 505–519.

Koste, W., & Shiel, R. J. (1980). Preliminary remarks on the characteristics of the Rotifer fauna of Australia (Notogaea). *Hydrobiologia, 73*, 221–227.

Koul, A. K. (1979). Seasons of Kashmir valley. *Geographical Review of India, 41*(2), 123–130.

Krecker, F. H. (1939). A comparative study of the animal population of certain submerged aquatic plants. *Ecology, 20*, 553–562.

Krishnamoorthi, K. P., & Visvesvara, G. (1966). Hydrobiological studies in Gandhisagar (Jammu Tank). Seasonal variations in plankton (1961–62). *Hydrobiologia, 27*, 501–514.

Krishnamurthy, S. R., & Bharati, S. G. (1996). A study on distribution of diatoms in the river Kali, around Dandeli, (North Kanara District) Karnataka-India. *Pollution Research, 15*(4), 321–324.

Kumar, A. (1995a). Some limnological aspects of freshwater tropical wetland of Santhal Pargana (Bihar) India. *Journal of Environment and Pollution, 2*(3), 137–141.

Kumar, A. (1995b). Studies on pollution in river Mayurakshi in South Bihar. *Indian Journal of Environment and Pollution, 2*(1), 21–26.

Kumar, A. (1996a). Comparative studies on diel variations of abiotic factors in lentic and lotic fresh water ecosystems of Santhal Parganas (Bihar). *Indian Journal of Environment and Pollution, 3*(1), 53–56.

Kumar, A. (1996b). Impact of organic pollution on macro-zoobenthos of the river Mayurakshi of Bihar. *Pollution Research, 15*(1), 85–88.

Kumar, A. (1997). Comparative hydrological studies of tropical waterbodies with special reference to sewage pollution in South Bihar. *Journal of Ecobiology, 9*(4), 255–267.

Kumar, A. (1998). Diversification of phytoplankton and its correlation with certain physico-chemical characteristics of the river Mayurakshi in tribal zone of Bihar. *International Journal of Ecology and Environmental Sciences, 24*, 215–224.

Kumar, B. A., Choudhury, A., & Mitra, A. (2003). Seasonal distribution of nutrients its biological importance in upper stretch of Gangetic West Bengal. In A. Kumar (Ed.), *Aquatic ecosystems* (pp. 29–36). New Delhi: A.P.H. Publishing Corporation.

Kumar, K. S., & Sarma, T. V. S. (1979). Faunal (phytal) affinity in two fresh water ponds with diverse ecological stresses in Vizianagarama. *India Geobios, 6*, 347–348.

Kumar, A., & Sharma, R. K. (1999). Physico-chemical characteristics of river Sengar in Distt.-Etawah (UP). *Indian Journal of Environmental Science, 3*(2), 201–203.

Kumar, A., & Singh, D. K. (1996). Some aspects of classification, ecology and conservation of fresh-water wetlands and their socio-economic importance in Santhal Pargana (Bihar). *Indian Journal of Environment and Pollution, 3*(2), 83–89.

Kundangar, M. R. D., & Ahmad, M. R. (1991). *Environmental features and plankton communities of two semi-urban lakes of Kashmir*. In Proceedings of national conference on aquatic sciences In India. New Delhi.

Kundangar, M. R., & Zutshi, D. P. (1985). Environmental features and plankton communities of two Himalayan rural lakes. Quoted from Sarwar, S. G. and Rifat, M. (1991).

Lakes and Waterways Development Authority, LAWDA. (2004). Government of Jammu and Kashmir. Archived from the original report.

Lakshminarayan, J. S. S. (1965). Studies of the phytoplankton of river Ganges, Varanasi, India. Part-I. Physico-chemical characteristics of river Ganga. *Hydrobiologia, 25*, 119–137.

Lang, C. (1989). Eutrophication of lake Neuchatel indicated by Oligochaeto communities. *Hydrobiologia, 174*, 57–65.

Lawrence, W. R. (1895). *The valley of Kashmir*. Srinagar: Chinar Publishing House.

Learner, M. A., Williams, R., Harcup, M., & Hughes, B. D. (1971). A survey macro-fauna of the river Cynon, a polluted fributary of the river Taff. (South Wales). Freshwater Birl, 1, 339–367.

Love, G. J., & Godwin, M. H. (1959). The effects of natural winter temperatures on the development of Anopheles quadrimaculatus in South-Western Georgia. *Ecology, 40*(2), 198–205.

Lower, R. L. (1974). *Environmental requirements and pollution tolerance of freshwater diatoms*. Cincinnati: Environmental Protection Agency.

Lund, J. W. G. (1951). A sedimentation technique for counting algae and other organisms. *Hydrobiologia, 3*, 390–394.
Lund, J. W. G. (1965). The ecosystem of freshwater phytoplankton. *Biological Reviews, 40*, 281–293.
Mac Crimmon, H. R., & Kelso, J. R. M. (1970). Seasonal variations in selected nutrients of river system. *Journal of the Fisheries Research Board of Canada , 27*(5), 837–846.
Macan, T. T. (1963). *Freshwater ecology* (p. 1338). London: Longmans Green and Co. Ltd.
Mackereth, F. J. H. (1963). Some methods of water analysis for limnologistics. *Freshwater Biology Association Science Publication* (England), 21, 70p.
Maemets, A. (1983). Rotifers as indicators of lake types in Estonia. *Hydrobiologia, 104*, 357–361.
Mahadevan, A., & Krishnaswamy, S. (1983). A quality profile of river Vaigai (S. India). *Indian Journal of Environmental Health, 25*(4), 288–299.
Mairs, D. F. (1966). A total alkalinity atlas for marine lake waters. *Limnology and Oceanography, 11*, 68–72.
Mallick, A. K., & Padhi, S. B. (1999). Seasonal fluctuation of diatom flora of Gopal pur coast in Orrisa. *Indian Journal of Environment and Ecoplanning, 2*(3), 289–293.
Mandal, B. K., & Moitra, S. K. (1975). Seasonal variations of benthos and bottom soil edaphic factors in a freshwater fish pond at Burdwan, West Bengal. *Tropical Ecology, 16*, 43–48.
Manikya, R. P. (1984). *Ecological studies in the river Tungabhadra (AP) with special reference to the effect of effluents on the river*. Ph.D. thesis. Hyderabad: Osmania University.
Mariappan, P., & Vasudevan (2002). Correlation coefficients of some physico-chemical parameters of drinking water ponds in eastern part of Sivagangai district, Tamil Nadu. *Pollution Research, 21*(4), 403–407.
Marshall, S. M., & Orr, A. P. (1972). *The biology of marine copepod*. Berlin: Springer.
Maruthi, Y., Avasn, S., Rao, M. V., Krishna, R., & Rao, S. (2000). Pollution status of river Sarada at Anakapalli, Andhra Pradesh. *Indian Journal of Environment and Ecoplanning, 3*(1), 45–48.
Mason, C. F. (1981). *Ecology of freshwater pollution*. London: Longman.
Mathuthu, A. S., Zaranyika, F. M., & Jannalagadde, S. B. (1993). Monitoring of water quality in upper Kukuvisi river in Harare, Zimbabwe. *Environment International, 19*(1), 51–61.
McMcClelland, W. T., & Brusven, M. A. (1980). Effects of sedimentation on the behaviour and distribution of riffle insects in a laboratory stream. *Aquatic Insects, 2*, 161–169.
Mc Murry, G., & Olive, J. H. (1975). Summer phytoplankton photosynthesis in a north eastern Ohio glacial lake. *Ohio Journal of Science, 75*, 238–250.
Mehra, N. K. (1986). Studies of primary productivity in a sub-tropical comparison between experimental and periodicity values. *Indian Journal of Experimental Biology, 24*, 189–192.
Mellanby, H. (1963). *Animal life in fresh water* (6th ed.). London: Chapman and Hall Ltd.
Mengastou, S., & Fernando, C. H. (1991). Seasonally and abundance of some dominant crustacean zooplankton in lake Awasa, a tropical rift valley lake in Ethiopia. *Hydrobiologia, 226*, 137–152.
Meybeek, M., Friedrich, G., Thomas, R., & Chapman, D. (1992). Rivers. In D Chapman (Ed.), *Water quality assesssments* (pp. 239–316). London/New York/Tokyo: Chapman and Hall.
Michael, R. G. (1964) *Limnological investigations on pond plankton, macrofauna and chemical constituents of water and their bearing on fish population*. Ph.D. thesis. West Bengal: Calcutta University.
Michael, R. G. (1968). Studies on bottom fauna in a tropical freshwater pond. *Hydrobiologia, 31*, 203–230.
Michael, R. G. (1969). Seasonal trends in physico-chemical factors and plankton of a freshwater fish pond and their role in fish culture. *Hydrobiologia, 33*(1), 144–160.
Michael, R. G., & Sharma, B. K. (1988). *Fauna of India and Adjacent countries: Indian cladocera*. Calcutta: Zoological Survey of India. 265 pages.
Mir, G. R. (1995). *Impact of floating gardens on the limnological features of Dal lake*. M.Phil dissertation of University of Kashmir

Mir, A. M., & Kachroo, P. (1982). Limnology of Kashmir lakes VII. The ecology of *Bacillariophyceae* in two lakes in Srinagar. *Proceedings of Indian National Science Academy, 48*(2), 378–390.

Mishra, K. K. (1988). Nutrient management in fresh water lakes. *I.J.E.P., 9*(5), 358–361.

Mishra, S. R., & Saxena, D. N. (1993). Phytoplanktonic composition of sewage polluted Morar (Kalpi) river in Gwalior, Madhya Pradesh. *Environment and Ecology, 11*, 625–629.

Mohanty, S. K. (1975). Some observation on the physico-chemical feature of the outer channel of the Chilka lake during 1971–73. *Bulletin of the Department of Marine Science, 8*(1), 69–89.

Moore, W. J. (1980). Zooplankton, and related phytoplankton cycles, in a eutrophic lake. *Hydrobiologia, 74*, 99–104.

Moyle, J. B. (1946). Some chemical factors influencing the distribution of aquatic plants in Minnesota. *American Midland Naturalist, 34*, 402–420.

Mukherjee, S. K. (1921). *The Dal lake (Kashmir) A study in biotic succession*. In Proceedings of 8th Indian Science Congress, Abstract, pp. 185.

Mukherjee, S. K. (1925). *The ecology of the Dal lake region Kashmir*. In Proceedings of 12th Indian Science Congress, Abstracts, pp. 188.

Mukherjee, S. K. (1926). *Aquatic and Marshland vegetation of the Dal lake of Kashmir*. In Proceedings of Linnean Society London, Abstract. p. 55.

Mukherjee, S. K. (1932). *Bathymeteric survey of Dal lake of Kashmir with special reference to the penetration of active rays to the different depths of water and effect on the incidence of vegetation*. In Proceedings of 19th Indian Sciences Congress, Abstract, pp. 328.

Munawar, M. (1970a). Limnological studies on fresh water ponds of Hyderabad. I. The biotope. *Hydrobiologia, 35*, 127–162.

Munawar, M. (1970b). Limnological studies on freshwater ponds of Hyderabad. II. Biocenose. *Hydrobiologia., 36*(1), 105–128.

Munawar, M. (1972). Limnological studies on Euglenineae in certain polluted and unpolluted environments. *Hydrobiologia, 39*, 307–320.

Munawar, M. (1974a). Limnological studies on freshwater ponds of Hyderabad (India). *Hydrobiologia, 44*, 13–27.

Munawar, M. (1974b). Limnological studies of freshwater ponds of Hyderabad. IV. Biocenose, Periodicity and species composition of unicellular and ecological phytoplankton in polluted and unpolluted environments. *Hydrobiologia, 45*, 1–32.

Murphy, J., & Riley, J. (1962). A modified single solution method for the determination of phosphate in natural waters. *Analitica Chimica Acta, 27*, 31.

Nanazato, T., & Yasuno, M. (1985). Population dynamics and production of cladoceran zooplankton in the highly eutrophic lake Kasumigaura. *Hydrobiologia, 124*, 13–22.

Nandan, S. N., & Aher, N. H. (2002). Limnological study of algae of the river Mausam in Maharashtra. *Indian Journal of Environment and Ecoplanning, 6*(3), 605–608.

Narain, S., & Chauhan, R. (2000). Water quality status of river complex Yamuna at Panchnada (Distt.: Etawah, U.P., India) I: An integrated management approach. *Pollution Research, 19*(3), 357–364.

Nassar, S. A. K. (1975). Seasonal variations in the physico-chemical and biological properties of the tropical shallow pond. *Japanese Journal of Ecology, 24*(4), 255–259.

Naumann, E. (1917). Beitraze Jur Kenntnis des Teichnanno- plankton II Uber das Neuston das subwassers. *Biol. zbl., 37*, 98–106.

Naumann, E. (1919). Beitraze Jur kenntnis des teichnanno- plankton II Uber das Neuston das subwassers. *Biol.zbl, 37*, 98–106.

Needham, J. G., & Llyod, J. J. (1916). *The life in inland waters*. New York: The Comstock Publishing Company.

Nogrady, T. (1983). Succession of planktonic rotifer populations in some lakes of eastern rift valley, Kenya. *Hydrobiologia, 98*, 45–54.

O' Brien, J. W. (1979). The predator-prey interaction of planktivorous fish and zooplankton. *American Scientist, 67*, 572–581.

Odum, E. P. (1971). *Fundamental of ecology* (2 and 3rd ed.). Philadephia: Saunders Co.

Ohle, W. (1934). Chemische and physikalishe under such urgen norddeutscher seen. *Archives of Hydrobiology, 26*(384–464), 584–658.
Orcutt, J. D., & Pace, M. L. (1984). Seasonal dynamics of rotifer and crustacean zooplankton populations in a eutrophic, monomictic lake with a note on rotifer sampling techniques. *Hydrobiologia, 119*, 73–80.
Oslen, S. (1950). Aquatic plants and hydrospheric factors. *Svensk Bot. Tidskr.,* 44 pp.
Oswald, W. J. (1960). Light conversion efficiency of algal growth in sewage. *Journal of the Sanitary Engineering Division, American Society of Civil Engineers, 86*, 71.
Ownbey, C. R., & Kee, D. A. (1967). Chloride in lake Eric. *Proceedings of Congregation Great Lakes, 1*, 382–389.
Pagioro, T. A., Roberto, M. C., & Lansac-Toha, F. A. (1997). Comparative lminological analysis of two lagoons on the floodplain of the upper Parana river Brazil. *International Journal of Ecology and Environmental Sciences, 23*, 229–239.
Pahwa, D. V., & Mehrotra, S. V. (1966). Observations on fluctuations in the abundance of plankton in relation to certain hydrobiological condition of river Ganga. *Proceedings of the National Academy of Sciences, 36*(2), 157–189.
Palharya, J.P., & Malviya, S. (1988). Pollution of Narmada river at Hoshangabad in M.P. and suggested measure for control. In R. K. Trivedy (Ed.) *Ecology and pollution of Indian rivers* (pp. 55–85). New Delhi: Asian Publishing House.
Palmer, C. M. (1969). A composite rating of algae tolerating organic pollution. *Journal of Phycology, 5*, 78–82.
Palmer, C. M. (1980). *Algae and water pollution* (p. 123). London: Castle House Publication.
Pandit, A. K. (1980). *Biotic factors and food chain structure in some typcial wetlands of Kashmir*. Ph.D. thesis. Srinagar: Kashmir University.
Pandit, A. K. (1984). Role of macrophytes in aquatic ecosystems and management of fresh water resources. *Journal of Environmental Management, 18*, 73–88.
Pandit, A. K. (2002a). Conservation of lakes in Kashmir Himalaya. In A. K. Pandit (Ed.), *Natural resources of Western Himalaya* (pp. 291–328). Srinagar: Valley Book House. ISBN 81-86592-11-3.
Pandit, A. K. (2002b). Current research trends in limnological studies on periphyton. *Journal of Research and Developement, 2*, 149–209.
Pandit, A. K., & Kaul, V. (1981). Trophic structure of some typical wetlands. In B. Gopal, R. E. Turner, R. G. Wetzel, & D. F. Whigham (Eds.), *Wetlands: Ecology and management part-II* (pp. 55–82). Jaipur: International Scientific Publication and National Institute of Ecology.
Pandey, A. K., Dutta, S., & Sharma, K. C. (2002). Physico-chemical characteristics of Hamor pond of Kishangarh, Ajmer (Rajasthan). *Indian Journal of Environment and Ecoplanning, 6*(1), 167–169.
Paramavsivam, M., & Sreenivasan, A. (1981). Changes in algae flora due to pollution in Cauvery river. *Indian Journal of Environmental Health, 23*(3), 222–238.
Patil, S. G., Harshey, D. K., & Singh, D. F. (1985). Limnological studies of a tropical fresh water fish tank at Jabalpur, M.P. *Geobios New Reports, 4*(2), 143–148.
Patlas, K. (1972). Crustacean plankton and the eutrophication of the St. Lawrence Great Lakes. *Journal of Fisheries Research Board of Canada, 29*, 1451–1462.
Patnaik, S. (1973). Observations on the seasonal fluctuations of plankton in the Chilka lake. *Indian Journal of Fisheries, 20*(1), 43–55.
Patra, A. K., & Nayak, L. D. (1982). Limnobiotic survey of river Mahanadi during winter. *Geobios New Reports., 1*(1), 20–23.
Patra, A. K., & Nayak, L. D. (1983). Diurnal rhythm in certain physico- chemical factors in reservior Hirakud. *Geobios New Reports., 2*(1), 87–89.
Payne, A. I. (1987). *The ecology of tropical lakes and rivers* (p. 301). Chichester: Wiley.
Pearsall, W. H. (1930). Phytoplankton in the English lakes. I. The proportion in the water of some dissolved substances of biological importance. *Journal of Ecology, 18*, 306–320.

Pearsall, W. H. (1932). Phytoplankton in the English lakes 2. The composition of the phytoplankton ion relation to dissolved substances. *Journal of Ecology, 20*, 241–262.

Pearsall, W. H., & Lind, E. M. (1942). The distribution of phytoplankton in some North-West Irish loughs. *Proceedings of the Royal Irish Academy, 70B*, 1–24.

Pejler, B. (1957). Taxonomical and ecological studies on planktonic rotatoria from Central Seveeden, K. *Sevenska Vetensk. Akad. Handle. Ser., 4*, 1–168.

Penfound, W. T. (1952). An outline for ecological life histories of herbaceous vascular hydrophytes. *Ecology, 33*, 123–128.

Pennak, R. W. (1978). *Freshwater invertebrates of the U.S. 2nd*, New York: Wiley.

Petts, G. E., & Greenwood, M. (1985). Channel changes and invertebrate faunas below Nant-Y-Mach dam, river Rheidol, Wales, U.K. *Hydrobiologia, 122*, 65–80.

Philipose, M. T. (1960). Freshwater phytoplankton of inland fishers. In *Proceedings of symposium on algology* (pp. 272–291). New Delhi: ICAR.

Philpose, M. T. (1967). *Chlorococcales*. New Delhi: Indian Council of Agricultural Research.

Pillai, A., Pandey, P., & Shukla, A. V. (1999). Physico-chemical studies of drinking water of drug municipality. *Pollution Research, 18*(1), 49–51.

Ponji, Y., Karpati, I., & Szemes, G. (1985). Water pollution. In M. Kovacs (Ed.), *Pollution control and conservation* (pp. 122–203). Chichester: Ellis Horwood Ltd.

Prakash, V. (1990). Seasonal variations in abiotic factors at some freshwater impoundments of Jodhpur. *India Journal of Current Biosciences, 3*, 95–97.

Prasannakumari, A. A., Arathy, M. S., & Gangadevi, T. (2000). Bio-geo-chemical studies of a temple pond with special reference to Macroflora. *Pollution Research, 19*(4), 623–631.

Prasannakumari, A. A., Devi, G. T., & Sukeskumar, C. P. (2003). Surface water quality of river Neyyar_Thiruvanan tha puram, Kerala. *Pollution Research, 22*(4), 515–525.

Prescott, G. W. (1978). *How to know fresh water algae* (3rd ed., pp. 17–280). Dubuque: Wm. C. Brown company.

Pulle, J. S., & Khan, A. M. (2003). studies on zooplanktonic community of Isapur Dam water, India. *Pollution Research, 22*(3), 451–455.

Purdy, W. C. (1922). A study of the pollution and natural purification of the Ohio river-I. The plankton and related organisms. *U. S. Public Health Bulletin., 131*, 1–78.

Qadri, M. Y., & Yousuf, A. R. (1978). Seasonal variations in the physico- chemical factors of a sub- tropical lake of Kashmir. *Journal of the Inland Fisheries Society of India, 10*(2), 89–96.

Qadri, M. Y., & Yousuf, A. R. (1980a). Limnological studies on lake Malpur Sar I. *The Biotope. Geobios, 7*(3), 117–119.

Qadri, M. Y., & Yousuf, A. R. (1980b). Influence of physico-chemical factors on the seasonality of cladocera in lake Manasbal. *Geobios, 7*(6), 273–276.

Radwan, S. (1976). Planktonic rotifers as indicators of lake trophy Annales Universitates. Marie Curie Sklodoroxsk. *Lublinpolnia., 31*(21), 227–235.

Rai, L. C. (1978). Ecological studies of algal communities of Ganges river at Varanasi. *Indian Journal of Ecology, 5*(1), 1–6.

Rai, D. N., & Duttamunshi, J. S. (1979). The influence of thick floating vegetation (Eichornia cressipes) on the physico-chemical environment of a freshwater wetland. *Hydrobiologia, 65*, 65–69.

Rai, L. C., & Kumar, H. D. (1977). Study on the seasonal variations in the algal communities of a pond receiving effluent of a fertilizers factory. *Indian Journal of Ecology, 4*, 124.

Rajmankova, E. (1975). Comparison of *Lemna gibba* and *Lemna minor* from the production ecological viewpoint. *Aquatic Botany, I*, 423–427.

Ramakrishnaiah, M., & Sarkar, S. K. (1982). Plankton productivity in relation to certain hydrobiological factors in Konar reservoir (Bihar). *Journal of the Inland Fisheries Society of India, 14*(1), 58–68.

Ramakrishnaiah, M., & Sarkar, S. K. (1986). Plankton productivity in relation to certain hydrobiological factors in Konar reservoir (India). *Journal of the Inland Fisheries Society of India, 14*(1), 58–68.

Ramanathan, K. R. (1964). *Ulotrichales*. New Delhi: Indian Council of Agricultural Research.

Rana, B. C., & Palria, S. (1988). Assessment, evaluation and abatement studies of a polluted river Bandi (Rajasthan). In R. K. Trivedy (Ed.), *Ecology and pollution of India rivers* (pp. 345–359). New Delhi: Arian Publishing House.

Randhawa, M. S. (1959). *Zygnemales*. New Delhi: Indian Council of Agricultural Research.

Rao, C. B. (1955). On the distribution of algae in a group of six small ponds. II. Algae periodicity. *Journal of Ecology, 43*, 291–308.

Rao, K. K. C., Rao, N. K., Rao, P. L. K. M., & Rao, S. C. (1991). Studies on groundwater quality in the industrial areas in and around Hyderabad, A. P., India (Part-II). *Pollution Research, 10*(2), 75–91.

Rather, S. A., Bhat, S. A., & Pandit, A. K. (2001). Water quality of Hokarsar, a typical wetland of Kashmir. *Journal of Reserch and Development, 1*, 38–43.

Raunkier, C. (1934). *The life forms of plants and statistical plant geography*. Oxford: Clarendon.

Rawson, D. S. (1939a). Problem of lake biology (Vol. 10, pp. 9–26). Publication for the American Association for the Advancement of Science by the Science Press.

Rawson, D. A. (1939b). Physical and chemical studies. Plankton and bottom fauna of Okanagan lake, B.C. in 1935 with appended data from adjacent smaller lakes Bull. *Journal of Fisheries Research Board of Canada, 56*, 3–70.

Rawson, D. S. (1955). Morphometry as dominent factor in the productivity of large lakes. *Verh. Int. Limnol., 12*, 291–308.

Ray, P., Singh, S. B., & Sehgal, K. L. (1966). A study of some aspects of ecology of rivers Ganga and Yamuna at Allahabad (U.P.) in 1958–59. *Proceedings of National Academy of Sciences India, 36*(3), 235–272.

Reddy, P. M., Venkateshwarlu, V., & Raj Kumar, R. (1991). Phytoplankton in the river Tungabhadra (A.P.) and the impact of paper mill effluent on the flora. *Indian Journal of Microbial Ecology, 2*, 71–77.

Reid, G. K. (1961). *Ecology of inland water and estuaries* (p. 375). New York: Reinhold Publishing Corporation.

Reid, G. K., & Wood, R. D. (1976). *Ecology of inland waters and estuaries* (2nd ed.). New York: Van Nostrand Reinhold Publ, Co.

Rice, C. H. (1938). Studies in the phytoplankton of the river Thames, I and II. *Annals of Botany, 2*, 539–581.

Richardson, J. L. (1968). Diatoms and lake topology in east and central Africa. *Internationale Revue der gesamten Hydrobiologia Hydrographie, 53*, 299–338.

Rishi, V., & Kachroo, P. (1981). Effects of sewage and fertilizers on phytoplankton of Doodhganga. *Geobios, 8*(4), 160–162.

Rodhe, W. (1948). Environmental requirements of fresh water plankton algae. Experimental studies in the ecology of phytoplankton. *Symbols Botany Upsal, 10*, 1–149.

Round, F. E. (1973). The ecology of benthic algae. In D. F. Jackson (Ed.), *Algae and man* (pp. 138–184). New York: Plenum Press.

Roy, P. N. (2000). Studies on hydrological status of a stream in Santal Parganas (South Bihar) with special reference to pollution. *Indian Journal of Environment and Ecoplanning, 3*(1), 127–130.

Roy, S. P., & Sharma, U. P. (1983). Dipteran larval population in a freshwater wetland at Bhagalpur (Bihar) in relation to physico-chemical parameters. *Indian Journal of Ecology, 10*(1), 129–136.

Rutter, F. (1953). *Fundamentals of limnology*. Toronto: University of Toronto Press.

Ruttner, F. (1968). *Fundamentals of limnology*. Toronto: University of Toronto Press.

Sabater, B., & Isabel, M. (1990). Successional dynamics of the phytoplankton in the lower part of the river Ebro. *Journal of Plant Research, 12*, 573–592.

Saha, L. C., & Chowdhary, S. K. (1985). Phytoplankton density in relation to aerobic factors of a pond at Bhagalpur. *Consevation of Physiology and Ecology, 10*(2), 58–100.

Saha, L. C., & Pandit, B. (1988). Density of zooplankton between lentic and lotic ecosystem. *Geobios. New Reports, 7*(1), 31–35.

Saha, T., Ghosh, P. B., & Bandyopadhyay, T. S. (2000). Waste water characteristics of some canals of Calcutta. *Journal of Environmental Research, 10*(2), 97–101.

Sajeev, K. I. D. K. (1999). *Analytical studies on the ecosystem of the Ropar headworks reservoir*. Ph.D. thesis. Patiala: Punjabi University.

Sandhu, J. S., Sehgal, H. S., & Toor, H. S. (1984). Rotifer species diversity in manured experimental fish tanks. *Indian Journal of Ecology, 3*(2), 303–308.

Saradhamani, N., & Siwakumar, N. N. (1995). A Preliminary study on ecology of the river Bhavani at the Kampatti, Tamilnadu. *Journal of Environment and Pollution, 2*(1), 35–37.

Sarwar, S. G. (1985). *Community, species distribution and seasonal variations in Aufwuchs (periphyton) population on natural and artificial substrates*. Ph.D. thesis. University of Kashmir.

Sarwar, S. G. (1987). Species composition and seasonal variation of periphyton on *Ceratophyllum demersum* in Waskur lake, Kashmir. *Geobios New Reports., 6*(2), 114–115.

Sarwar, S. G. (1989). Impact of developmental activities on Dal lake. *Development without destruction* pp. 106.

Sarwar, S. G. (1991a). *Trophic status of Dal lake (Kashmir)*. In Proceedings national conference on aquatic sciences in India. New Delhi.

Sarwar, S. G. (1991b). Studies on periphyton communities of marsh lands around Srinagar. *Geobios New Report, 10*, 22–27.

Sarwar, S. G. (1999). Water quality and periphytic algal component of Anchar lake, Kashmir. In K. Vijay Kumar (Ed.), *Freshwater ecosystem of India* (pp. 237–250). Delhi: Daya publishing House.

Sarwar, S. G., & Majid, I. (1997). Abiotic features and diatom population of Wular lake, Kashmir. *Ecology Environment and Conservation, 3*(3), 121–127.

Sarwar, S. G., & Parveen, A. (1996). Community structure and population density of zooplankton in two interconnected lakes of Srinagar, Kashmir. *Pollution Research, 15*(1), 35–38.

Sarwar, S. G., & Rifat, M. (1991). Abiotie features of a fresh water lotic ecosystem of Kashmir. (Doodhganga river). *Pollution Research, 10*(2), 69–73.

Sarwar, S. G., & Wazir, A. M. (1988). Abiotic environment of a freshwater lentic ecosystem of Kashmir. *Geobios, 15*(6), 282–284.

Sarwar, S. G., & Zutshi, D. P. (1987a). Primary productivity of periphyton. *Geobios, 14*, 127–129.

Sarwar, S. G., & Zutshi, D. P. (1987b). Studies on periphyton population of Himalayan lake I. Species composition an community structure on natural and artificial substrates. *Proceedings of Indian National Science Academy, 52*(3), 239–243.

Sarwar, S. G., & Zutshi, D. P. (1988). Species distribution and community structure of periphytom algae on artificial substrata. *Tropical Ecology, 29*(2), 116–120.

Sarwar, S. G., & Zutshi, D. P. (1989). Periphytic algal flora of Phragmites communis trin. *Journal of the Indian Institute of Science, 69*, 275–283.

Sarwar, S. G., Naqshi, A. R., & Mir, G. R. (1996). Impact of floating gardens on the limnological features of dal lake. *Pollution Research, 15*(3), 217–221.

Sawyer, C. N. (1947). Fertilization of lakes by agricultural and urban drainage. *Journal of New Water Lakes Association, 51*, 109–127.

Sawyer, C. N. (1960). *Chemistry for sanitary engineers*. New York: Mc Graw Hill Book Co.

Saxsena, K. L., Chakrabarti, R. N., Khan, A. Q., Chattopadhya, S. N., & Chandra, H. (1966). Pollution studies of river Ganges near Kanpur. *Environmental Health, 8*(4), 270–285.

Schneider, R. F. (1962). Seasonal succession of certain invertebrates in northwestern Florida lake. *Quarterly Journal of Florida Academy of Science, 25*, 127–141.

Scott, D. (1966). The substrate cover-fraction concept. In K. W. Cummins, C. A. Tryon, R. T. Hartman (Eds.), Organism-substrate relationships in streams (Special Publication No. 4). Pymatuning Laboratory of Ecology, University of Pittsbury, 145 pp.

Scullion, J. C., Parish, A., & Edwards, R. W. (1982). Comparison of benthic macro invertebrate fauna and substratum composition in riffles and pools in the impounded river Elan and the unregulated river Wye, Mid-Wales. *Fresh Water Biology, 12*, 579–595.

Sedamkar, E., & Angadi, S. B. (2003). Physico-chemical parameters of two fresh waterbodies of Gulbarga – India, with special reference to Phytoplankton. *Pollution Research, 22*(3), 411–422.

Seenayya, G. (1971). Ecological studies in the plankton of certain fresh water ponds of Hyderabad. India. I. Physico-chemical complexes. *Hydrobiologia, 37*, 7–31.

Sengar, R. M. S., & Singh, S. (1986). Pollution status of Keetham lake (Soor sarovar) at Agra. *Geobios, 13*, 56–61.

Serruya, C. (1978). *Lake kinneret*. (Monographiae Biologicae, Vol 32, pp. 501). The Hague: W. Junk.

Serruya, C., & Pollingher, U. (1983). *Lakes of the warm belt* (p. 569). Cambridge: Cambridge University Press.

Sewell, R. B. S. (1934). A study of fauna of salt lakes Calcutta. *Records of Indian Muslims, 36*, 45–121.

Sexena, D. N., & Mishra, S. R. (1990). On the planktonic fauna of industrial waste water from Birla nagar industrial complex, Gwaliar, India. *Geobios New Reports., 9*, 186–188.

Shah, A. R. (1988). Physico-chemical aspects of pollution in river Jhelum (Kashmir) during 1981-83. In R. K. Trivedy (Ed.), *Ecology and pollution of Indian rivers* (pp. 163–207). New Delhi: Asian Publishing House.

Sharma, B. K. (1980). Contributions to the Rotifer fauna of Orissa, India. *Hydrobiologia, 70*, 225–233.

Sharma, S. (1986). Effect of physico-chemical factors on benthic fauna of Bhagirathi river, Garhwal Himalayas. *Indian Journal of Ecology, 13*(1), 133–137.

Sharma, S. (1988). *Potamological studies of Morar (Kalpi) river*. Gwalior. M. Phil. thesis. Gwalior: Jiwaji University.

Sharma, V. P., & Dhaneshwar, R. S. (1986). Significance of pH in the environment. *I. J. E. P., 6*(4), 272–274.

Sharma, B. K., & Michael, R. G. (1987). Review of taxonomic studies on freshwater cladocera from India with remarks on biogeography. *Hydrobiologia, 145*, 29–33.

Sharma, A. L. N., & Patnaik, P. K. (1985). Ecological studies on zooplankton of freshwater ponds in and around Bhuvaneshwar. *Journal of Environmental Biology, 6*(4), 245–256.

Sharma, R. K., & Rathore, V. (2000). Pollution ecology with reference to commercially important fisheries prospect in a rural based water body: The lake Sarsai Nawar, Etawah in U.P. *Pollution Research, 19*(4), 641–644.

Sharma, J. P., & Srivastava, J. B. (1986). Ecological observations on rotifer fauna of some fresh water ponds of Jammu (J and K) India. *Geobios New Reports., 5*(11), 6–10.

Sharma, M. R., & Verma, P. S. (2003). Water quality of springs in Hamirpur area of outer Himalayas. *Pollution Research, 22*(3), 369–372.

Sharma, K. D., Lal, N., & Pathak, P. D. (1981). Water quality of sewage rains entering Yamuna at Agra. *Indian Journal of Environmental Health, 23*(2), 118–122.

Sharma, S., Kaushik, S., & Mathur, R. (1989). Pollution studies of Saank, Asun and Kuari rivers in M.P. In V. P. Agrawal, B. N. Desai, & S. A. H. Abidi (Eds.), *Management of aquatic ecosystems* (pp. 211–218). Muzzaffarnagar: Society of Bioscience.

Sharma, M. S., Liyaquat, F., Barbar, D., & Chishty, N. (2000). Biodiversity of freshwater zooplankton in relation to heavy metal pollution. *Pollution Research, 19*(1), 147–157.

Shastree, N. K., Islam, M. S., Pathak, S., & Afshan, M. (1991). Studies on the physico-chemical dimensions of the lentic hydrosphere of Ravindra Sarovar (Gaya). In N. K. Shastree (Ed.), *Current trends in limnology I* (pp. 133–152). Delhi: Narendra Publishing House.

Shetty, H. P. C., Saha, S. D., & Glosh, B. B. (1961). Observation on the distribution and fluctuation of plankton in the Hooghly Maltash eastuarine system with notes on their relation to commercial fish landings. *Indian Journal of Fisheries, 8*(2), 326–363.

Shivanikar, S. V., Patil, P. M., Vaidya, D. P., & Bandela, N. N. (1999). Environmental temperature fluctuation determine dissolved oxygen level in Godavari river water. *Pollution Research, 18*(4), 415–418.

Shrivastava, H. N. (1962). Aquatic fauna as indicator of pollution. *Environmental Health, 4*, 106–113.

Shukla, A. (1995). Population dynamics of rotifers fauna at Govindgarh lake, Rewa (M.P.) India. *Journal of Environmental Pollution, 2*(4), 177–181.

Shukla, A., Singh, A. K., Chaturvedi, S. and Tripathi, N. P. (1995) Population dynamics of rotifers fauna at Govindgarh lake, Rewa (M.P.) India. *Journal of Environmental Pollution, 2*(4), 177–181.

Shyamsunder (1988). Monitoring the water quality of river Jhelum, Kashmir. In R. K. Trivedy (Ed.), *Ecology and pollution of Indian rivers* (pp. 131–161). New Delhi: Asian Publishing House.

Singh, V. P. (1960). Phytoplankton ecology of inland waters of U.P. In *Proceedings of symposium on algology* (pp. 243–271). New Delhi: ICAR.

Singh, R. N. (1965). Limnological relations of Indian inland waters with special reference to water blooms. *Proceedings of the International Association of Limnology, 12*, 831–883.

Singh, R. S. (1982). Systematics and ecology of aquatic crustacea of Rewa. *Geobios New Reports., 1*(11), 53–54.

Singh, B. B. (1995). Pollution status of Rapti river at Gorakpur. *Journal of Environment and Pollution, 2*(3), 117–120.

Singh, S. K. (1996). *Correlation among different water quality parameters of groundwater of Jhunjhunu district, Rajasthan.* 28th annual convention of Indian water works ssociation. (pp. 89–98). Jodhpur.

Singh, J. P., & Roy, S. P. (1990). Investigations of limnological profiles of Kawar lake (Bagusarai, Bihar). In V. P. Agrawal & P. Das (Eds.), *Recent trends in limnology* (pp. 457–467). Muzzaffarnagar: Society of Biosciences.

Singh, S. B., & Sahai, R. (1978). Seasonal fluctuation of zooplankton population in relation to certain physico-chemical factors of a pond. *Geobios, 5*(15), 228–230.

Singh, S. R., & Srivastava, V. K. (1991). Investigation on the periodicity of phytoplankton in relation to certain hydrological conditions in the stretch of Ganga river between Buxer and Baffia. *Pollution Research, 10*(2), 93–101.

Singh, V. P., Saxena, P. N., Tiwari, A., Lonsane, B. K., Khan, M. A., & Arora, I. (1969). Algal flora of sewage Uttar Pradesh in relation to physicochemical variables. *Environmental Health, 11*, 208–219.

Singh, H. R., Badola, S. P., & Dobriyal, A. K. (1982). Ecology of the river Nayar of Garhwal Himalaya, Uttar Pradesh. *Journal of Zoology, 2*, 72–76.

Sinha, R. K. (1987). *Evaluation of pollution, water quality and carbon cyeling from hydrobiological parameters in river Ganges in and around Patna*. Ph. D. thesis. Patna: Patna Univ.

Sinha, R. K.,& Prasad, K. (1988). *Ganga basin reseach project. BuxarBarh.* Final technical reports. Patna: Patna Univ.

Sinha, A. K., Srivastava, R. K., Diwivedi, A., & Pande, D. P. (1991). *Sediment characteristics of river Ganga at Shuklagani (Unnao), Gagasso(Rae-Bareli) and Phakhama (Allahabad).* In Proceedings national conference on aquatic science in India. New Delhi.

Sivakumar, R., Mohanraj, R., & Azeez, P. A. (2000). Physico-chemical analysis of water sources of Ooty, South India. *Pollution Research, 19*(1), 143–146.

Siver, P. A. (1977). Comparison of attached diatom communities on natural and artificial substrates. *Journal of Phycology, 13*, 402–406.

Skuja, H. (1964). Grundzuge der Algenflora and Algenvegetation der Fjeldgegenden Um Abisko in Sehwechisch Lappland. *Nova Acta. Royal Society of Scientists. Upsal.* No. 4, 18, 465 pp.

Sladecek, J. (1961). A guide to limno-saprobiai organisms. *Sc. Pop. Institute Technic Program Technology. Water, 7*, 543–562.

Sladecek, V. (1983). Rotifers as indicators of water quality. *Hydrobiologia, 100*, 169–201.

Smith, G. M. (1950). *Freshwater algae of United States*. New York: McGraw Hill.

Spence, D. H. N. (1964). The macrophytic vegetation on freshwater lochs, swamps and associated fern. In J. H. Burnell (Ed.), *The vegetation of Scotland*. London: Oliver & Boyol, Edinburgn.
Spencer, D. F., Young, H. Y., & Greene, R. W. (1983). Alterations in the zooplankton community of a fly ash treated lake. *Hydrobiologia, 107*, 123–130.
Sreenivasan, A. (1970). Limnology of tropical impoundments. A comparative study of major reservoirs in Madras state (India). *Hydrobiologia, 36*(3–4), 443–469.
Stein, M. A. (1896). Notes on the Ancient Topography of the Pīr Pantsāl Route. *Journal of the Asiatic Society of Bengal*, Vol. LXIV, Part I, No. 4, 1895. Calcutta 1896.
Stevenson, R. J. (1984). Procedures for mounting algae in a syrup medium. *Transactions of the American Microscopical Society, 103*, 320–321.
Stockner, J. G., & Benson, W. W. (1971). The succession of diatom assemblages in the recent sediments of lake Washington. *Limnology and Oceanography, 12*, 413–532.
Subba Rao, N. V. (1989). *Handbook of fresh water molluscs of India*. Calcutta: Zoological Survey of India. 289 pages.
Suduzuki, M., Waternabe, K., Suzuki, K., & Narita, K. (1983). Occurrence of rotifera in the field under natural and intertionally - changed conditions. *Hydrobiologia, 104*, 34–37.
Sugunan, V. V. (1995). *Reservoir fishes of India. FAO fisheries technical paper 345*. Rome: Food and Agriculture organization of the United Nations. 423 pages.
Sukumaran, P. K., Raghavan, S. L., & Rahman, M. F. (1984). Observations on physico-chemical conditions of water and plankton of two lakes in Maland region of Karnataka. *Journal of the Inland Fisheries Society of India, 16*(1-2), 53–55.
Suryanarayanan, N. (1996). Environmental State of the art of Indian lakes/reservoirs pages 19–34. In M. Ando (Ed.), *Conservation and management of lakes/reservoirs in India*. Shiga: International lake Environment Committee Foundation.
Suvarna, C. A., & Somashekar, R. K. (2000). Study on the phytoplankton constituents of river Cauvery and its tributaries. *Indian Journal of the Environment and Ecoplanning, 3*(1), 49–56.
Swarup, K., & Singh, S. R. (1979). Limnological studies of Suraha lake (Ballia). I. Variations in the water quality. *Journal of the Inland Fisheries Society of India, 2*(1), 22–23.
Syal, J. (1996). *Present ecological status of river Satluj in the region of Punjab*. Ph.D. thesis. Patiala: Punjabi University.
Sylvester, R. O. (1961). Nutrient content of drainage water from forested, urban and agricullural areas. *Technical report of Taft Sanitary Engineering Center*, W 61-3, 80–88.
Thienemann, A. (1925). Die Binnengewasser Mittleleuropas: Eine Limnologische Einfuthrung. *Die. Binnengewasser., 1*, 1–255.
Thomas, S., & Azis, P. K. A. (1996). Spatial and temporal distribution of nutrients in the Peppara reservoir – A man made ecosystem on the western Ghats, South India. *Pollution Research, 15*(1), 5–10.
Thomas, S., Harikrishan, K., Sanil, G., Paulmurugan, R., & Das, M. R. (2001). Studies on the water quality of Kuttanand wetland ecosystem of Kerala. *Pollution Research, 20*(1), 59–66.
Thomosn, L. M. (1952). *Soils and soil fertility*. New York/London: Mc Graw Hill.
Thorat, S. R., & Sultana, M. (2000). Pollution status of Salim Ali lake, Aurangabad (M.S.). *Pollution Research, 19*(2), 307–309.
Threlkeld, S. T. (1979). The mid summer dynamics of two Daphnia sp. in Wintergreen lake. *Ecology, 60*, 165–179.
Thresh, J. C., Suckling, E. V., & Beale, J. E. (1944). *The examination of Water and water supplies* (6th ed.). London: E.W. Taylor.
Tilzer, M. M. (1973). Diurnal periodicity in the phytoplankton assemblage of a high mountain lake. *Limnology and Oceanography, 18*, 15–30.
Tonopi, G. T. (1980). *Fresh water animals of India – An ecological approach*. New Delhi: Oxford and IBH Publishing Co.
Tripathi, B. D., Sikandar, M., & Shukla, S. C. (1991). Physico-chemical characterization of city sewage discharged into river Ganga at Varanasi, India. *Environment International, 17*(5), 469–478.

Trivedy, R. K., & Goel, P. K. (1986). *Chemical and biological methods for water pollution studies*. Karad: Environmental Publications.

Unni, K. S. (1985). Comparative limnology of several reservoirs in central India. *Internationale Revue der Gesamten Hydrobiologie, 70*, 845–856.

Unni, K. S., & Naik, L. (1997). Distribution and ecology of zooplankton in the headwaters of the tropical river Narmada. *International Journal of Ecology and Environmental Science, 23*, 1–16.

Upadhyaya, N. P., & Roy, N. N. (1982). Studies in river pollution in Kathmandu valley. *Indian Journal of Evironmental Health, 24*(2), 124–135.

Valarmathi, S., Azariah, J., & Govindasamy, C. (2002). Seasonal variation of physico-chemical parameters in estuarine and coastal waters samples of Madras coast, Bay of Bengal, India. *Pollution Research, 21*(2), 177–185.

Valecha, V., & Bhatnagar, G. P. (1988). Seasonal changes of phytoplankton in relation to some physicochemical factors in lower lake of Bhopal. *Geobios, 15*, 170–173.

Vareethiah, K., & Haniffa, M. A. (1998). Phytoplankton community organisation and species succession in a Bar-built estuary. *Journal of Environment and Pollution, 5*(3), 209–214.

Vashisht, H. S., & Sra, G. S. (1979). *The biological characteristics of Chandigarh waste in relation to the physico-chemical factors*. In Proceedings of the symposium on environmental biololgy. (pp. 429–440).

Vashist, H. S., & Monga, K. S. (1981). The hydrobiology of Sukhna main line canal (Narwana Branch). *Indian Ecology, 8*(2), 313–314.

Vasisht, H. S., & Kapoor, N. K. (1981). The ecology of polluted waters of Yamunanagar (Haryana). *Indian Journal of Ecology, 8*(1), 134–135.

Vasisht, H. S., & Sharma, K. (1975). Ecology of a typical urban pond in Ambala city of Haryana. *Indian Journal of Ecology, 2*(1), 79–86.

Vass, K. K. (1973). *Studies on hydrobiology and primary productivity of some Kashmir lakes*. Ph.D. thesis. Srinagar: Kashmir University.

Vass, K. K., Bhanot, K. K., & Ghosh, A. N. (1978). Periphyton production in 2 ponds of Brackish water farm at Kakdwip West Bengal, India. *Journal of Indian Fisheries Society of India, 10*, 32–38.

Venkataraman, G. S. (1961). *Vaucheriaceae*. New Delhi: Indian Council of Agricultural Research.

Venkateshwarlu, V. (1969). An ecological study of the algae of the river Mossi, Hyderabad. (India) with special reference to pollution, III The algal periodicity. *Hydrobiologia, 34*(3-4), 533–560.

Venkateshwarlu, V., & Reddy, M. P. (1985). Algae as biomonitors in river ecology. *Symposium Biomonitoring state environment*, 183–189.

Venkateshwarlu, V., & Joyanti, T. V. (1968). Hydrobiological studies of river Sabarmati to evaluate quality. *Hydrobiologia, 31*, 442–448.

Venkateshwarlu, V., Kumar, P. T. S., & Kumari, J. N. (1990). Ecology of algae of the river Moori, Hyderabad - A comprehensive study. *Journal of Environmental Biology, 11*, 79–91.

Varma, C. M. (1998). Chemical and biological evaluation of an industrially polluted river. *Journal of Environment and Pollution, 5*(2), 181–187.

Verma, M. C. (2003). Chemical and biological evaluation of an industrially polluted river. In A. Kumar (Ed.), *Aquatic ecosystems* (pp. 113–119). New Delhi: A.P.H. Publishing corporation.

Verma, S. R., & Dalela, R. C. (1975). Studies on the pollution of the Kalinadi by industrial wastes near Mansurpur Part-II. Biological index of pollution and biological characteristics of the river. *Acta. Hydrachemical Hydrobiology, 3*(3), 259–274.

Verma, S. R., & Shukla, G. R. (1968). Hydrobiological studies of a temple tank, Devikund in Deoband (U. P.) India. *Environmental Health, 10*, 177–188.

Verma, S. R., & Shukla, G. R. (1969). Pollution in a perennial stream, Khala Laksar, Shaharanpur, India. *Environmental Health, 11*, 145–162.

Verma, S. R., & Shukla, G. R. (1970). The physico-chemical conditions of Kamla Nehru tank, Muzzaffarnagar (U.P.) in relation to the biological productivity. *Environmental Health, 12*, 110–128.

Verma, S. R., Tyagi, A. K., & Dalela, R. C. (1978). Physico-chemical and biological characteristics of Kadarabad in U. P. *Indian Journal of Environmental Health, 20*(1), 1–13.

Vijver, V. D., Bart, L., & Beyens, L. (1997). Freshwater diatoms from some islands in the maritime Antarctic region. *Antarctic Science, 9*(4), 418–425.

Vollenweider, R. A. (1972). Input output models with special reference to the phosphorus loading concept in Limnology (Workshop conference). Kestanienbawn, Switzerland.

Wanganeo, A. (1984). Primary production characteristics of a Himalayan lake in Kashmir. *Internationale Revue der Gesamten Hydrobiologie, 69*(1), 79–90.

Wanganeo, A., Wangneo, R., & Zutshi, D. P. (1996). Limnological studies on a dimictic Himalayan Lake. In S. Nath (Ed.), *Recent advances in fish ecology limnology and eco-conservation* (pp. 37–54). Delhi. 110035: Daya Publishing House.

Ward, H. B., & Whipple, G. C. (1959). *Fresh water biology* (2nd ed.). New York: Wiley.

Water, T. F. (1957). The effects of lime application to acid bog lakes in northern Michigan. *Transactions of an American Fisheries Society, 86*, 329.

Weiler, R. R. (1988). Chemical limnology of Georgian Bay and the north channel between 1974 and 1980. *Hydrobiologia, 163*, 77–83.

Weilgolaski, F. E. (1975). Biological indicators of pollution. *Urbana. Ecol., 1*, 63–79.

Welch, P. S. (1948). *Limnological methods*. Philadelphia: The Blaskiston Company. 381 pp.

Welch, P. S. (1952). *Limnology* (2nd ed.). New York: Mc Graw Hill Book Co.

Westlake, D. F. (1965) Some basic data for investigation of the productivity of aquatic macrophytes. In C. R. Goldman (Ed.), Primary productivity in aquatic environments. Memoir Institute of Hydrobiology, 17(suppl). Berkeley: University of California Press.

Wetzel, R. G. (1975). *Limnology*. Philadelphia/London: W. B. Saunder's Company.

Wetzel, R. G., & Likens, G. E. (2000). *Limnological analysis* (IIIth ed.). New York/Berlin: Springer.

Whipple, G. C., Fair, G. M., & Whipple, M. C. (1927). *The microscopy of drinking water*. New York: Wiley.

Wilcox (1955) Distribution of aquatic plants in Minnesota Svensk Bot. Tidskr.

Wilkens, H. (1972). Under suchungen uben des vorkoman planktischer rotatorien in stadtgewassern and ithre Beziechung zur saprobiuo. *Mitt. Hanbung Zool. Mus. Inst., 68*, 1–20.

Willem, H. O., Desonet, F., & Evans, M. J. C. (1972). A hydrobiological study of the polluted river Lieve (Gheut, Belgium). *Hydrobiologia, 39*(1), 91–134.

Williams, D. D., & Feltmate, B. W. (1992). *Aquatic insects*. Oxon: C.A.B. International Walling Ford.

Williams, W. D. (1969). Eutrophication of lakes. *Proceedings. Royal Society of Victoria, 83*, 17–26.

Winkler, L. W. (1888). The determination of dissolved oxygen in water. *Berlin. Deut. Chem. Ges., 21*, 2843.

Winner, J. M., Patrick, P. H., & Wallen, D. G. (1980). Zooplankton species diversity in lake St. Clair, Ontario, Canada. *Hydrobiologia, 75*, 57–63.

Woodword, G. M. (1984). Pollution control. in the Humber estuary. *Water Pollution Control, 83*(1), 82–90.

Yadava, Y. S., Singh, R. K., Choudhary, M., & Kolchhar, V. (1987). Limnology and productivity of Dighali Beel Assam. *Tropical Ecology, 28*, 137–146.

Yoshimura, S. (1933). Rapid eutrophication within recent years of lake Haruna, Gunmen, Japan. *Journal of Geography and Geology, 11*, 31–41.

Yousuf, A. R. (1996). Fishery resources of Kashmir. In A. H. Khan & A. K. Pandit (Eds.), *Ecology, environment and energy* (pp. 87–128).

Yousuf, A. R., & Qadri, M. Y. (1985). Seasonal fluctuations of zooplankton in lake Manasbal (Kashmir). *Indian Journal of Ecology, 12*(2), 354–359.

Yousuf, A. R., & Qadri, M. Y. (1986). Some limnological aspects of Mirgund wetland. *Geobios New Reports, 5*(1), 27–30.

Yousuf, A. R., Shah, G. M., & Qadri, M. Y. (1986). Some limnological aspects of Mirgund wetland. *Geobios New Reports., 5*(1), 27–30.

Yousuf, A. R., Gazala, F., & Kousar, J. P. (2002). Ecology and feeding biology of commercially important cyprinid fishes of Anchar lake, Kashmir, with a note on their conservation. In A. K. Pandit (Ed.), *Natural resources of Western Himalaya* (pp. 243–272). Srinagar: Valley Book House Hazratbal.

Zafar, A. R. (1955). *On The periodicity and distribution of algae in certain fish ponds in vicinity of Hyderabad, India*. Ph.D. thesis. Hyderabad: Osmania University.

Zafar, A. R. (1959a). Taxonomy of lakes. *Hydrobiologia, 13*(3), 287–299.

Zafar, A. R. (1959b). Two years observation on the periodicity of Euglenaceae in two fish breeding ponds. *Journal of the Indian Botanical Society, 34*, 430–450.

Zafar, A. R. (1964). On the ecology of algae in certain fish ponds of Hyderabad. India I. Physico-chemical complexes. *Hydrobiologia., 23*, 179–195.

Zafar, A. R. (1967). On the ecology of algae in certain fish ponds of Hyderabad, India - III. The periodicity. *Hydrobiologia, 30*(1), 96–112.

Zutshi, D. P. (1968). *Ecology of some Kashmir lakes*. Ph. D. thesis. Srinagar: J & K University.

Zutshi, D. P., & Khan, M.A. (1978). On lake typology of Kashmir. *Environment Physiological Ecology*, 465–472.

Zutshi, D. P., & Vass, K. K. (1971). Ecology and production of *Salvinia natans* in Kashmir. *Hydrobiologia, 38*, 303–320.

Zutshi, D. P., & Vass, K. K. (1978). Limnological studies on Dal lake, II- Chemical features. *Indian Journal of Ecology, 5*, 90–97.

Zutshi, D. P., & Vass, K. K. (1982). Limnological studies on Dal lake, Srinagar-III. Biological features. *Proceedings of Indian National Science Academy B, 48*(2), 234–241.

Zutshi, D. P., Subla, B. A., Khan, M. A., & Wanganeo, A. (1980). Comparative limnology of nine lakes of Jammu and Kashmir Himalayas. *Hydrobiologia, 72*(1–2), 101–112.

Zutshi, D. P., & Wanganeo, A. (1984). The phytoplankton and primary productivity of a high altitude subtropical lake. *Verh. Internat. Verein Limnol., 22*, 1168–1172.

MIX
Papier aus verantwortungsvollen Quellen
Paper from responsible sources
FSC® C105338

If you have any concerns about our products,
you can contact us on
ProductSafety@springernature.com

In case Publisher is established outside the EU,
the EU authorized representative is:
Springer Nature Customer Service Center GmbH
Europaplatz 3, 69115 Heidelberg, Germany

Printed by Libri Plureos GmbH
in Hamburg, Germany